Stochastic Processes

Stochastic Processes

Theory and Applications

Special Issue Editors

Alexander Zeifman
Victor Korolev
Alexander Sipin

MDPI • Basel • Beijing • Wuhan • Barcelona • Belgrade

Special Issue Editors
Alexander Zeifman
Vologda State University
Russia

Victor Korolev
Lomonosov Moscow State University
Russia

Alexander Sipin
Vologda State University
Russia

Editorial Office
MDPI
St. Alban-Anlage 66
4052 Basel, Switzerland

This is a reprint of articles from the Special Issue published online in the open access journal *Mathematics* (ISSN 2227-7390) in 2019 (available at: https://www.mdpi.com/journal/mathematics/special_issues/Stochastic_Processes_Theory_Applications).

For citation purposes, cite each article independently as indicated on the article page online and as indicated below:

LastName, A.A.; LastName, B.B.; LastName, C.C. Article Title. *Journal Name* **Year**, *Article Number*, Page Range.

ISBN 978-3-03921-962-9 (Pbk)
ISBN 978-3-03921-963-6 (PDF)

Contents

About the Special Issue Editors

Alexander I. Zeifman—Professor, Head of Department of Applied Mathematics, Vologda State University, Vologda, Russia; Senior Researcher, Institute of Informatics Problems, Federal Research Center "Computer Sciences and Control" of the Russian Academy of Sciences, Russia; Chief Researcher, Vologda Research Center of the Russian Academy of Sciences, Russia. Graduate of Vologda State Pedagogical Institute, 1976. Candidate of Science in Physics and Mathematics (PhD), 1981. Doctor of Science in Physics and Mathematics (1994, Institute of Control Sciences, Russian Academy of Sciences). Main research interests: stochastic models, continuous-time Markov chains, bounds on the rate of convergence, perturbation bounds, queueing models, biological models, queueing theory.

Victor Yu. Korolev—Professor, Head of Department of Mathematical Statistics, Faculty of Computational Mathematics and Cybernetics, Lomonosov Moscow State University, Moscow, Russia; Leading researcher, Institute of Informatics Problems, Federal Research Center "Computer Sciences and Control" of the Russian Academy of Sciences, Moscow, Russia; Professor, Hangzhou Dianzi University, Hangzhou, China. Graduate of Faculty of Computational Mathematics, Lomonosov Moscow State University, 1977. Candidate of Science in Physics and Mathematics (PhD), 1981. Doctor of Science in Physics and Mathematics (1994, Lomonosov Moscow State University). Main research interests: Limit theorems of probability theory and their applications in distribution theory, statistics, risk theory, reliability theory. Probability models of real processes in physics, meteorology, financial mathematics and other fields.

Alexander S. Sipin—Professor at the Department of Applied Mathematics, Vologda State University, Institute of Mathematics, Natural and Computer Sciences, Russia. Graduate of Faculty of Mathematics and Mechanics, Leningrad State University, now St.Petersburg State University, Russia in 1975. Candidate of Science in Physics and Mathematics (PhD), 1979. Doctor of Science in Physics and Mathematics (2016, St. Petersburg State University, Russia). Research interests: Monte Carlo and quasi-Monte Carlo methods, Markov chains, meshless numerical methods for solving boundary value problems.

Preface to "Stochastic Processes"

The aim of this Special Issue was to publish original research articles that cover recent advances in the theory and application of stochastic processes. There is particular focus on applications of stochastic processes as models of dynamic phenomena in various research areas, such as queuing theory, physics, biology, economics, medicine, reliability theory, and financial mathematics.

After a thorough review of the submitted papers, 15 works were selected and included in the final collection. These papers were written by scientists representing Belorussia, China, Finland, Iran, Italy, Korea and Russia.

Most papers concern queueing theory and Markov processes. There are also papers dealing with problems of risk theory, Monte Carlo methods and probability models of meteorological phenomena.

The Editors hope that this Special issue will be of interest to specialists in probability theory and its applications, as well as to specialists in related fields of science.

Alexander Zeifman, Victor Korolev, Alexander Sipin
Special Issue Editors

mathematics

MDPI

Article

A Note on a Generalized Gerber–Shiu Discounted Penalty Function for a Compound Poisson Risk Model

Jiechang Ruan [1], Wenguang Yu [2,*], Ke Song [2], Yihan Sun [2], Yujuan Huang [3] and Xinliang Yu [2]

1 Department of Humanities and Social Sciences, Yibin Vocational & Technical College, Yibin 644003, China; rjcjie@163.com
2 School of Insurance, Shandong University of Finance and Economics, Jinan 250014, China; sk19982013@163.com (K.S.); sunyihanhan12345@163.com (Y.S.); 20163433@sdufe.edu.cn (X.Y.)
3 School of Science, Shandong Jiaotong University, Jinan 250357, China; 211018@sdjtu.edu.cn
* Correspondence: yuwg@sdufe.edu.cn

Received: 13 August 2019; Accepted: 20 September 2019; Published: 24 September 2019

Abstract: In this paper, we propose a new generalized Gerber–Shiu discounted penalty function for a compound Poisson risk model, which can be used to study the moments of the ruin time. First, by taking derivatives with respect to the original Gerber–Shiu discounted penalty function, we construct a relation between the original Gerber–Shiu discounted penalty function and our new generalized Gerber–Shiu discounted penalty function. Next, we use Laplace transform to derive a defective renewal equation for the generalized Gerber–Shiu discounted penalty function, and give a recursive method for solving the equation. Finally, when the claim amounts obey the exponential distribution, we give some explicit expressions for the generalized Gerber–Shiu discounted penalty function. Numerical illustrations are also given to study the effect of the parameters on the generalized Gerber–Shiu discounted penalty function.

Keywords: compound Poisson risk model; generalized Gerber–Shiu discounted penalty function; Laplace transform; Dickson–Hipp operator; recursive formula

MSC: 91B30; 91B70; 60G55

---

1. Introduction

The classical compound Poisson risk process $\{U(t)\}_{t\geq 0}$ is defined by

$$U(t) = u + ct - \sum_{i=1}^{N(t)} X_i, \quad t \geq 0, \tag{1}$$

where u is the non-negative amount of initial reserves, and $c > 0$ denotes the constant premium rate per unit time. The counting process $\{N(t)\}_{t\geq 0}$, representing the total claim numbers up to time t , is a homogeneous Poisson processes with intensity λ. $\{X_i\}_{i\geq 1}$ is a sequence of independent and identically distributed non-negative random variables, where X_i is the i-th claim amount. Let $f(x)$ denote the density function of X, and let $E[X]$ and $\hat{f}(s) = \int_0^\infty e^{-sx} f(x)dx$ denote the expectation and Laplace transform of X, respectively. To avoid ruin from being a certain event, we assume $c > \lambda E[X]$.

We say that ruin occurs whenever $U(t)$ becomes negative. The time to ruin of the insurance company is defined as

$$\tau = \inf\{t : U(t) < 0\}, \tag{2}$$

where $\tau = \infty$ if for all $t \geq 0, U(t) \geq 0$. For the initial reserves $U(0) = u$, the probability of ruin is defined as

$$\psi(u) = P(\tau < \infty | U(0) = u), \quad u \geq 0. \tag{3}$$

The probability of ruin is an important risk measure in the study of ruin in risk theory. It has been widely studied in actuarial science. In 1998, famous actuarial scholars Hans Gerber and Elias Shiu first proposed an expected discounted penalty function, which is also called Gerber–Shiu discounted penalty function, to study ruin related problems. Recently, it has become a powerful risk measurement tool in ruin theory. Given the initial surplus $U(0) = u$, we define the classical Gerber–Shiu discounted penalty function as follows:

$$\Phi(u, \delta) = E[e^{-\delta\tau} W(U(\tau-), |U(\tau)|) I(\tau < \infty) | U(0) = u], \quad u \geq 0, \tag{4}$$

where $\delta \geq 0$ is the force of interest, $W(x, y)$ is a non-negative measurable penalty function, and $I(\cdot)$ is the indicator function. It is clear that the Gerber–Shiu discounted penalty function becomes the probability of ruin when $\delta = 0, W(x, y) = 1$. For the recent literature on the Gerber–Shiu discounted penalty function, we can refer to work by Lin et al. [1], Zhang et al. [2], Yu [3,4], Wang et al. [5], Avram et al. [6], Zhang [7], Chi [8], Peng and Wang [9], Li et al. [10], Huang et al. [11], Preischl and Thonhauser [12], Zeng et al. [13,14], Yu et al. [15], Dickson and Qazvini [16], Zhang and Su [17,18], Li et al. [19], and Zhao and Yin [20], among others.

In recent years, Gerber–Shiu discounted penalty function has been extended by many actuarial scholars, so that the new risk measures can be used to study more related quantities. For example, Cai et al. [21] studied the ruin-related Gerber–Shiu discounted penalty risk measures by bringing in the conception of path consumption. Cheung [22] extended the Gerber–Shiu discounted penalty function by introducing the penultimate claim before ruin under a Sparre–Andersen renewal risk model. Chueng [23], Cheung and Woo [24] proposed a new Gerber–Shiu type function by incorporating the total claims up to ruin. Chueng and Feng [25] studied a new kind of generalized Gerber–Shiu discounted penalty function under the Markov arrival process. Wang and Li [26] extended the discount rate from constant to a random variable for the Gerber–Shiu discounted penalty function in the classical risk model. Wang and Zhang [27] provided a smooth extension of the Gerber–Shiu discounted penalty function by introducing an auxiliary function. As is known to all, the ruin time is also an important random variable in the study of risk theory. We can study the Laplace transform of the ruin time by the Gerber–Shiu discounted penalty function, while other mathematical characteristics associated with the ruin time cannot be directly studied through Gerber–Shiu discounted penalty function. In recent years, many actuarial scholars have paid attention to the moment of the ruin time. For instance, Egidio dos Reis [28], Lin and Willmot [29] and Drekic and Willmot [30] studied the moment of the ruin time under the compound Poisson risk model. Pitts and Politis [31] proposed an approximation approach of the moment of ruin time. Yu et al. [32] studied the moment of ruin time under the Markov arrival risk model. The moment of ruin time was introduced in the Gerber–Shiu discounted penalty function by Lee and Willmot [33]; then, they studied this new Gerber–Shiu discounted penalty function Sparre-Andersen risk model [34]. Schmidli [35] considered a new Gerber–Shiu discounted penalty function, which is modified with an additional penalty for reaching a level above the initial capital. Deng et al. [36] studied a generalized Gerber–Shiu discounted penalty function, in which the interest rates follow a Markov chain with finite state space. Li and Lu [37] studied the generalized expected discounted penalty function in a risk process with credit and debit interests.

In this paper, we introduce a new generalized Gerber–Shiu type function. For non-negative integer n, define

$$\Phi_n(u, \delta) = E[\tau^n e^{-\delta\tau} W(U(\tau-), |U(\tau)|) I(\tau < \infty) | U(0) = u], \quad u \geq 0. \tag{5}$$

We call $\Phi_n(u,\delta)$ a generalized Gerber–Shiu discounted penalty function. It is obvious that $\Phi_0(u,\delta) = \Phi(u,\delta)$. When $W(x,y) = 1$, $\Phi_n(u,\delta)$ is the discounted nth moment of ruin time, then we can use it to study the expectation and variance of ruin time. Note that the generalized Gerber–Shiu discounted penalty function defined by formula (5) is different from the function studied in Lee and Willmot [34] since we bring in the surplus before ruin.

In this paper, we mainly discuss the calculation method of the generalized Gerber–Shiu discounted penalty function $\Phi_n(u,\delta)$. In Section 2, we propose a recursion method by Laplace transform to calculate $\Phi_n(u,\delta)$. In Section 3, we present an exact expression of $\Phi_n(u,\delta)$ when claim amounts are exponentially distributed. Numerical examples are also given to explain the effect of the related parameters. Finally, conclusions are given in Section 4.

2. Recursion Calculation of $\Phi_n(u,\delta)$

First, we define the Laplace transform of $\Phi_n(u,\delta)$ by

$$\hat{\Phi}_n(u,\delta) = \int_0^{\infty} e^{-su}\Phi_n(u,\delta)du, \quad Re(s) \geq 0.$$

For convenience, we introduce the Dickson–Hipp operator T_s, which, for any integral function h on $(0,+\infty)$, is defined as

$$T_s h(x) = \int_x^{\infty} e^{-s(y-x)}h(y)dy = \int_0^{\infty} e^{-sy}h(x+y)dy, \quad x \geq 0.$$

It is easily seen that $T_s h(0) = \int_0^{\infty} e^{-sy}h(y)dy = \hat{h}(s)$. The Dickson–Hipp operator has interchangeability, that is to say, for $s \neq r$,

$$T_s T_r h(x) = T_r T_s h(x) = \frac{T_s h(x) - T_r h(x)}{r-s}.$$

For for more properties of the Dickson–Hipp operator, we refer interested readers to Dickson and Hipp [38] and Li and Garrido [39].

When $n = 0$, it follows from Gerber and Shiu [35] and Laplace transform that $\hat{\Phi}_0(s,\delta)$ satisfies the following equation:

$$\{cs - \lambda(1-\hat{f}(s)) - \delta\}\hat{\Phi}_0(s,\delta) = \lambda\{\hat{\omega}(\rho(\delta)) - \hat{\omega}(s)\}, \tag{6}$$

where

$$\omega(u) = \int_u^{\infty} W(u, x-u)f(x)dx,$$

$$\hat{\omega}(s) = \int_0^{\infty} e^{-su}\omega(u)du,$$

and $\rho(\delta)$ is the positive root of the following equation

$$cs - \lambda(1-\hat{f}(s)) - \delta = 0. \tag{7}$$

By Equation (6), we can obtain the renewal equation satisfied by Gerber–Shiu discounted penalty function $\Phi_0(u,\delta)$, and we can further derive the analytic expression of $\Phi_0(u,\delta)$. Now, we consider the case $n \geq 1$. The derivative of $\Phi_n(u,\delta)$ with respect to δ is given by

$$\Phi_n(u,\delta) = (-1)^n \frac{d^n}{d\delta^n}\Phi_0(u,\delta). \tag{8}$$

Applying Laplace transform on both sides of Equation (8) gives

$$\hat{\Phi}_n(s,\delta) = (-1)^n \frac{d^n}{d\delta^n}\hat{\Phi}_0(s,\delta). \tag{9}$$

Then, taking a derivative on both sides of Equation (6) with respect to δ yields

$$\{cs - \lambda(1-\hat{f}(s)) - \delta\}\frac{d^n}{d\delta^n}\hat{\Phi}_0(s,\delta) - n\frac{d^{n-1}}{d\delta^{n-1}}\hat{\Phi}_0(s,\delta) = \lambda\frac{d^n}{d\delta^n}\hat{\omega}(\rho(\delta)).$$

In addition, by Equation(9), we have

$$\{cs - \lambda(1-\hat{f}(s)) - \delta\}\hat{\Phi}_n(s,\delta) + n\hat{\Phi}_{n-1}(s,\delta) = (-1)^n\lambda\frac{d^n}{d\delta^n}\hat{\omega}(\rho(\delta)). \tag{10}$$

Setting $s = \rho(\delta)$ in Equation (10) yields

$$(-1)^n\lambda\frac{d^n}{d\delta^n}\hat{\omega}(\rho(\delta)) = n\hat{\Phi}_{n-1}(\rho(\delta),\delta).$$

Substituting the above result back into Equation (10) gives

$$\{cs - \lambda(1-\hat{f}(s)) - \delta\}\hat{\Phi}_n(s,\delta) = n\hat{\Phi}_{n-1}(\rho(\delta),\delta) - n\hat{\Phi}_{n-1}(s,\delta). \tag{11}$$

Since $\rho(\delta)$ is the root of Equation (7), we have

$$\begin{aligned} cs - \lambda(1-\hat{f}(s)) - \delta &= cs - \lambda(1-\hat{f}(s)) - \delta - \{c\rho(\delta) - \lambda[1-\hat{f}(\rho(\delta))] - \delta\} \\ &= c(s-\rho(\delta)) + \lambda[\hat{f}(s) - \hat{f}(\rho(\delta))] \\ &= (s-\rho(\delta))\left[c + \lambda\frac{\hat{f}(s)-\hat{f}(\rho(\delta))}{s-\rho(\delta)}\right] \\ &= (s-\rho(\delta))[c - \lambda T_s T_{\rho(\delta)} f(0)]. \end{aligned} \tag{12}$$

Plugging Equation (12) back into Equation (11) gives

$$(s-\rho(\delta))[c - \lambda T_s T_{\rho(\delta)} f(0)]\hat{\Phi}_n(s,\delta) = n\hat{\Phi}_{n-1}(\rho(\delta),\delta) - n\hat{\Phi}_{n-1}(s,\delta).$$

We can rewrite the above equation to obtain

$$\left[1 - \frac{\lambda}{c}T_s T_{\rho(\delta)} f(0)\right]\hat{\Phi}_n(s,\delta) = \frac{n}{c}\frac{\hat{\Phi}_{n-1}(\rho(\delta),\delta) - \hat{\Phi}_{n-1}(s,\delta)}{s-\rho(\delta)} = \frac{n}{c}T_s T_{\rho(\delta)}\Phi_{n-1}(0,\delta). \tag{13}$$

Applying Laplace transform on both sides of Equation (13) gives

$$\Phi_n(u,\delta) = \int_0^\infty \frac{\lambda}{c}T_{\rho(\delta)} f(x)\Phi_n(u-x,\delta)dx + \frac{n}{c}T_{\rho(\delta)}\Phi_{n-1}(u,\delta), \tag{14}$$

where $g(x) = \frac{\lambda}{c}T_{\rho(\delta)} f(x), x \geq 0$. Since by Gerber and Shiu [40] we have $\int_0^\infty g(x)dx < 1$, then Equation (14) is a defective renewal equation.

Define

$$H(u) = \sum_{n=1}^{\infty} g^{*n}(u), \quad u \geq 0,$$

where g^{*n} denotes the nth convolution of g. Then, we can express the solution of Equation (14) as follows:

$$\Phi_n(u,\delta) = \frac{n}{c} T_{\rho(\delta)} \Phi_{n-1}(u,\delta) + \frac{n}{c} \int_0^u H(u-x) T_{\rho(\delta)} \Phi_{n-1}(x,\delta) dx. \tag{15}$$

The above equation gives a recursion algorithm for computing $\Phi_n(u,\delta)$, where the initial value is given by $\Phi_0(u,\delta)$.

3. Explicit Expressions for Exponential Claim Distribution and Numerical Examples

In this section, we suppose that the claim amounts are exponentially distributed, and the density function is given by

$$f(x) = \alpha e^{-\alpha x}, \ \ \alpha, x > 0.$$

To obtain some explicit results, we assume the penalty function $W(x,y) = W_1(y)$. Then, we have

$$\omega(u) = \int_u^\infty W_1(x-u) \alpha e^{-\alpha x} dx = \beta e^{-\alpha u}, \tag{16}$$

where $\beta = \int_0^\infty W_1(x) \alpha e^{-\alpha x} dx$. Solving Equation (7), we obtain

$$\rho(\delta) = \frac{\lambda + \delta - c\alpha + \sqrt{(\lambda + \delta - c\alpha)^2 + 4c\alpha\delta}}{2c}. \tag{17}$$

We denote the other root of Equation (7) by $-R(\delta)$; then, it is easy to obtain

$$R(\delta) = -\frac{\lambda + \delta - c\alpha - \sqrt{(\lambda + \delta - c\alpha)^2 + 4c\alpha\delta}}{2c}. \tag{18}$$

Now, we derive an expression for $\Phi_n(u,\delta)$ by Laplace inverse transform. First, by Equation (6), we have

$$\frac{c}{s+\alpha}(s - \rho(\delta))(s + R(\delta)) \hat{\Phi}_0(s,\delta) = \frac{\lambda\beta}{\alpha + \rho(\delta)} - \frac{\lambda\beta}{\alpha + s}.$$

Then, we obtain

$$\hat{\Phi}_0(s,\delta) = \frac{\lambda\beta}{c(\alpha + \rho(\delta))(s + R(\delta))}.$$

Applying Laplace inverse transform in the above equation gives

$$\Phi_0(u,\delta) = \frac{\lambda\beta}{c(\alpha + \rho(\delta))} e^{-R(\delta)u}, \ \ u \geq 0. \tag{19}$$

Next, we consider $n \geq 1$. Combining the derivative of formula (19) w.r.t δ and Equation (8), we find that $\Phi_n(u,\delta)$ has the following expression:

$$\Phi_n(u,\delta) = \sum_{k=0}^n A_{n,k} \frac{u^k}{k!} e^{-R(\delta)u}, \ \ u \geq 0. \tag{20}$$

Finally, we discuss how to determine the coefficients $A_{n,k}$ in formula (20).
When $n = 0$, comparing Equations (19) and (20), we obtain

$$A_{0,0} = \frac{\lambda\beta}{c[\alpha + \rho(\delta)]}. \tag{21}$$

Taking the Laplace transform of formula (20) yields

$$\hat{\Phi}_n(s,\delta) = \sum_{k=0}^{n} \frac{A_{n,k}}{[s+R(\delta)]^{k+1}}. \tag{22}$$

By formulas (21) and (22), we have

$$\begin{aligned} \hat{\Phi}_n(s,\delta) &= \frac{n[\hat{\Phi}_{n-1}(\rho(\delta),\delta) - \hat{\Phi}_{n-1}(s,\delta)]}{cs - \lambda(1-\hat{f}(s)) - \delta} \\ &= \frac{n(s+\alpha)}{c(s-\rho(\delta))(s+R(\delta))} \sum_{k=0}^{n-1} A_{n-1,k} \cdot \left[\frac{1}{(\rho(\delta)+R(\delta))^{k+1}} - \frac{1}{(s+R(\delta))^{k+1}} \right] \\ &= \frac{L_n(s)}{[s+R(\delta)]^{n+1}}, \end{aligned} \tag{23}$$

where

$$L_n(s) = \frac{n(s+\alpha)}{c(s-\rho(\delta))} \sum_{k=0}^{n-1} A_{n-1,k} \cdot [s+R(\delta)]^{n-k-1} \frac{[s+R(\delta)]^{k+1} - [\rho(\delta)+R(\delta)]^{k+1}}{[\rho(\delta)+R(\delta)]^{k+1}} \tag{24}$$

is an n-order polynomial.

By partial fraction expansion of formula (23), we obtain

$$A_{n,k} = \frac{1}{(n-k)!} \frac{d^{n-k}}{ds^{n-k}} L_n(s)|_{s=-R(\delta)}, \ k = 0,1,2,......,n. \tag{25}$$

Noting that the polynomial $L_n(s)$ only depends on $A_{n-1,k}$, we can calculate $A_{n,k}$ recursively from $A_{0,k}$.

Without losing generality, we give the explicit expressions of $\Phi_n(u,\delta)$ for $n = 1,2$.

For $n = 1$, from formula (20), we have

$$\Phi_1(u,\delta) = A_{1,0} \cdot e^{-R(\delta)u} + A_{1,1} \cdot ue^{-R(\delta)u}, \ u \geq 0. \tag{26}$$

From formula (24), we have

$$L_1(s) = \frac{A_{0,0} \cdot (s+\alpha)}{c[\rho(\delta)+R(\delta)]}. \tag{27}$$

By formulas (25) and (27), we have

$$\begin{aligned} A_{1,0} &= \frac{d}{ds} L_1(s)|_{s=-R(\delta)} = \frac{A_{0,0}}{c[\rho(\delta)+R(\delta)]}, \\ A_{1,1} &= L_1(-R(\delta)) = \frac{A_{0,0} \cdot (\alpha - R(\delta))}{c[\rho(\delta)+R(\delta)]}. \end{aligned}$$

Then, by formula (26), we get

$$\Phi_1(u,\delta) = \frac{A_{0,0}}{c[\rho(\delta)+R(\delta)]} e^{-R(\delta)u} + \frac{A_{0,0} \cdot (\alpha - R(\delta))}{c[\rho(\delta)+R(\delta)]} e^{-R(\delta)u}, \ u \geq 0. \tag{28}$$

For $n = 2$, from formula (20), we have

$$\Phi_2(u,\delta) = A_{2,0} \cdot e^{-R(\delta)u} + A_{2,1} \cdot ue^{-R(\delta)u} + \frac{1}{2} A_{2,2} \cdot u^2 e^{-R(\delta)u}, \ u \geq 0. \tag{29}$$

From formula (24), we have

$$L_2(s) = \frac{2A_{1,0}}{c[\rho(\delta)+R(\delta)]}(s+\alpha)(s+R(\delta)) + \frac{2A_{1,1}}{c[\rho(\delta)+R(\delta)]^2}(s+\alpha)(s+\rho(\delta)+R(\delta)), \quad (30)$$

$$L_2'(s) = \frac{2A_{1,0}\cdot[2s+R(\delta)+\alpha]}{c[\rho(\delta)+R(\delta)]} + \frac{2A_{1,1}\cdot[2s+2R(\delta)+\alpha+\rho(\delta)]}{c[\rho(\delta)+R(\delta)]^2}, \quad (31)$$

$$L_2''(s) = \frac{4A_{1,0}}{c[\rho(\delta)+R(\delta)]} + \frac{4A_{1,1}}{c[\rho(\delta)+R(\delta)]^2}. \quad (32)$$

By formulas (25) and (30)–(32), we have

$$\begin{aligned} A_{2,0} &= \frac{1}{2}\frac{d^2}{ds^2}L_2(s)|_{s=-R(\delta)} = \frac{2A_{1,0}}{c[\rho(\delta)+R(\delta)]} + \frac{2A_{1,1}}{c[\rho(\delta)+R(\delta)]^2}, \\ A_{2,1} &= \frac{d}{ds}L_2(s)|_{s=-R(\delta)} = \frac{2A_{1,0}\cdot[\alpha-R(\delta)]}{c[\rho(\delta)+R(\delta)]} + \frac{2A_{1,1}\cdot[\alpha+\rho(\delta)]}{c[\rho(\delta)+R(\delta)]^2}, \\ A_{2,2} &= L_2(s)|_{s=-R(\delta)} = \frac{2A_{1,1}\cdot[\alpha-R(\delta)]}{c[\rho(\delta)+R(\delta)]}. \end{aligned}$$

Then, by formula (29), we get

$$\begin{aligned} \Phi_2(u,\delta) &= \{\frac{2A_{1,0}}{c[\rho(\delta)+R(\delta)]} + \frac{2A_{1,1}}{c[\rho(\delta)+R(\delta)]^2}\}e^{-R(\delta)u} \\ &+ \frac{2A_{1,0}\cdot[\alpha-R(\delta)]}{c[\rho(\delta)+R(\delta)]} + \frac{2A_{1,1}\cdot[\alpha+\rho(\delta)]}{c[\rho(\delta)+R(\delta)]^2}ue^{-R(\delta)u} \\ &+ \frac{A_{1,1}\cdot[\alpha-R(\delta)]}{c[\rho(\delta)+R(\delta)]}u^2e^{-R(\delta)u}, \quad u\geq 0. \end{aligned} \quad (33)$$

Next, we give the numerical simulation of $\Phi_0(u,\delta)$, $\Phi_1(u,\delta)$ and $\Phi_2(u,\delta)$ to illustrate the effect of the related parameters on the generalized Gerber–Shiu discounted penalty function by Matlab (Version: matlab2016a; Manufacturer: The MathWorks, Inc.; Natick, Massachusetts 01760 USA)

Example 1. *Suppose $W(x,y) = 1$, then $\beta = 1$. We give the influence of the relevant parameters on the function $\Phi_0(u,\delta)$, $\Phi_1(u,\delta)$ and $\Phi_2(u,\delta)$. See Figures 1–4.*

It is easy to see that the images we get from Figures 1–4 are the opposite of the images some scholars get from the traditional classical risk model. For example, in Figure 1, $\Phi_0(u,\delta)$ is the Laplace transform of the ruin time. $\Phi_0(u,\delta)$ goes down as c increases. This means that the higher the premium income rate c is, the smaller the function $\Phi_0(u,\delta)$ is. The reason is that $e^{-\delta\tau}$ is a decreasing function of τ. Increased premiums mean greater ruin time τ, which in turn leads to smaller functions $e^{-\delta\tau}$. Similarly, $\tau e^{-\delta\tau}$ and $\tau^2 e^{-\delta\tau}$ are also a subtractive function of τ when τ is large. Thus, $\Phi_1(u,\delta)$ and $\Phi_2(u,\delta)$ go down as c increases. The same conclusion appears in Figures 2–4. We are not explain it one more time in the following Figures 5–8.

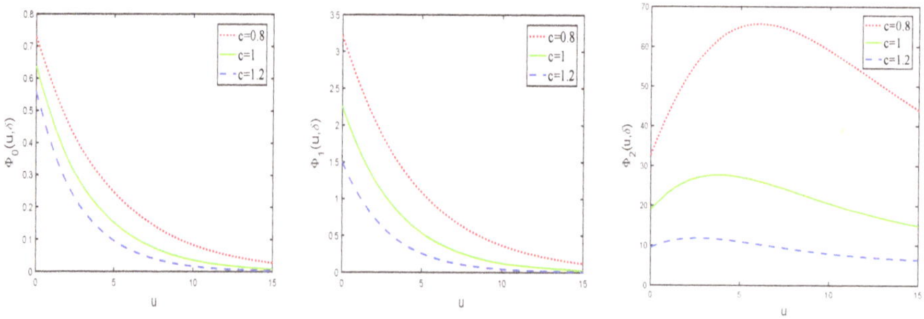

(a) (b) (c)

Figure 1. Set $\lambda = 0.6, \alpha = 0.8, \delta = 0.05$. (**a**) The influence of the parameter c on the function $\Phi_0(u, \delta)$. (**b**) The influence of the parameter c on the function $\Phi_1(u, \delta)$. (**c**) The influence of the parameter c on the function $\Phi_2(u, \delta)$.

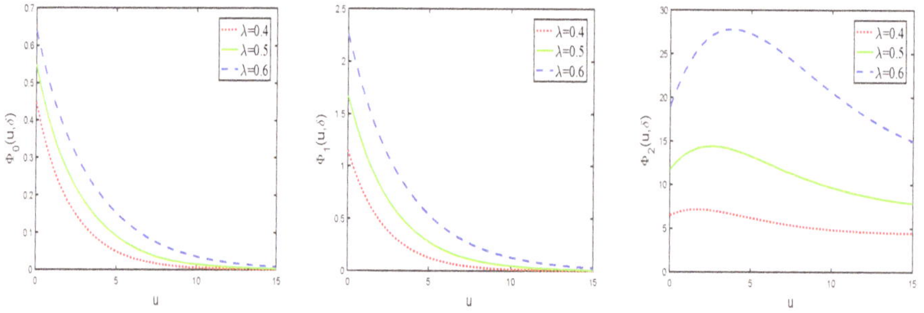

(a) (b) (c)

Figure 2. Set $c = 1, \alpha = 0.8, \delta = 0.05$. (**a**) The influence of the parameter λ on the function $\Phi_0(u, \delta)$. (**b**) The influence of the parameter λ on the function $\Phi_1(u, \delta)$. (**c**) The influence of the parameter λ on the function $\Phi_2(u, \delta)$.

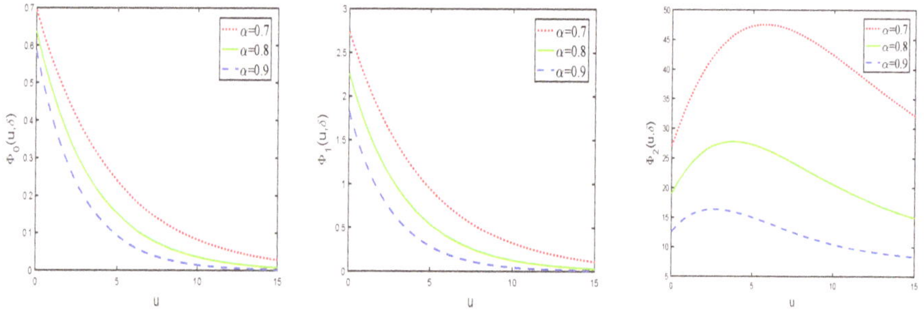

(a) (b) (c)

Figure 3. Set $c = 1, \lambda = 0.6, \delta = 0.05$. (**a**) The influence of the parameter α on the function $\Phi_0(u, \delta)$. (**b**) The influence of the parameter α on the function $\Phi_1(u, \delta)$. (**c**) The influence of the parameter α on the function $\Phi_2(u, \delta)$.

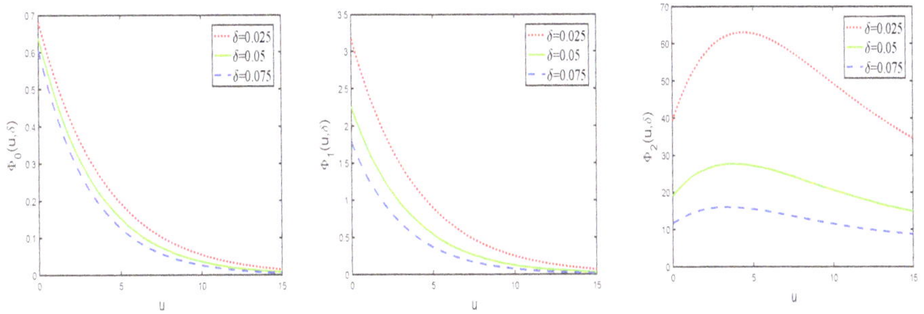

(**a**) (**b**) (**c**)

Figure 4. Set $c = 1, \lambda = 0.6, \alpha = 0.8$. (**a**) The influence of the parameter δ on the function $\Phi_0(u,\delta)$. (**b**) The influence of the parameter δ on the function $\Phi_1(u,\delta)$. (**c**) The influence of the parameter δ on the function $\Phi_2(u,\delta)$.

Example 2. *Suppose* $W(x,y) = y$, *then* $\beta = \dfrac{1}{\alpha}$. *We give the influence of the relevant parameters on the function* $\Phi_0(u,\delta)$, $\Phi_1(u,\delta)$ *and* $\Phi_2(u,\delta)$. *See Figures 5–8.*

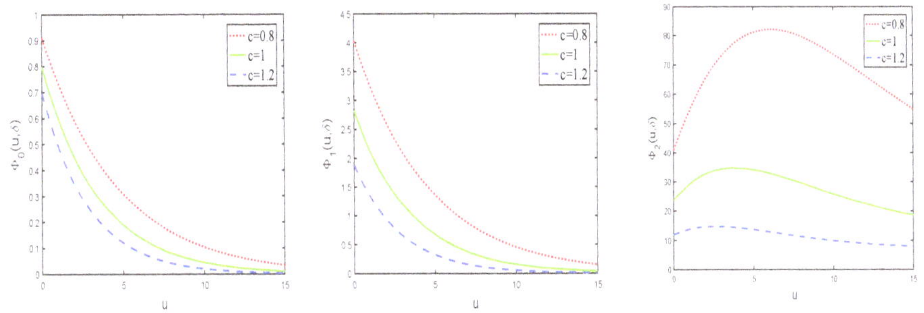

(**a**) (**b**) (**c**)

Figure 5. Set $\lambda = 0.6, \alpha = 0.8, \delta = 0.05$. (**a**) The influence of the parameter c on the function $\Phi_0(u,\delta)$. (**b**) The influence of the parameter c on the function $\Phi_1(u,\delta)$. (**c**) The influence of the parameter c on the function $\Phi_2(u,\delta)$.

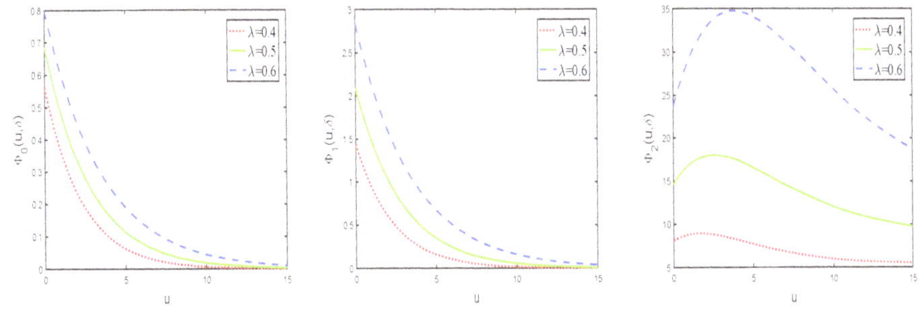

(**a**) (**b**) (**c**)

Figure 6. Set $c = 1, \alpha = 0.8, \delta = 0.05$. (**a**) The influence of the parameter λ on the function $\Phi_0(u,\delta)$. (**b**) The influence of the parameter λ on the function $\Phi_1(u,\delta)$. (**c**) The influence of the parameter λ on the function $\Phi_2(u,\delta)$.

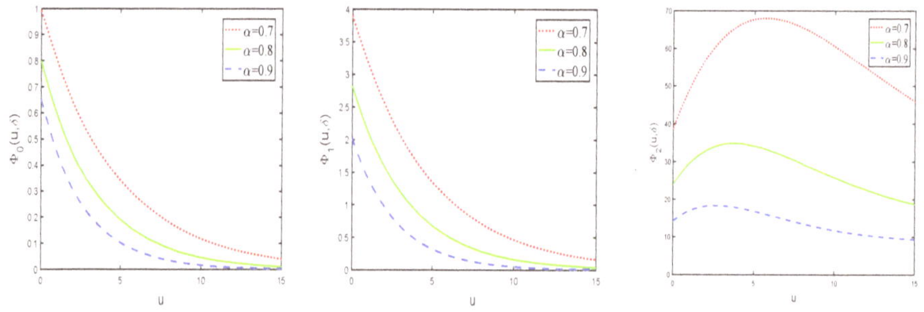

(a) (b) (c)

Figure 7. Set $c = 1, \lambda = 0.6, \delta = 0.05$. (**a**) The influence of the parameter α on the function $\Phi_0(u, \delta)$. (**b**) The influence of the parameter α on the function $\Phi_1(u, \delta)$. (**c**) The influence of the parameter α on the function $\Phi_2(u, \delta)$.

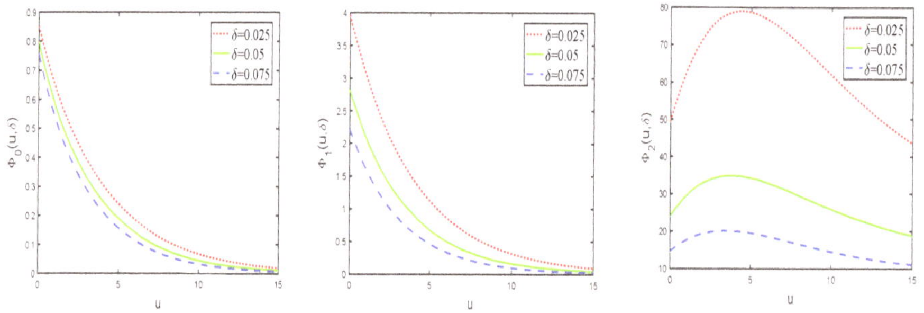

(a) (b) (c)

Figure 8. Set $c = 1, \lambda = 0.6, \alpha = 0.8$. (**a**) The influence of the parameter δ on the function $\Phi_0(u, \delta)$. (**b**) The influence of the parameter δ on the function $\Phi_1(u, \delta)$. (**c**) The influence of the parameter δ on the function $\Phi_2(u, \delta)$.

4. Conclusions

In this paper, we discuss a generalized Gerber–Shiu discounted penalty function, which relies on the moment of the time to ruin under the compound Poisson risk model. We present a recursion algorithm for calculating the generalized Gerber–Shiu discounted penalty function by Laplace transform and renewal theory when the claim amounts are subject to an exponential distribution. Furthermore, we derive some explicit expressions of the generalized Gerber–Shiu discounted penalty function for $n = 0, 1, 2$. In addition, we also give numerical examples to explain the effects of parameters c, λ, α and δ on the generalized Gerber–Shiu discounted penalty function $\Phi_0(u, \delta), \Phi_1(u, \delta)$ and $\Phi_2(u, \delta)$. It is very easy to see the effects of these parameters on the generalized Gerber–Shiu discounted penalty function from Figures 1–8. The insurance company can bring the real claim data into the model for numerical simulation and obtain relevant parameters, so that the ruin probability, the Laplace transform of the ruin time and the discounted expected time to ruin can be calculated. The acquisition of these actuarial quantities will effectively improve the operating level of the insurance company.

Author Contributions: Data curation, W.Y., Y.S., Y.H. and X.Y.; Methodology, J.R. and W.Y.; Software, W.Y., K.S. and Y.S.; Writing—original draft, J.R. and W.Y.

Funding: This research is partially supported by the National Social Science Foundation of China (Grant No. 15BJY007), the National Natural Science Foundation of China (Grant Nos. 11301303, 71804090), the Taishan Scholars Program of Shandong Province (Grant No. tsqn20161041), the Humanities and Social Sciences Project of the Ministry Education of China (Grant No. 16YJC630070, Grant No. 19YJA910002), the Natural Science Foundation of Shandong Province (Grant No. ZR2018MG002), the Fostering Project of Dominant Discipline and Talent Team of Shandong Province Higher Education Institutions (Grant No. 1716009), the 1251 Talent Cultivation Project of Shandong Jiaotong University, the Risk Management and Insurance Research Team of Shandong University of Finance and Economics, the Shandong Jiaotong University 'Climbing' Research Innovation Team Program, and the Collaborative Innovation Center Project of the Transformation of New and Old Kinetic Energy and Government Financial Allocation.

Acknowledgments: The authors would like to thank the four anonymous referees for their helpful comments and suggestions, which improved an earlier version of the paper.

Conflicts of Interest: The authors declare no conflict of interest.

References

1. Lin, X.S.; Willmot, G.E.; Drekic, S. The classical risk model with a constant dividend barrier: Analysis of the Gerber–Shiu discounted penalty function. *Insur. Math. Econ.* **2003**, *33*, 391–408.
2. Zhang, H.Y.; Zhou, M.; Guo, J.Y. The Gerber–Shiu discounted penalty function for classical risk model with a two-step premium rate. *Stat. Probab. Lett.* **2006**, *76*, 1211–1218. [CrossRef]
3. Yu, W.G. Some results on absolute ruin in the perturbed insurance risk model with investment and debit interests. *Econ. Model.* **2013**, 31, 625–634. [CrossRef]
4. Yu, W.G. On the expected discounted penalty function for a Markov regime switching risk model with stochastic premium income. *Discret. Dyn. Nat. Soc.* **2013**, *2013*, 1–9. [CrossRef]
5. Wang, Y.Y.; Yu W.G.; Huang, Y.J., Yu X.L.; Fan H.L. Estimating the expected discounted penalty function in a compound Poisson insurance risk model with mixed premium income. *Mathematics* **2019**, *7*, 305. [CrossRef]
6. Avram, F.; Palmowski, Z.; Pistorius, M.R. On Gerber–Shiu functions and optimal dividend distribution for a Lévy risk process in the presence of a penalty function. *Ann. Appl. Probab.* **2015**, *25*, 1868–1935. [CrossRef]
7. Zhang, Z.M. Estimating the Gerber–Shiu function by Fourier-Sinc series expansion. *Scand. Actuar. J.* **2017**, *2017*, 898–919. [CrossRef]
8. Chi, Y.C. Analysis of expected discounted penalty function for a general jump diffusion risk model and applications in finance. *Insur. Math. Econ.* **2010**, *46*, 385–396. [CrossRef]
9. Peng, J.Y.; Wang, D.C. Uniform asymptotics for ruin probabilities in a dependent renewal risk model with stochastic return on investments. *Stochastics* **2018**, *90*, 432–471. [CrossRef]
10. Li, S.M.; Lu, Y.; Sendova, K.P. The expected discounted penalty function: From infinite time to finite time. *Scand. Actuar. J.* **2019**, *2019*, 336–354. [CrossRef]
11. Huang, Y.J.; Yu, W.G.; Pan, Y.; Cui, C.R. Estimating the Gerber–Shiu expected discounted penalty function for Lévy risk model. *Discrete Dyn. Nat. Soc.* **2019**, *2019*, 1–15. [CrossRef]
12. Preischl, M.; Thonhauser, S. Optimal reinsurance for Gerber–Shiu functions in the Cramér-Lundberg model. *Insur. Math. Econ.* **2019**, *87*, 82–91. [CrossRef]
13. Zeng, Y.; Li, D.P.; Chen, Z.; Yang, Z. Ambiguity aversion and optimal derivative-based pension investment with stochastic income and volatility. *J. Econ. Dyn. Control* **2018**, *88*, 70–103. [CrossRef]
14. Zeng, Y.; Li, D.P.; Gu, A.L. Robust equilibrium reinsurance–investment strategy for a mean–variance insurer in a model with jumps. *Insur. Math. Econ.* **2016**, *66*, 138–152. [CrossRef]
15. Yu, W.G.; Huang, Y.J.; Cui, C.R. The absolute ruin insurance risk model with a threshold dividend strategy. *Symmetry* **2018**, *10*, 377. [CrossRef]
16. Dickson, D.C.M.; Qazvini, M. Gerber–Shiu analysis of a risk model with capital injections. *Eur. Actuar. J.* **2016**, *6*, 409–440. [CrossRef]
17. Zhang, Z.M.; Su, W. A new efficient method for estimating the Gerber–Shiu function in the classical risk model. *Scand. Actuar. J.* **2018**, *2018*, 426–449. [CrossRef]
18. Zhang, Z.M.; Su, W. Estimating the Gerber–Shiu function in a Lévy risk model by Laguerre series expansion. *J. Comput. Appl. Math.* **2019**, *346*, 133–149. [CrossRef]
19. Li, J.C.; Dickson, D.C.M.; Li, S.M. Some ruin problems for the MAP risk model. *Insur. Math. Econ.* **2015**, *65*, 1–8. [CrossRef]

20. Zhao, Y.X.; Yin, C.C. The expected discounted penalty function under a renewal risk model with stochastic income. *Appl. Math. Comput.* **2012**, *218*, 6144–6154. [CrossRef]
21. Cai, J.; Feng, R.; Willmot, G.E. On the total discounted operating costs up to default and its applications. *Adv. Appl. Probab.* **2009**, *41*, 495–522. [CrossRef]
22. Cheung, E.C.K. A generalized penalty function in Sparre–Andersen risk models with surplus-dependent premium. *Insur. Math. Econ.* **2011**, *48*, 384–397. [CrossRef]
23. Cheung, E.C.K. Moments of discounted aggregate claim costs until ruin in a Sparre–Andersen risk model with general interclaim times. *Insur. Math. Econ.* **2013**, *53*, 343–354. [CrossRef]
24. Cheung, E.C.K.; Woo, J.K. On the discounted aggregate claim costs until ruin in dependent Sparre–Andersen risk processes. *Scand. Actuar. J.* **2016**, *2016*, 63–91. [CrossRef]
25. Cheung, E.C.K.; Feng, R. A unified analysis of claim costs up to ruin in a Markovian arrival risk process. *Insur. Math. Econ.* **2013**, *53*, 98–109. [CrossRef]
26. Wang, H.C.; Li, N.X. On the Gerber–Shiu function with random discount rate. *Commun. Stat. Theory Methods* **2017**, *46*, 210–220. [CrossRef]
27. Wang, W.Y.; Zhang, Z.M. Computing the Gerber–Shiu function by frame duality projection. *Scand. Actuar. J.* **2019**, *2019*, 291–307. [CrossRef]
28. Egidio dos Reis, A.D. On the moments of ruin and recovery times. *Insur. Math. Econ.* **2000**, *27*, 331–343. [CrossRef]
29. Lin, X.S.; Willmot, G.E. The moments of the time of ruin, the surplus before ruin, and the deficit at ruin. *Insur. Math. Econ.* **2000**, *27*, 19–44. [CrossRef]
30. Drekic, S.; Willmot, G.E. On the moments of the time of ruin with applications to phase-type claims. *N. Am. Actuar. J.* **2005**, *9*, 17–30. [CrossRef]
31. Pitts, S.M.; Politis, K. Approximations for the moments of ruin time in the compound Poisson model. *Insur. Math. Econ.* **2008**, *42*, 668–679. [CrossRef]
32. Yu, K.; Ren, J.; Stanford, D.A. The moments of the time of ruin in Markovian risk models. *N. Am. Actuar. J.* **2010**, *14*, 464–471. [CrossRef]
33. Lee, W.Y.; Willmot, G.E. On the moments of the time to ruin in dependent Sparre–Andersen models with emphasis on Coxian interclaim times. *Insur. Math. Econ.* **2014**, *59*, 1–10. [CrossRef]
34. Lee, W.Y.; Willmot, G.E. The moments of the time to ruin in dependent Sparre–Andersen models with Coxian claim sizes. *Scand. Actuar. J.* **2016**, *2016*, 550–564. [CrossRef]
35. Schmidli, H. Extended Gerber–Shiu functions in a risk model with interest. *Insur. Math. Econ.* **2015**, *61*, 271–275. [CrossRef]
36. Deng, Y.C.; Liu, J.; Huang, Y.; Li, M.; Zhou, J.M. On a discrete interaction risk model with delayed claims and stochastic incomes under random discount rates. *Commun. Stat. Theory Methods* **2018**, *47*, 5867–5883. [CrossRef]
37. Li, S.M.; Lu, Y. On the generalized Gerber–Shiu function for surplus processes with interest. *Insur. Math. Econ.* **2013**, *52*, 127–134. [CrossRef]
38. Dickson, D.C.M.; Hipp, C. On the time to ruin for Erlang (2) risk processes. *Insur. Math. Econ.* **2001**, *29*, 333–344. [CrossRef]
39. Li, S.M.; Garrido, J. On ruin for the Erlang(n) risk process. *Insur. Math. Econ.* **2004**, *34*, 391–408. [CrossRef]
40. Gerber, H. U.; Shiu, E.S.W. On the time value of ruin. *N. Am. Actuar. J.* **1998**, *2*, 48–72. [CrossRef]

mathematics

MDPI

Article

Valuing Guaranteed Minimum Death Benefits by Cosine Series Expansion

Wenguang Yu [1,*], Yaodi Yong [2], Guofeng Guan [1], Yujuan Huang [3,*], Wen Su [2] and Chaoran Cui [4]

1 School of Insurance, Shandong University of Finance and Economics, Jinan 250014, China; guofeng1997012@163.com
2 College of Mathematics and Statistics, Chongqing University, Chongqing 401331, China; yaodiyong@cqu.edu.cn (Y.Y.); wensu@cqu.edu.cn (W.S.)
3 School of Science, Shandong Jiaotong University, Jinan 250357, China
4 School of Computer Science & Technology, Shandong University of Finance and Economics, Jinan 250014, China; crcui@sdufe.edu.cn
* Correspondence: yuwg@sdufe.edu.cn (W.Y.); yujuanh518@163.com (Y.H.)

Received: 2 August 2019; Accepted: 5 September 2019; Published: 10 September 2019

Abstract: Recently, the valuation of variable annuity products has become a hot topic in actuarial science. In this paper, we use the Fourier cosine series expansion (COS) method to value the guaranteed minimum death benefit (GMDB) products. We first express the value of GMDB by the discounted density function approach, then we use the COS method to approximate the valuation Equations. When the distribution of the time-until-death random variable is approximated by a combination of exponential distributions and the price of the fund is modeled by an exponential Lévy process, explicit equations for the cosine coefficients are given. Some numerical experiments are also made to illustrate the efficiency of our method.

Keywords: equity-linked death benefits; Fourier cosine series expansion; guaranteed minimum death benefit; option; valuation; Lévy process

MSC: 91B30; 60G99; 62P05; 62M05

---

1. Introduction

The guaranteed minimum death benefit (GMDB) is a common rider embedded in variable annuity products that promises a minimum payout upon the death of the insured. In this product, policyholders first pay premiums to the insurance company, and then, investment accounts are established for capital investment. When the insured dies, a payment shall be given to the designated beneficiary, and the payout amount depends on the performance of the policyholder's account. This mechanism not only provides insurance guarantee for policyholders, but also has the opportunity to benefit from the financial market, which appeals to customers.

For $t \geq 0$, let $S(t) = S(0)e^{X(t)}$ denote the price of a stock fund or mutual fund at time t. For a person currently aged x, let T_x denote the remaining lifetime, called the time-until-death random variable hereafter. Moreover, we assume T_x is independent with the asset price process $S(t)$ throughout this paper. Consider a GMDB rider that guarantees a payment of $b(S(T_x))$ to the beneficiary when the insured dies, where $b(\cdot)$ is an equity-linked death benefit function. For a constant force of interest $\delta \geq 0$, we are interested in valuing the following expectation:

$$V_x := E[e^{-\delta T_x} b(S(T_x))], \tag{1}$$

which represents a fair price for an equity-linked life-contingent payment at T_x. Since most contracts have a finite expiry date, we can modify (1) by introducing an expiry date T and consider:

$$V_{x,T} := E[e^{-\delta T_x} b(S(T_x)) \mathbf{I}(T_x \le T)], \tag{2}$$

where $\mathbf{I}(A)$ denotes the indicator function of event A.

We are also interested in the case when the death benefit amount depends on two stocks (or stock funds). Let $\{S_1(t), S_2(t)\}_{t\ge 0}$ denote the price process of two stocks, and let:

$$X_i(t) = \ln(S_i(t)/S_i(0)), \qquad i = 1, 2, \ t \ge 0.$$

In the sequel, it is also assumed that T_x is independent with both $S_1(t)$ and $S_2(t)$. Accordingly, we are interested in evaluating the following expectations:

$$\overline{V}_x := E[e^{-\delta T_x} b(S_1(T_x), S_2(T_x))], \qquad \overline{V}_{x,T} := E[e^{-\delta T_x} b(S_1(T_x), S_2(T_x)) \mathbf{I}(T_x \le T)].$$

Recently, the valuation of the GMDB product has drawn many researchers' attention. Under an exponential mortality law, Milevsky and Posner [1] proposed a risk-neutral framework to derive valuation equations for GMDB contracts. Besides, they conducted a numerical study under the Gompertz-Makeham law. Later, in Bauer et al. [2], they supposed that the death should only occur at the policy anniversary date, which facilitates a discrete numerical valuation approach for fairly valuing varieties of guaranteed riders, including GMDB. Gerber et al. [3] proposed a discounted density approach to value GMDB in a Brownian motion risk model, and their results were extended by Gerber et al. [4] and Siu et al. [5] to the jump diffusion model and the regime-switching jump diffusion model, respectively. Based on the fact that the combination of exponential distributions is weakly dense on the set of probability functions defined on $[0, \infty)$ (see Dufresne [6] and Ko and Ng [7]), we notice that the density function of the time-until-death random variable was approximated by a linear combination of exponential distributions in Gerber et al. [3,4] and Siu et al. [5]. More recently, Zhang and Yong [8] and Zhang et al. [9] used two different methods to value GMDB products. Recent related literature can be found in Dai et al. [10], Bélanger et al. [11], Kang and Ziveyi [12], Asmussen et al. [13], and Zhou and Wu [14].

When the density function of the time-until-death random variable is approximated by a linear sum of exponential distributions, Gerber et al. [3,4,15] derived explicit valuation equations for GMDB contracts under various payoffs. The simplicity of using a combination of exponential distributions is excellent, but they are not representative of reality. A direct way to calibrate this is to use life table data. In Ulm [16,17], he emphasized the valuation of GMDB products under mortality laws, such as the De Moivre law of mortality and the Makeham law of mortality. A similar consideration could be found in Liang et al. [18]. They novelly introduced the piecewise constant forces of the mortality assumption to describe the time-until-death variable, then decomposed the valuation problem and presented explicit valuation equations for GMDB.

Except the aforementioned assumptions on the time-until-death random variable, the modeling of the asset process has attracted the attention of scholars. Brownian motion was widely used to model the log-asset price process, which was adopted in Milevsky and Posner [1], Bauer et al. [2] (Section 4), Gerber et al. [3], and Liang et al. [18]. In the field of financial markets, this case is basically a Black-Scholes framework. However, more processes could be also implemented. In Gerber et al. [4], Kou's jump model was used as the log-asset process. A counterpart study of Gerber et al. [3] was referred to by Gerber et al. [15], in which a random walk exponentially generates the price process. To study the valuation issue in a different perspective, scholars turned to the regime-switching model, which was used to investigate the performance of an object subject to the economic changes in financial markets. In the literature, interest readers can refer to Fan et al. [19], Siu et al. [5], Ignatieva et al. [20], and Hieber [21].

In Fang and Oosterlee [22], a highly-efficient option pricing method was proposed to price European options. The method was based on the Fourier cosine series expansions, and it is now called the cosine series expansion (COS) method in the literature. The COS method is quite easy to implement to approximate an integrable function as long as the objective function has a closed-form Fourier transform. The one-dimensional COS (1D COS) method in [22] was extended to the two-dimensional COS method (2D COS) by Ruijter and Oosterlee [23] to price financial options in two-dimensional asset price processes. Leitao et al. [24] proposed a data-driven COS method. Except for option pricing, this method has been adopted in insurance ruin theory. For example, Chau et al. [25,26] used the 1D COS method to compute the ruin probability and the expected discounted penalty function; Zhang [27] approximated the density function of the time to ruin by both 1D and 2D COS methods; Yang et al. [28] proposed a nonparametric estimator for the deficit at ruin by the 2D COS method; Wang et al. [29] and Huang et al. [30] used the 1D COS method to estimate the expected discounted penalty function under some risk models with stochastic premium income. The COS method has also been used by some authors to value variable annuities. For example, Deng et al. [31] used the 1D COS method for equity-indexed annuity products under general exponential Lévy models; Alonso-García et al. [32] extended the 1D COS method to the pricing and hedging of variable annuities embedded with guaranteed minimum withdraw benefit riders. The latest research on Fourier transform was given by Zhang et al. [33], Chan [34], Zhang and Liu [35], Have and Oosterlee [36], Shimizu and Zhang [37], Tour [38], Zhang [39], and Wang et al. [40].

The discounted density method proposed by Gerber et al. [3,4] can be successfully used to value GMDB products when the log-return process is the Brownian motion or Kou's jump diffusion model. Under these two models, the density functions have some closed-form expressions. However, in practice, we cannot obtain the closed-from expression for the discounted density function. In this paper, we use the COS method to approximate the discounted density function, which is applicable since most of the widely-used Lévy processes have explicit characteristic functions. To the best of our knowledge, this is the first paper exploring the COS method on numerical valuation of the GMDB product under the general exponential Lévy models. In particular, there are few papers on GMDB valuation dependent on two stock prices in the literature, and this is the first paper dealing with this problem.

This paper aims to value the GMDB contracts in a risk-neutral framework, mainly by numerically solving Equations (1) and (2). In subsequent sections, we first briefly recall the COS method in Fang and Oosterlee [22], then adopt the method to approximate V_x and $V_{x,T}$ in the one-dimensional framework. We define auxiliary functions to simplify deductions and display equations under different payoffs in Section 2. Motivated by Gerber et al. [3], we consider the situation where the density function of T_x is approximated by a linear combination of exponential distributions, and calculate cosine coefficients in Section 4.1. Under the multi-dimensional case, we shed light on a two-dimensional framework in Section 3. Finally, numerical examples are presented in Section 4, in which we display tables and figures to illustrate the performance of our proposed approach.

2. 1D COS Approximation

In the section, we shall use the 1D COS method to compute V_x and $V_{x,T}$. The idea of the 1D COS method is that every absolutely integrable function f can be approximated on a truncated domain $[a_1, a_2]$ by a truncated Fourier cosine series with N terms,

$$f(y) \approx \sum_{k=0}^{N-1}{}' \mathcal{A}_k(f, a_1, a_2) \cos\left(k\pi \frac{y - a_1}{a_2 - a_1}\right), \tag{3}$$

where $\sum'$ means that the first term in the summation has half weight, and the cosine coefficients are given by:

$$\mathcal{A}_k(f, a_1, a_2) = \frac{2}{a_2 - a_1} \mathrm{Re}\left\{\mathcal{F}f\left(\frac{k\pi}{a_2 - a_1}\right) \exp\left(-i\frac{ka_1\pi}{a_2 - a_1}\right)\right\}. \tag{4}$$

Here, $\mathcal{F}f(s) = \int e^{isy} f(y)dy$, $s \in \mathbb{R}$, is the Fourier transform of f, and $\text{Re}(\cdot)$ means taking the real part.

Let $f_{T_x}(\cdot)$ denote the probability density function of T_x, and for $t > 0$, let $f_{X(t)}(\cdot)$ be the probability density function of $X(t)$. By changing the order of integrals, we can obtain:

$$\begin{aligned} V_x &= \int_0^\infty E[e^{-\delta t} b(S(t))] f_{T_x}(t)dt \\ &= \int_0^\infty e^{-\delta t} \int_{-\infty}^\infty b(S(0)e^y) f_{X(t)}(y)dy f_{T_x}(t)dt \\ &= \int_{-\infty}^\infty b(S(0)e^y) f^\delta_{X(T_x)}(y)dy, \end{aligned} \tag{5}$$

where:

$$f^\delta_{X(T_x)}(y) = \int_0^\infty e^{-\delta t} f_{X(t)}(y) f_{T_x}(t)dt \tag{6}$$

is the discounted density function of the random variable $X(T_x)$. Similarly, for the T-year life-contingent option, we have:

$$V_{x,T} = \int_{-\infty}^\infty b(S(0)e^y) f^\delta_{X(T_x),T}(y)dy, \tag{7}$$

where:

$$f^\delta_{X(T_x),T}(y) = \int_0^T e^{-\delta t} f_{X(t)}(y) f_{T_x}(t)dt. \tag{8}$$

We shall implement the 1D COS method to compute the integrals in (5) and (7). Instead of expanding the discounted densities $f^\delta_{X(T_x)}$ and $f^\delta_{X(T_x),T}$ via Fourier cosine series, we shall consider the following auxiliary functions:

$$g^\delta_n(y) = e^{ny} f^\delta_{X(T_x)}(y), \quad g^\delta_{n,T}(y) = e^{ny} f^\delta_{X(T_x),T}(y), \qquad n \geq 0.$$

Suppose that both g^δ_n and $g^\delta_{n,T}$ belong to $L^1(\mathbb{R})$, then by Equations (3) and (4), we have:

$$g^\delta_n(y) \approx \sum_{k=0}^{N-1}{}' \mathcal{A}_k(g^\delta_n, a_1, a_2) \cos\left(k\pi \frac{y - a_1}{a_2 - a_1}\right) \tag{9}$$

and:

$$g^\delta_{n,T}(y) \approx \sum_{k=0}^{N-1}{}' \mathcal{A}_k(g^\delta_{n,T}, a_1, a_2) \cos\left(k\pi \frac{y - a_1}{a_2 - a_1}\right). \tag{10}$$

Remark 1. *The cosine coefficients $\mathcal{A}_k(g^\delta_n, a_1, a_2)$ and $\mathcal{A}_k(g^\delta_{n,T}, a_1, a_2)$ can be explicitly computed when $\{X(t)\}_{t\geq 0}$ is a Lévy process and f_{T_x} is a combination of exponential density function. To this end, it suffices to specify the Fourier transforms $\mathcal{F}g^\delta_n$ and $\mathcal{F}g^\delta_{n,T}$. Suppose that $\{X(t)\}_{t\geq 0}$ (with $X(0) = 0$) is a Lévy process with characteristic function:*

$$\phi_{X(t)}(s) = E[e^{isX(t)}] = e^{t\Psi_X(s)}, \quad s \in \mathbb{R}, \tag{11}$$

where $\Psi_X(s) = \ln(E[e^{isX(1)}])$ is called the characteristic exponent. Furthermore, suppose that:

$$f_{T_x}(t) = \sum_{j=1}^m A_j \alpha_j e^{-\alpha_j t}, \qquad t > 0, \tag{12}$$

where $\alpha_j > 0$ and $\sum_{j=1}^m A_j = 1$. Under these assumptions, one easily obtains:

$$\mathcal{F}g^\delta_n(s) = \sum_{j=1}^m \frac{A_j \alpha_j}{\delta + \alpha_j - \Psi_X(s - in)} \tag{13}$$

and:

$$\mathcal{F}g_{n,T}^{\delta}(s) = \sum_{j=1}^{m} A_j\alpha_j \frac{1-e^{-(\delta+\alpha_j-\Psi_X(s-in))T}}{\delta+\alpha_j-\Psi_X(s-in)}. \tag{14}$$

In the sequel, we shall consider three payoff functions.

Case 1. $b(s) = s$.

In this case, Equations (5) and (7) become:

$$V_x = S(0)\int_{-\infty}^{\infty} g_1^{\delta}(y)dy, \qquad V_{x,T} = S(0)\int_{-\infty}^{\infty} g_{1,T}^{\delta}(y)dy.$$

For small number c_1 and large number c_2, using Equation (9), we can approximate V_x as follows,

$$\begin{aligned} V_x &\approx S(0)\int_{c_1}^{c_2} g_1^{\delta}(y)dy \\ &\approx S(0)\int_{c_1}^{c_2} \sum_{k=0}^{N-1}{}' \mathcal{A}_k(g_1^{\delta},a_1,a_2) cos\left(k\pi\frac{y-a_1}{a_2-a_1}\right)dy \\ &= S(0)\sum_{k=0}^{N-1}{}' \mathcal{A}_k(g_1^{\delta},a_1,a_2)\cdot\chi_k(a_1,a_2,c_1,c_2), \end{aligned} \tag{15}$$

where:

$$\chi_k(a_1,a_2,c_1,c_2) = \begin{cases} \dfrac{a_2-a_1}{k\pi}\left[\sin\left(k\pi\dfrac{c_2-a_1}{a_2-a_1}\right)-\sin\left(k\pi\dfrac{c_1-a_1}{a_2-a_1}\right)\right], & k\neq 0, \\ c_2-c_1, & k=0. \end{cases}$$

Similarly, $V_{x,T}$ can be approximated as follows,

$$V_{x,T} \approx S(0)\sum_{k=0}^{N-1}{}' \mathcal{A}_k(g_{1,T}^{\delta},a_1,a_2)\cdot\chi_k(a_1,a_2,c_1,c_2). \tag{16}$$

Case 2. $b(s) = s^n\mathbf{I}_{(s>K)}$ with $n \geq 0$ and $K > 0$.

Here, the positive constant K denotes the strike price. Put $\kappa = \ln(K/S(0))$. It follows from Equation (5) that:

$$V_x = \int_{\kappa}^{\infty}[S(0)]^n e^{ny} f_{X(T_x)}^{\delta}(y)dy = \int_{\kappa}^{\infty}[S(0)]^n g_n^{\delta}(y)dy.$$

Then, using the 1D COS method, we obtain for large number c_2,

$$V_x \approx [S(0)]^n\sum_{k=0}^{N-1}{}' \mathcal{A}_k(g_n^{\delta},a_1,a_2)\cdot\chi_k(a_1,a_2,\kappa,c_2). \tag{17}$$

Similarly, for $V_{x,T}$, we have:

$$V_{x,T} \approx [S(0)]^n\sum_{k=0}^{N-1}{}' \mathcal{A}_k(g_{n,T}^{\delta},a_1,a_2)\cdot\chi_k(a_1,a_2,\kappa,c_2). \tag{18}$$

Remark 2. *For the call option, the payoff is given by:*

$$b(s) = (s-K)_+ = s\mathbf{I}_{(s>K)} - K\mathbf{I}_{(s>K)}.$$

By applying Equations (5) and (17), we obtain:

$$\begin{aligned} V_x &= \int_{\kappa}^{\infty} S(0)e^y f^{\delta}_{X(T_x)}(y)dy - \int_{\kappa}^{\infty} K f^{\delta}_{X(T_x)}(y)dy \\ &= S(0)\int_{\kappa}^{\infty} g^{\delta}_1(y)dy - K\int_{\kappa}^{\infty} g^{\delta}_0(y)dy \\ &\approx S(0)\sum_{k=0}^{N-1}{}' \mathcal{A}_k(g^{\delta}_1, a_1, a_2)\cdot \chi_k(a_1,a_2,\kappa,c_2) - K\sum_{k=0}^{N-1}{}' \mathcal{A}_k(g^{\delta}_0, a_1, a_2)\cdot \chi_k(a_1,a_2,\kappa,c_2). \end{aligned} \tag{19}$$

For the T-year life-contingent option, we have:

$$V_{x,T} \approx S(0)\sum_{k=0}^{N-1}{}' \mathcal{A}_k(g^{\delta}_{1,T}, a_1, a_2)\cdot \chi_k(a_1,a_2,\kappa,c_2) - K\sum_{k=0}^{N-1}{}' \mathcal{A}_k(g^{\delta}_{0,T}, a_1, a_2)\cdot \chi_k(a_1,a_2,\kappa,c_2). \tag{20}$$

Case 3. $b(s) = s^n \mathbf{I}_{(s<K)}$ with $n \geq 0$ and $K > 0$.

By Equation (5) and the 1D COS method, we have for small number c_1,

$$\begin{aligned} V_x &= \int_{\infty}^{\kappa} [S(0)]^n g^{\delta}_n(y)dy \approx [S(0)]^n \int_{c_1}^{\kappa} g^{\delta}_n(y)dy \\ &\approx [S(0)]^n \sum_{k=0}^{N-1}{}' \mathcal{A}_k(g^{\delta}_n, a_1, a_2)\cdot \chi_k(a_1,a_2,c_1,\kappa). \end{aligned} \tag{21}$$

Similarly, for $V_{x,T}$, we have:

$$V_{x,T} \approx [S(0)]^n \sum_{k=0}^{N-1}{}' \mathcal{A}_k(g^{\delta}_{n,T}, a_1, a_2)\cdot \chi_k(a_1,a_2,c_1,\kappa). \tag{22}$$

Remark 3. *For the put option, the payoff function is given by:*

$$b(s) = (K-s)_+ = K\mathbf{I}_{(s<K)} - s\mathbf{I}_{(s<K)},$$

which together with Equations (21) and (22) gives:

$$V_x \approx K\sum_{k=0}^{N-1}{}' \mathcal{A}_k(g^{\delta}_0, a_1, a_2)\cdot \chi_k(a_1,a_2,c_1,\kappa) - S(0)\sum_{k=0}^{N-1}{}' \mathcal{A}_k(g^{\delta}_1, a_1, a_2)\cdot \chi_k(a_1,a_2,c_1,\kappa) \tag{23}$$

and:

$$V_{x,T} \approx K\sum_{k=0}^{N-1}{}' \mathcal{A}(g^{\delta}_{0,T}, a_1, a_2)\cdot \chi_k(a_1,a_2,c_1,\kappa) - S(0)\sum_{k=0}^{N-1}{}' \mathcal{A}(g^{\delta}_{1,T}, a_1, a_2)\cdot \chi_k(a_1,a_2,c_1,\kappa). \tag{24}$$

3. 2D COS Approximation

In this section, we use the 2D COS method to compute $\overline{V}_x$ and $\overline{V}_{x,T}$. For a bivariate integrable function f, we denote its Fourier transform by:

$$\mathcal{F}f(s_1,s_2) = \int\int e^{is_1 y + is_2 z} f(y,z)dydz, \qquad s_1, s_2 \in \mathbb{R}.$$

It follows from Ruijter and Oosterlee [23] that f can be approximated on a truncated domain $[a_1, a_2] \times [b_1, b_2]$ by truncated Fourier cosine series expansions with $N_1 \times N_2$ terms,

$$f(y,z) \approx \sum_{k_1=0}^{N_1-1}{}' \sum_{k_2=0}^{N_2-1}{}' \mathcal{B}_{k_1,k_2}(f) \cos\left(k_1\pi \frac{y-a_1}{a_2-a_1}\right) \cos\left(k_2\pi \frac{z-b_1}{b_2-b_1}\right), \tag{25}$$

where the cosine coefficients are given by:

$$\mathcal{B}_{k_1,k_2}(f) = \frac{1}{2}\left(\mathcal{B}^{+}_{k_1,k_2}(f) + \mathcal{B}^{-}_{k_1,k_2}(f)\right)$$

with:

$$\mathcal{B}^{\pm}_{k_1,k_2}(f) = \frac{2}{a_2-a_1}\frac{2}{b_2-b_1}\mathrm{Re}\left\{\mathcal{F}f\left(\frac{k_1\pi}{a_2-a_1}, \pm\frac{k_2\pi}{b_2-b_1}\right)\exp\left(-ik_1\pi\frac{a_1}{a_2-a_1} \mp ik_2\pi\frac{b_1}{b_2-b_1}\right)\right\}. \tag{26}$$

For $t > 0$, we denote the joint probability density function of $(X_1(t), X_2(t))$ by $f_{X_1(t),X_2(t)}(y,z)$, and for $\delta \geq 0$, define the following discounted density functions:

$$\begin{aligned} f^{\delta}_{X_1(T_x),X_2(T_x)}(y,z) &= \int_0^{\infty} e^{-\delta t} f_{X_1(t),X_2(t)}(y,z) f_{T_x}(t)dt, \\ f^{\delta}_{X_1(T_x),X_2(T_x),T}(y,z) &= \int_0^{T} e^{-\delta t} f_{X_1(t),X_2(t)}(y,z) f_{T_x}(t)dt. \end{aligned}$$

For $\overline{V}_x$, by changing the order of integrals, we have:

$$\begin{aligned} \overline{V}_x &= \int_0^{\infty} E[e^{-\delta t} b(S_1(t), S_2(t))] f_{T_x}(t)dt \\ &= \int_0^{\infty} e^{-\delta t} \int_{-\infty}^{\infty}\int_{-\infty}^{\infty} b(S_1(0)e^y, S_2(0)e^z) f_{X_1(t),X_2(t)}(y,z)dydz f_{T_x}(t)dt \\ &= \int_{-\infty}^{\infty}\int_{-\infty}^{\infty} b(S_1(0)e^y, S_2(0)e^z) f^{\delta}_{X_1(T_x),X_2(T_x)}(y,z)dydz. \end{aligned} \tag{27}$$

For $\overline{V}_{x,T}$, we have:

$$\overline{V}_{x,T} = \int_{-\infty}^{\infty}\int_{-\infty}^{\infty} b(S_1(0)e^y, S_2(0)e^z) f^{\delta}_{X_1(T_x),X_2(T_x),T}(y,z)dydz. \tag{28}$$

As in Section 2, we shall pay attention to the following auxiliary functions,

$$g^{\delta}_{m,n}(y,z) = e^{my+nz} f^{\delta}_{X_1(T_x),X_2(T_x)}(y,z), \qquad g^{\delta}_{m,n,T}(y,z) = e^{my+nz} f^{\delta}_{X_1(T_x),X_2(T_x),T}(y,z).$$

Suppose that both $g^{\delta}_{m,n}$ and $g^{\delta}_{m,n,T}$ are absolutely integrable. Then, by Equation (25), we obtain:

$$g^{\delta}_{m,n}(y,z) \approx \sum_{k_1=0}^{N_1-1}{}' \sum_{k_2=0}^{N_2-1}{}' \mathcal{B}_{k_1,k_2}(g^{\delta}_{m,n}) \cos\left(k_1\pi \frac{y-a_1}{a_2-a_1}\right) \cos\left(k_2\pi \frac{z-b_1}{b_2-b_1}\right) \tag{29}$$

and:

$$g^{\delta}_{m,n,T}(y,z) \approx \sum_{k_1=0}^{N_1-1}{}' \sum_{k_2=0}^{N_2-1}{}' \mathcal{B}_{k_1,k_2}(g^{\delta}_{m,n,T}) \cos\left(k_1\pi \frac{y-a_1}{a_2-a_1}\right) \cos\left(k_2\pi \frac{z-b_1}{b_2-b_1}\right). \tag{30}$$

To calculate cosine coefficients in Equations (29) and (30), we suppose $(X_1(t), X_2(t))$ is a two-dimensional Lévy process with $X_1(0) = X_2(0) = 0$. The characteristic exponent is defined by:

$$\Psi_{X_1,X_2}(s_1, s_2) = \ln(E[e^{is_1X_1(1)+is_2X_2(1)}]).$$

Furthermore, suppose that f_{T_x} is a combination of the exponential density function given by Equation (12). By Fubini theorem, we have:

$$\begin{aligned} \mathcal{F}g^{\delta}_{m,n}(s_1,s_2) &= \int\int e^{is_1y+is_2z}e^{my+nz}f^{\delta}_{X_1(T_x),X_2(T_x)}(y,z)dydz \\ &= \int\int e^{(is_1+m)y+(is_2+n)z}\int_0^{\infty}e^{-\delta t}f_{X_1(t),X_2(t)}(y,z)f_{T_x}(t)dtdydz \\ &= \int_0^{\infty}e^{-(\delta-\Psi_{X_1,X_2}(s_1-mi,s_2-ni))t}f_{T_x}(t)dt \\ &= \sum_{j=1}^{m}\frac{A_j\alpha_j}{\delta+\alpha_j-\Psi_{X_1,X_2}(s_1-mi,s_2-ni)} \end{aligned} \tag{31}$$

and:

$$\begin{aligned} \mathcal{F}g^{\delta}_{m,n,T}(s_1,s_2) &= \int_0^{T}e^{-(\delta-\Psi_{X_1,X_2}(s_1-mi,s_2-ni))t}f_{T_x}(t)dt \\ &= \sum_{j=1}^{m}A_j\alpha_j\frac{1-e^{-(\delta+\alpha_j-\Psi_{X_1,X_2}(s_1-mi,s_2-ni))T}}{\delta+\alpha_j-\Psi_{X_1,X_2}(s_1-mi,s_2-ni)}. \end{aligned} \tag{32}$$

In the sequel, we study the numerical approximation of the value of life-contingent two-asset options. We shall consider the Margrabe option, maximum/minimum option, and geometric option. First, the contingent Margrabe option, also called the exchange option, is considered. Note that its payoff function is:

$$[S_1(T_x)-S_2(T_x)]_+.$$

Then, we have:

$$\begin{aligned} &E[e^{-\delta T_x}[S_1(T_x)-S_2(T_x)]_+\ |S_1(0)<S_2(0)] \\ &= \int_0^{\infty}\int_{-\infty}^{\infty}\int_{-\infty}^{\infty}e^{-\delta t}[S_1(0)e^y-S_2(0)e^z]_+\cdot f_{X_1(t),X_2(t)}(y,z)f_{T_x}(t)dydzdt \\ &= \int_{-\infty}^{\infty}\int_{-\infty}^{\infty}[S_1(0)e^y-S_2(0)e^z]_+\cdot f^{\delta}_{X_1(T_x),X_2(T_x)}(y,z)dydz \\ &= \int\int_{(y,z)\in\mathcal{D}_1}(S_1(0)e^y-S_2(0)e^z)f^{\delta}_{X_1(T_x),X_2(T_x)}(y,z)dydz \\ &= S_1(0)\int\int_{(y,z)\in\mathcal{D}_1}e^yf^{\delta}_{X_1(T_x),X_2(T_x)}(y,z)dydz-S_2(0)\int\int_{(y,z)\in\mathcal{D}_1}e^zf^{\delta}_{X_1(T_x),X_2(T_x)}(y,z)dydz \\ &= S_1(0)\int\int_{(y,z)\in\mathcal{D}_1}g^{\delta}_{1,0}(y,z)dydz-S_2(0)\int\int_{(y,z)\in\mathcal{D}_1}g^{\delta}_{0,1}(y,z)dydz, \end{aligned} \tag{33}$$

where $\mathcal{D}_1$ denotes the region $\{(y,z):y-z>\ln\frac{S_2(0)}{S_1(0)}\}$. Set $\mathcal{D}_1'=\mathcal{D}_1\cap([a_1,a_2]\times[b_1,b_2])$. Utilizing Equation (29), we obtain the following approximation formula:

$$\begin{aligned} &E[e^{-\delta T_x}[S_1(T_x)-S_2(T_x)]_+\ |S_1(0)<S_2(0)] \\ &\approx S_1(0)\sum_{k_1=0}^{N_1-1}{}'\sum_{k_2=0}^{N_2-1}{}'\mathcal{B}_{k_1,k_2}(g^{\delta}_{1,0})\int\int_{(y,z)\in\mathcal{D}_1'}\cos\left(k_1\pi\frac{y-a_1}{a_2-a_1}\right)\cos\left(k_2\pi\frac{z-b_1}{b_2-b_1}\right)dydz \\ &\quad -S_2(0)\sum_{k_1=0}^{N_1-1}{}'\sum_{k_2=0}^{N_2-1}{}'\mathcal{B}_{k_1,k_2}(g^{\delta}_{0,1})\int\int_{(y,z)\in\mathcal{D}_1'}\cos\left(k_1\pi\frac{y-a_1}{a_2-a_1}\right)\cos\left(k_2\pi\frac{z-b_1}{b_2-b_1}\right)dydz, \end{aligned}$$

where the double integrals in both summations can be analytically computed.

What follows next are the approximation procedures for maximum/minimum options. The payoff functions for maximum and minimum options are given by:

$$\max\{S_1(T_x),S_2(T_x)\},\quad \min\{S_1(T_x),S_2(T_x)\}.$$

With a trivial mathematical change, we can turn them into the following forms:

$$\begin{aligned} \max\{S_1(T_x),S_2(T_x)\} &= S_2(T_x)+[S_1(T_x)-S_2(T_x)]_+, \\ \min\{S_1(T_x),S_2(T_x)\} &= S_1(T_x)-[S_1(T_x)-S_2(T_x)]_+. \end{aligned}$$

Hence, the valuation equations can be obtained by taking discounted expectations on both sides of the above equations,

$$\begin{aligned} E[e^{-\delta T_x}\max\{S_1(T_x),S_2(T_x)\}] &= E[e^{-\delta T_x}S_2(T_x)] + E[e^{-\delta T_x}[S_1(T_x)-S_2(T_x)]_+], \\ E[e^{-\delta T_x}\min\{S_1(T_x),S_2(T_x)\}] &= E[e^{-\delta T_x}S_1(T_x)] - E[e^{-\delta T_x}[S_1(T_x)-S_2(T_x)]_+], \end{aligned}$$

which imply that we can take advantage of the deductions of exchange option and basic option taking the form $b(s) = s$ to compute the above expectations.

Finally, we pay attention to the geometric option. The payoff function with strike price K is given by:

$$\left[\sqrt{S_1(T_x)S_2(T_x)} - K\right]_+ .$$

When condition $\sqrt{S_1(0)S_2(0)} < K$ holds, we can develop the following approximation.

$$\begin{aligned} &E[e^{-\delta T_x}[\sqrt{S_1(T_x)S_2(T_x)} - K]_+ \mid \sqrt{S_1(0)S_2(0)} < K] \\ &= \int_{-\infty}^{+\infty}\int_{-\infty}^{+\infty}[\sqrt{S_1(T_x)S_2(T_x)} - K]_+ \cdot f^{\delta}_{X_1(T_x),X_2(T_x)}(y,z)dydz \\ &= \sqrt{S_1(0)S_2(0)}\int\int_{(y,z)\in\mathcal{D}_2} e^{\frac{1}{2}(y+z)} f^{\delta}_{X_1(T_x),X_2(T_x)}(y,z)dydz - K\int\int_{(y,z)\in\mathcal{D}_2} f^{\delta}_{X_1(T_x),X_2(T_x)}(y,z)dydz \\ &= \sqrt{S_1(0)S_2(0)}\int\int_{(y,z)\in\mathcal{D}_2} g^{\delta}_{\frac{1}{2},\frac{1}{2}}(y,z)dydz - K\int\int_{(y,z)\in\mathcal{D}_2} g^{\delta}_{0,0}(y,z)dydz, \end{aligned} \tag{34}$$

where $\mathcal{D}_2$ denotes the region $\{(y,z) : y+z > \ln\frac{K^2}{S_1(0)S_2(0)}\}$. Set $\mathcal{D}_2' = \mathcal{D}_2 \cap ([a_1,a_2]\times[b_1,b_2])$. Then, Equation (34) can be approximated by:

$$\begin{aligned} &E[e^{-\delta T_x}[\sqrt{S_1(T_x)S_2(T_x)} - K]_+ \mid \sqrt{S_1(0)S_2(0)} < K] \\ &\approx \sqrt{S_1(0)S_2(0)}\sum_{k_1=0}^{N_1-1}{}'\sum_{k_2=0}^{N_2-1}{}'\mathcal{B}_{k_1,k_2}(g^{\delta}_{\frac{1}{2},\frac{1}{2}})\int\int_{(y,z)\in\mathcal{D}_2'}\cos\left(k_1\pi\frac{y-a_1}{a_2-a_1}\right)\cos\left(k_2\pi\frac{z-b_1}{b_2-b_1}\right)dydz \\ &- K\sum_{k_1=0}^{N_1-1}{}'\sum_{k_2=0}^{N_2-1}{}'\mathcal{B}_{k_1,k_2}(g^{\delta}_{0,0})\int\int_{(y,z)\in\mathcal{D}_2'}\cos\left(k_1\pi\frac{y-a_1}{a_2-a_1}\right)\cos\left(k_2\pi\frac{z-b_1}{b_2-b_1}\right)dydz, \end{aligned}$$

where analytical expressions of the double integrals in both summations exist.

Remark 4. *When we study $\overline{V}_{x,T}$, the approximation Equations are similar to those of $\overline{V}_x$, and the only modification is replacing $\mathcal{F}g^{\delta}_{m,n}$ by $\mathcal{F}g^{\delta}_{m,n,T}$.*

4. Numerical Illustrations

In this section, we present some numerical examples to show the performance of the proposed approach. For the linear combination of exponential distributions, which is used to approximate the density function of T_x, we use the case considered in Gerber et al. [3], which is given by:

$$f_{T_x}(t) = 3\times 0.08e^{-0.08t} - 2\times 0.12e^{-0.12t}, \qquad t > 0. \tag{35}$$

In the following subsections, we shall show that the COS method is very efficient for valuing GMDB.

4.1. The 1D COS Results

In this subsection, we use some Lévy processes to model the log asset process $X(t)$ (see Remark 1). Note that the probability law of the Lévy process $X(t)$ is determined by its characteristic exponent $\Psi_X(s)$. In this section, all computations were performed in MATLAB 2016b with Intel Core i7 at 3.4 GHz and RAM of 8 GB. We shall consider the following five models.

- Black-Scholes model (BS): $\Psi_X(s) = i\mu s - \frac{\sigma^2}{2}s^2$;
- Kou's jump diffusion model (Kou): $\Psi_X(s) = i\mu s - \frac{\sigma^2}{2}s^2 + \lambda_K \left(\frac{(1-p)\eta_2}{\eta_2 + is} + \frac{p\eta_1}{\eta_1 - is} - 1 \right)$;
- Merton's jump diffusion model (MJD): $\Psi_X(s) = i\mu s - \frac{\sigma^2}{2}s^2 + \lambda_J \left(e^{is\mu_J - \frac{\sigma_J^2}{2}s^2} - 1 \right)$;
- Variance Gammamodel (VG): $\Psi_X(s) = i\mu s - \frac{\sigma^2}{2}s^2 - \frac{1}{\nu}\ln\left(1 - i\nu\mu_\nu s + \nu\frac{\sigma_\nu^2}{2}s^2\right)$;
- Normal inverse Gaussian model (NIG): $\Psi_X(s) = i\mu s - \delta_{NIG}\left(\sqrt{\alpha^2 - (\beta + is)^2} - \sqrt{\alpha^2 - \beta^2}\right)$.

In the sequel, let $\delta = 0.05$ and μ be chosen to satisfy the risk-neutral requirement. Here, we need to assume that T_x is still independent with the stock price process even under the risk-neutral measure. Thus, μ is set such that $\Psi_X(-i) = \delta$. The values of other parameters are listed in Table 1. Furthermore, set $a_1 = -100$, $a_2 = 100$. For the death benefit function b, we consider the following two cases:

$$\text{call option}: \ b(s) = (s-K)_+; \quad \text{put option}: \ b(s) = (K-s)_+.$$

Table 1. Parameter setting for the log-asset process X.

Model	Abbreviation	Parameter Sets
Black-Scholes model	BS	$\sigma = 0.25,\ S(0) = 100;$
Kou's jump diffusion model	Kou	$\sigma = 0.25,\ S(0) = 100,\ \lambda_K = 0.6,\ \eta_1 = 4,\ \eta_2 = 1,\ p = 0.5;$
Merton's jump diffusion model	Merton	$\sigma = 0.25,\ S(0) = 100,\ \lambda_J = 0.6,\ \mu_J = 0.01,\ \sigma_J = 0.13;$
Variance Gamma model	VG	$\sigma = 0.25,\ S(0) = 100,\ \nu = 2,\ \mu_\nu = 0.01,\ \sigma_\nu = 0.05;$
Normal inverse Gaussian model	NIG	$S(0) = 100,\ \alpha = 2,\ \beta = 0.5,\ \delta_{NIG} = 0.05.$

First, in Tables 2 and 3, we display some results when X is the Black-Scholes model and Kou's jump diffusion model where the strike price $K = 80, 90, 110, 120$, respectively. Note that if X is Black-Scholes or Kou's jump diffusion, reference values can be obtained from Gerber et al. [3,4]. In Tables 2 and 3, we calculate the relative errors and average running time to show the performance of our method, where the average running time (in seconds) is reported based on 1000 operations. From a horizontal perspective of Tables 2 and 3, our approximation results performed better as the expansion term N increased. When $N = 2^{12}$, relative errors were around 10^{-8}. Furthermore, we present Monte Carlo simulation results with sample size 10^7. We found that the COS method can result in smaller relative errors for each case, and this method requires less time than MC. When X is Merton's jump diffusion, VG, and NIG, true reference values are not available. In these cases, we present MC simulation results with sample size 10^7. From the simulation results given in Table 4, we see that the differences between approximation results and simulation results were small and our method required less time than MC.

Next, we consider valuation with a finite expiry date. In Tables 5 and 6, we assume $T = 20$ and display some GMDB valuation results. In finite-time cases, reference values are available only when X is BS. For Kou's jump diffusion, we display MC results with sample size 10^7. With the call payoff function, it is observed that as K increases, the results decrease, which is opposite those with put payoff. Compared with MC, again, we found that the COS method required less time. Now, we further display the valuation results in Table 7 under different expiry dates in the BS model. As T grew from five to infinity, which denotes a whole life insurance type, valuation results increased as expected. If the expiry date is too far from the present, the probability that the insured dies becomes larger; hence, the insurance company is likely to make a payment. Besides, the dynamics of the asset is driven by a Brownian motion, with a positive drift; hence, the account accumulates from present to future. From a vertical perspective, it is also observed that relative errors decrease obviously as N increases.

Finally, to illustrate the accuracy of the proposed method, denary logarithms of relative errors are displayed in Figures 1 and 2 with different expansion terms and strike prices. For a fixed K, the curve tended to descend as expansion term N increased.

Table 2. Approximation results for guaranteed minimum death benefit (GMDB), where X is BS, $b(s) = (K - s)_+$.

K	Reference Value	BS						MC		
		$N = 2^4$		$N = 2^8$		$N = 2^{12}$				
		Relative Errors	Time	Relative Errors	Time	Relative Errors	Time	Relative Errors	Time	Std
80	3.6161	6.37	0.1009	3.10×10^{-3}	0.1010	1.41×10^{-8}	0.0977	1.02×10^{-3}	1215.7831	7.5637
90	4.9871	4.59	0.0986	1.24×10^{-2}	0.1002	4.54×10^{-8}	0.1021	8.96×10^{-4}	1215.2156	9.4440
110	8.4402	2.72	0.0970	1.54×10^{-2}	0.1041	3.13×10^{-8}	0.0987	7.48×10^{-4}	1215.4682	13.5673
120	10.4920	2.21	0.0973	1.37×10^{-2}	0.0999	1.66×10^{-8}	0.1018	7.31×10^{-4}	1215.6826	15.7678

Table 3. Approximation results for GMDB, where X is Kou, $b(s) = (K - s)_+$.

K	Reference Value	Kou						MC		
		$N = 2^4$		$N = 2^8$		$N = 2^{12}$				
		Relative Errors	Time	Relative Errors	Time	Relative Errors	Time	Relative Errors	Time	Std
80	18.0238	1.06	0.1106	3.34×10^{-4}	0.1217	3.45×10^{-9}	0.1184	4.81×10^{-4}	1043.5026	16.7433
90	20.9370	9.43×10^{-1}	0.1208	4.60×10^{-4}	0.1173	1.04×10^{-8}	0.1166	5.18×10^{-4}	1043.4856	18.9326
110	27.0526	7.74×10^{-1}	0.1157	1.31×10^{-3}	0.1208	9.50×10^{-9}	0.1139	5.89×10^{-4}	1043.8940	23.2865
120	30.2424	7.13×10^{-1}	0.1222	1.48×10^{-3}	0.1151	5.28×10^{-9}	0.1205	6.17×10^{-4}	1043.7561	25.4520

Table 4. Approximation results for GMDB, where X is Merton's jump diffusion model (MJD), the Variance Gamma model (VG), or the normal inverse Gaussian model (NIG), $b(s) = (K - s)_+$.

K	Merton (Time)		MC (Time)	VG (Time)		MC (Time)	NIG (Time)		MC (Time)
	$N = 2^8$	$N = 2^{12}$		$N = 2^8$	$N = 2^{12}$		$N = 2^8$	$N = 2^{12}$	
80	4.4349 (0.0877)	4.4514 (0.0863)	4.4508 (1027.4452)	3.8263 (0.0870)	3.8395 (0.0886)	3.8412 (1237.0955)	6.1072 (0.0891)	6.1399 (0.0879)	6.1435 (1695.8358)
90	5.9255 (0.0866)	5.9823 (0.0884)	5.9822 (1027.4338)	5.1947 (0.0853)	5.2556 (0.0890)	5.2574 (1237.0935)	7.9234 (0.0870)	7.9881 (0.0901)	7.9932 (1695.8093)
110	9.6156 (0.0854)	9.7228 (0.0861)	9.7250 (1027.4386)	8.6668 (0.0852)	8.7901 (0.0856)	8.7909 (1237.0956)	12.2390 (0.0876)	12.3349 (0.0879)	12.3326 (1695.8200)
120	11.7834 (0.0879)	11.8986 (0.0872)	11.9015 (1027.4401)	10.7420 (0.0863)	10.8770 (0.0885)	10.8771 (1237.1015)	14.6968 (0.0881)	14.7924 (0.0894)	14.8006 (1695.8699)

Table 5. Approximation results for finite GMDB, where X is BS or Kou, $b(s) = (s - K)_+$, $T = 20$.

K	BS (Relative Errors)			Kou (Time)			MC(Time)
	$N = 2^4$	$N = 2^8$	$N = 2^{12}$	$N = 2^4$	$N = 2^8$	$N = 2^{12}$	
80	19.1403 (0.414)	32.6564 (3.43×10^{-4})	32.6676 (1.56×10^{-9})	27.4259 (0.1194)	42.7130 (0.1173)	42.7070 (0.1173)	42.7070 (1090.9950)
90	17.0844 (0.436)	30.2620 (2.04×10^{-3})	30.3241 (7.46×10^{-9})	25.7974 (0.1194)	41.4205 (0.1175)	41.4301 (0.1183)	41.4293 (1091.1883)
110	13.2010 (0.497)	26.1378 (4.95×10^{-3})	26.2680 (1.00×10^{-8})	22.7491 (0.1183)	39.1093 (0.1174)	39.1448 (0.1197)	39.1418 (1091.1751)
120	11.3513 (0.537)	24.3848 (5.86×10^{-3})	24.5286 (7.08×10^{-9})	21.3090 (0.1176)	38.0807 (0.1172)	38.1253 (0.1184)	38.1210 (1091.1742)

Table 6. Approximation results for finite GMDB, where X is Merton, VG, or NIG, $b(s) = (s-K)_+$ and $T = 20$.

K	Merton (Time)		MC (Time)	VG (Time)		MC (Time)	NIG (Time)		MC (Time)
	$N = 2^8$	$N = 2^{12}$		$N = 2^8$	$N = 2^{12}$		$N = 2^8$	$N = 2^{12}$	
80	33.2207 (0.1128)	33.2371 (0.1112)	33.2582 (1033.1533)	32.8072 (0.1109)	32.8204 (0.1105)	32.8216 (1242.9794)	34.3088 (0.1145)	34.3415 (0.1140)	34.3638 (1701.5796)
90	30.9514 (0.1121)	31.0082 (0.1111)	31.0310 (1033.2315)	30.4485 (0.1126)	30.5094 (0.1104)	30.5123 (1242.8595)	32.2714 (0.1167)	32.3360 (0.1140)	32.3611 (1701.3483)
110	27.0436 (0.1103)	27.1508 (0.1133)	27.1778 (1033.0862)	26.3865 (0.1090)	26.5099 (0.1139)	26.5150 (1242.7859)	28.8046 (0.1141)	28.9006 (0.1160)	28.9307 (1701.5626)
120	25.3774 (0.1104)	25.4925 (0.1122)	25.5214 (1033.0618)	24.6586 (0.1095)	24.7936 (0.1129)	24.7996 (1242.9656)	27.3386 (0.1139)	27.4342 (0.1182)	27.4663 (1701.6632)

Table 7. Approximation results for finite GMDB, where X is BS, $b(s) = (s-K)_+$ and $K = 120$.

T	5	10	30	60	∞
$N = 2^8$	1.2988 (8.60×10^{-2})	7.0082 (2.01×10^{-2})	39.2337 (3.65×10^{-3})	55.9713 (2.56×10^{-3})	58.2215 (2.46×10^{-3})
$N = 2^{12}$	1.4211 (1.22×10^{-7})	7.1521 (2.42×10^{-8})	39.3774 (4.41×10^{-9})	56.1150 (3.09×10^{-9})	58.3653 (2.97×10^{-9})

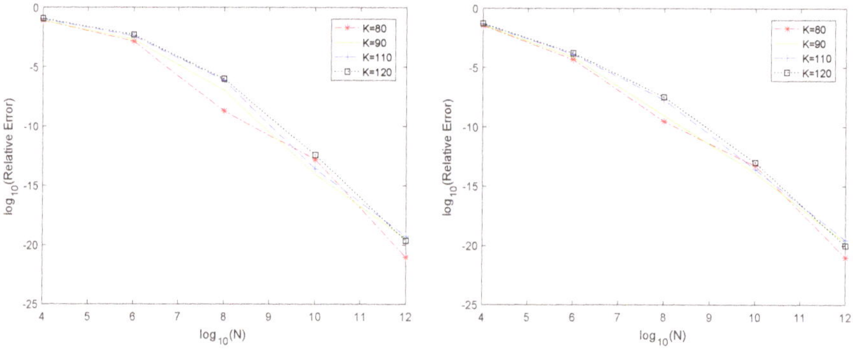

Figure 1. Relative errors with payoff $b(s) = (s-K)_+$: (**a**) BS model; (**b**) Kou model.

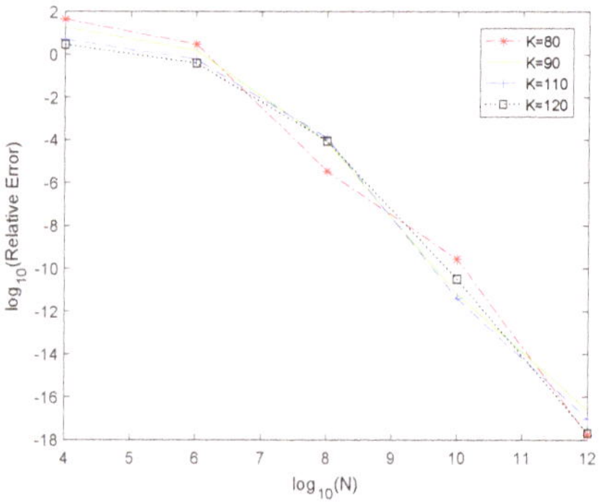

Figure 2. Relative errors for BS model with payoff $b(s) = (K-s)_+$ and finite expiry time $T = 20$.

4.2. The 2D COS Results

In this subsection, the valuation results involving two assets in a GMDB contract are presented. For illustrative purpose, we assumed that the density function of T_x is represented by the combination of exponential distribution Equation (35). Besides, we considered a bivariate normal distribution. In addition, suppose that $(S_1(t), S_2(t))$ is a bivariate geometric Brownian motion, where the log-asset price at time t is bivariate normally distributed:

$$(X_1(t), X_2(t)) \sim \mathcal{N}(\boldsymbol{\mu}, \boldsymbol{\Sigma} t),$$

where:

$$\boldsymbol{\mu} = [0.02, -0.005], \; \boldsymbol{\Sigma} = [0.04, 0.015; 0.015, 0.09].$$

Set $(S_1(0), S_2(0)) = (90, 110)$, and the strike price $K = 100$. We considered the cosine expansion over a symmetric interval $[-100, 100] \times [-100, 100]$.

In Gerber et al. [3], the valuation equations for GMDB with the exchange option type were explicitly obtained by the Esscher transform, from which we can compute reference values and relative errors. Motivated by the idea in Gerber et al. [3], we further considered some other option types. Approximation results are given in Table 8, and the relative errors confirmed the accuracy of the proposed procedure. It was observed that as N increased, the relative errors decreased.

Table 8. Approximation results for 2D GMDB.

Option Type	$N = 2^6$	$N = 2^8$	$N = 2^{10}$	$N = 2^{12}$
Exchange Option	140.8768	153.5866	153.6412	153.6411
	(8.31×10^{-2})	(3.55×10^{-4})	(6.38×10^{-7})	(6.74×10^{-10})
Geometric Option	111.4024	114.0135	114.0281	114.0281
	(2.30×10^{-2})	(1.29×10^{-4})	(1.54×10^{-7})	(1.27×10^{-10})
Maximum Option	470.8768	483.5866	483.6412	483.6411
	(2.64×10^{-2})	(1.12×10^{-4})	(2.03×10^{-7})	(2.14×10^{-10})
Minimum Option	129.1232	116.4134	116.3588	116.3589
	(1.10×10^{-1})	(4.68×10^{-4})	(8.43×10^{-7})	(8.90×10^{-10})

5. Conclusions

In this paper, we used the COS method to value GMDB products under a general Lévy framework. When the death benefit payment depended on only one stock fund, we used the 1D COS method to value the products; when it depended on two stock funds, the 2D COS method was used for valuation. Various numerical results illustrated the accuracy and efficiency of the proposed method.

Our COS method can only be used to value GMDB contracts that depend on the terminal value of the stock price; however, we note that some products are also dependent on the running maximum or minimum of the stock price; for example, life-contingent lookback options and barrier options. Hence, we have to develop the COS method or search for another numerical method to solve this problem. We leave this for future research.

Author Contributions: Data curation, G.G., W.S., and C.C.; formal analysis, W.Y., Y.Y., and Y.H.; methodology, Y.Y.; writing, original draft, W.Y., Y.H., and Y.Y.

Funding: This research received no external funding.

Acknowledgments: The authors would like to thank the four anonymous referees for their helpful comments and suggestions, which improved an earlier version of the paper. This research is partially supported by the National Natural Science Foundation of China (Grant No. 11301303), the National Social Science Foundation of China (Grant No. 15BJY007), the Taishan Scholars Program of Shandong Province (Grant No. tsqn20161041), the Humanities and Social Sciences Project of the Ministry Education of China (Grant Nos. 19YJA910002 and 16YJC630070), the Shandong Provincial Natural Science Foundation (Grant No. ZR2018MG002), the Fostering Project of Dominant Discipline and Talent Team of Shandong Province Higher Education Institutions (Grant No.

1716009), the Shandong Jiaotong University "Climbing" Research Innovation Team Program, the Risk Management and Insurance Research Team of Shandong University of Finance and Economics, the 1251 Talent Cultivation Project of Shandong Jiaotong University, the Collaborative Innovation Center Project of the Transformation of New and Old Kinetic Energy and Government Financial Allocation, and the Outstanding Talents of Shandong University of Finance and Economics.

Conflicts of Interest: The authors declare that they have no conflicts of interest.

References

1. Milevsky, M.A.; Posner, S.E. The titanic option: Valuation of the guaranteed minimum death benefit in variable annuities and mutual funds. *J. Risk Insur.* **2001**, *68*, 93–128. [CrossRef]
2. Bauer, D.; Kling A.; Russ, J. A universal pricing framework for guaranteed minimum benefits in variable annuities. *Astin Bull.* **2008**, *38*, 621–651. [CrossRef]
3. Gerber, H.U.; Shiu, E.S.W.; Yang, H.L. Valuing equity-linked death benefits and other contingent options: A discounted density approach. *Insur. Math. Econ.* **2012**, *51*, 73–92. [CrossRef]
4. Gerber, H.U.; Shiu, E.S.W.; Yang, H.L. Valuing equity-linked death benefits in jump diffusion models. *Insur. Math. Econ.* **2013**, *53*, 615–623. [CrossRef]
5. Siu, C.C.; Yam, S.C.P.; Yang, H.L. Valuing equity-linked death benefits in a regime-switching framework. *Astin Bull.* **2015**, *45*, 355–395. [CrossRef]
6. Dufresne, D. Stochastic life annuities. *N. Am. Actuar. J.* **2007**, *11*, 136–157. [CrossRef]
7. Ko, B.; Ng, A.C.Y. Discussion on "Stochastic Annuities" by Daniel Dufresne, January, 2007. *N. Am. Actuar. J.* **2007**, *11*, 170–171. [CrossRef]
8. Zhang, Z.M.; Yong, Y.D. Valuing guaranteed equity-linked contracts by Laguerre series expansion. *J. Comput. Appl. Math.* **2019**, *357*, 329–348. [CrossRef]
9. Zhang, Z.M.; Yong, Y.D.; Yu, W.G. Valuing equity-linked death benefits in general exponential Lévy models. *J. Comput. Appl. Math.* **2020**. [CrossRef]
10. Dai, M.; Kwok, Y.K.; Zong, J.P. Guaranteed minimum withdrawal benefit in variable annuities. *Math Financ.* **2008**, *18*, 493–667. [CrossRef]
11. Bélanger, A.C.; Forsyth, P.A.; Labahn, G. Valuing the guaranteed minimum death benefit clause with partial withdrawals. *Appl. Math. Financ.* **2009**, *16*, 451–496. [CrossRef]
12. Kang, B.; Ziveyi, J. Optimal surrender of guaranteed minimum maturity benefits under stochastic volatility and interest rates. *Insur. Math. Econ.* **2018**, *79*, 43–56. [CrossRef]
13. Asmussen, S.; Laub, P.J. ; Yang, H.L. Phase-type models in life insurance: Fitting and valuation of equity-linked benefits. *Risks* **2019**, *7*, 17. [CrossRef]
14. Zhou, J.; Wu, L. Valuing equity-linked death benefits with a threshold expense strategy. *Insur. Math. Econ.* **2015**, *62*, 79–90. [CrossRef]
15. Gerber, H.U.; Shiu, E.S.W.; Yang, H.L. Geometric stopping of a random walk and its applications to valuing equity-linked death benefits. *Insur. Math. Econ.* **2015**, *64*, 313–325. [CrossRef]
16. Ulm, E.R. Analytic solution for return of premium and rollup guaranteed minimum death benefit options under some simple mortality laws. *Astin Bull.* **2008**, *38*, 543–563. [CrossRef]
17. Ulm, E.R. Analytic solution for ratchet guaranteed minimum death benefit options under a variety of mortality laws. *Insur. Math. Econ.* **2014**, *58*, 14–23. [CrossRef]
18. Liang, X.Q.; Tsai, C.C.L.; Lu, Y. Valuing guaranteed equity-linked contracts under piecewise constant forces of mortality. *Insur. Math. Econ.* **2016**, *70*, 150–161. [CrossRef]
19. Fan, K.; Shen, Y.; Siu, T.K.; Wang, R.M. Pricing annuity guarantees under a double regime-switching model. *Insur. Math. Econ.* **2015**, *62*, 62–78. [CrossRef]
20. Ignatieva, K.; Song, A.; Ziveyi, J. Pricing and hedging of guaranteed minimum benefits under regime-switching and stochastic mortality. *Insur. Math. Econ.* **2016**, *70*, 286–300. [CrossRef]
21. Hieber, P. Cliquet-style return guarantees in a regime switching Lévy model. *Insur. Math. Econ.* **2017**, *72*, 138–147. [CrossRef]
22. Fang, F.H.; Oosterlee, C.W. A novel pricing method for European options based on Fourier-Cosine series expansions. *SIAM J. Sci. Comput.* **2008**, *31*, 826–848. [CrossRef]
23. Ruijter, M.J.; Oosterlee, C.W. Two-dimensional Fourier Cosine series expansion method for pricing financial options. *SIAM J. Sci. Comput.* **2012**, *34*, B642–B671. [CrossRef]

24. Leitao, Á.; Oosterlee, C.W.; Ortiz-Gracia, L.; Bohte, S.M. On the data-driven COS method. *Appl. Math. Comput.* **2018**, *317*, 68–84. [CrossRef]
25. Chau, K.W.; Yam S.C.P.; Yang, H. Fourier-cosine method for Gerber-Shiu functions. *Insur. Math. Econ.* **2015**, *61*, 170–180. [CrossRef]
26. Chau, K.W.; Yam S.C.P.; Yang, H. Fourier-cosine method for ruin probabilities. *J. Comput. Appl. Math.* **2015**, *281*, 94–106. [CrossRef]
27. Zhang, Z.M. Approximation the density of the time to ruin via Fourier-Cosine series expansion. *Astin Bull.* **2017**, *47*, 169–198. [CrossRef]
28. Yang, Y.; Su, W.; Zhang, Z.M. Estimating the discounted density of the deficit at ruin by Fourier cosine series expansion. *Stat. Probabil. Lett.* **2019**, *146*, 147–155. [CrossRef]
29. Wang, Y.Y.; Yu, W.G.; Huang, Y.J.; Yu, X.L.; Fan, H.L. Estimating the expected discounted penalty function in a compound Poisson insurance risk model with mixed premium income. *Mathematics* **2019**, *7*, 305. [CrossRef]
30. Huang, Y.J.; Yu, W.G.; Pan, Y.; Cui, C.R. Estimating the Gerber-Shiu expected discounted penalty function for Lévy risk model. *Discrete Dyn. Nat. Soc.* **2019**, *2019*. [CrossRef]
31. Deng, G.; Dulaney, T.; McCann, C.J.; Yan, M. Efficient valuation of equity-indexed annuities under Lévy processes using Fourier-cosine series. *J. Comput. Financ.* **2017**, *21*, 1–27.
32. Alonso-García, J.; Wood, O.; Ziveyi, J. Pricing and hedging guaranteed minimum withdrawal benefits under a general Lévy ramework using the COS method. *Quant. Financ.* **2017**, *18*, 1049–1075. [CrossRef]
33. Zhang, Z.M.; Yang, H.L.; Yang, H. On a nonparametric estimator for ruin probability in the classical risk model. *Scand. Actuar. J.* **2014**, *2014*, 309–338. [CrossRef]
34. Chan, T.L.R. Efficient computation of European option prices and their sensitivities with the complex Fourier series method. *N. Am. J. Econ. Financ.* **2019**, doi:10.1016/j.najef.2019.100984. [CrossRef]
35. Zhang, Z.M.; Liu, C.L. Nonparametric estimation for derivatives of compound distribution. *J. Korean Stat. Soc.* **2015**, *44*, 327–341. [CrossRef]
36. Have, Z.V.D.; Oosterlee, C.W. The COS method for option valuation under the SABR dynamics. *Int. J. Comput. Math.* **2018**, *95*, 444–464. [CrossRef]
37. Shimizu, Y.; Zhang, Z.M. Estimating Gerber-Shiu functions from discretely observed Lévy driven surplus. *Insur. Math. Econ.* **2017**, *74*, 84–98. [CrossRef]
38. Tour, G.; Thakoor, N.; Khaliq, A.Q.M.; Tangman, D.Y. COS method for option pricing under a regime-switching model with time-changed Lévy processes. *Quant. Financ.* **2018**, *18*, 673–692. [CrossRef]
39. Zhang, Z.M. Estimating the Gerber-Shiu function by Fourier-Sinc series expansion. *Scand. Actuar. J.* **2017**, *2017*, 898–919. [CrossRef]
40. Wang, Y.Y.; Yu, W.G.; Huang, Y.J. Estimating the Gerber-Shiu function in a compound Poisson risk model with stochastic premium income. *Discrete Dyn. Nat. Soc.* **2019**, *2019*. [CrossRef]

mathematics

MDPI

Article

On Two Interacting Markovian Queueing Systems

Valeriy A. Naumov [1], Yuliya V. Gaidamaka [2,3,*] and Konstantin E. Samouylov [2,3]

1 Service Innovation Research Institute, 8 A Annankatu, Helsinki 00120, Finland
2 Department of Applied Informatics and Probability, Peoples' Friendship University of Russia (RUDN University), Miklukho-Maklaya St. 6, Moscow 117198, Russian
3 Institute of Informatics Problems, Federal Research Center "Computer Science and Control" of the Russian Academy of Sciences, Vavilov St. 44-2, Moscow 119333, Russian
* Correspondence: gaydamaka-yuv@rudn.ru

Received: 29 July 2019; Accepted: 21 August 2019; Published: 1 September 2019

Abstract: In this paper, we study a Markovian queuing system consisting of two subsystems of an arbitrary structure. Each subsystem generates a multi-class Markovian arrival process of customers arriving to the other subsystem. We derive the necessary and sufficient conditions for the stationary distribution to be of product form and consider some particular cases of the subsystem interaction for which these conditions can be easily verified.

Keywords: Markovian arrival process; multi-class arrival processes; product form

1. Introduction

The product form of the stationary distribution greatly simplifies the analysis of complex queueing systems. In such systems, the stationary distribution of the network can be written as the product of the distributions of its nodes with the arrival rates modified to reflect the routing of customers in the network. Jackson [1] showed that a specific class of open queueing networks has a product-form stationary distribution. Gordon and Newell [2] introduced product-form Markovian models for closed queueing networks. Kobayashi [3] and Towsley [4] developed forms of state-dependent routing in a queueing network, allowing a product-form solution. Baskett et al. [5] found product-form solutions for open, closed or mixed queueing networks with multiple classes of customers and various service disciplines and service time distributions. Surveys by Disney and König [6], Nelson [7] and Balsamo [8], as well as books by Kelly [9], Ross [10], Whittle [11] and Walrand [12] cover various aspects of product-form solutions for conventional queueing networks.

Gelenbe [13,14] proposed models of a queueing network with positive and negative customers: the G-networks. These models, which include G-networks with signals [15], resets [16], and multiple customer classes [17], radically extend the class of Markovian queueing systems with the product-form stationary distributions. The complexity of such systems continues to increase, with the introduction of new extensions of G-networks being an essential area of research [18,19].

In conventional queueing networks, each node in isolation can be represented by a birth and death process. Nodes in a Markov network [20] are Markovian queueing systems whose behavior can be represented by general discrete-state Markov processes. The details of the nodes' internal structure can be ignored. Each node of a Markov network can have three types of state changes: arrival, departure and internal transitions, which are distinguished only by the rates or probabilities at which they occur. A transition of the network as a whole involves changes at only one or two nodes. The former case corresponds to a network transition consisting of an internal change at one node. The latter consists of a departure transition at one node that triggers an arrival transition at another node determined by a routing probability.

Naumov [21] obtained the necessary and sufficient conditions for a product-form solution for a Markov network consisting of two nodes, the first of which generates a Markovian arrival process (MAP) of customers arriving to the second node. Chao et al. [20] obtained the necessary and sufficient conditions for a general Markov network to be of product form. Chao and Miyazawa [22] extended the notion of quasi-reversibility to Markov networks and applied it to the study of networks with triggered movements and positive and negative signals. In [23], Chao provided an overview of product-form Markov networks.

The procedure for establishing the existence of a product-form stationary distribution for Markov networks includes a solution of a system of nonlinear equations [20]. The objective of this paper is to simplify this procedure for some Markov networks so that it can be easily applied in practice. We consider two-node Markov networks and multi-class Markovian arrival processes (MMAP). In Section 2, we formulate the basic properties of MMAP. In Section 3, we derive a matrix formulation of the product-form conditions and develop a simple procedure to check whether a Markov network with two customer classes is of product form. Examples given in Section 4 illustrate the theory developed in the paper.

The following notation conventions are used throughout the article. Bold lowercase letters denote vectors and bold capitalized letters denote matrices. Inequality $\mathbf{x} \le \mathbf{y}$ represents $x_m \le y_m$ for all vectors $\mathbf{x}$ and $\mathbf{y}$; $\delta(i,j) = 1$ if $i = j$ and $\delta(i,j) = 0$ otherwise; $\mathbf{I} = [\delta(i,j)]$ denotes the identity matrix; $\mathbf{u}$ is the column vector of all ones; the i-th component of vector $\mathbf{e}_i$ is equal to one, and the others are equal to zero.

2. Multi-Class Markovian Arrival Process

Consider a multi-class arrival process $\mathcal{T} = ((\tau_n, \sigma_n), n = 1, 2, \ldots)$, where $0 \le \tau_1 \le \tau_2 \le \ldots$ are the arrival times and $\sigma_n \in \{1, 2, \ldots, k\}$ is the class of the n-th customer. Denote $N_v(t) = \sum\limits_{\tau_n \le t} \delta(v, \sigma_n)$ the number of class v customers arrived during time t and $\mathbf{N}(t) = (N_1(t), N_2(t), \ldots, N_k(t))$. The process $\mathcal{T}$ is a MMAP if for some random process $X(t)$ with a finite set of states $\mathcal{X}$ the process $(X(t), \mathbf{N}(t))$ is a homogeneous Markov process and the following conditions are satisfied:

$$\begin{aligned}
\mathbb{P}\{X(t+\varepsilon) = j, \sum_{v=1}^{k} N_v(t+\varepsilon) > \sum_{v=1}^{k} n_v + 1 | X(t) = i, \mathbf{N}(t) = \mathbf{n}\} &= o(\varepsilon), \\
\mathbb{P}\{X(t+\varepsilon) = j, \mathbf{N}(t+\varepsilon) = \mathbf{n} | X(t) = i, \mathbf{N}(t) = \mathbf{n}\} &= \delta(i,j) + A_0(i,j) + o(\varepsilon), \\
\mathbb{P}\{X(t+\varepsilon) = j, \mathbf{N}(t+\varepsilon) = \mathbf{n} + \mathbf{e}_v | X(t) = i, \mathbf{N}(t) = \mathbf{n}\} &= A_v(i,j) + o(\varepsilon),\ v = 1, 2, \ldots, k,
\end{aligned} \tag{1}$$

for all $i, j \in \mathcal{X}$, $t, \varepsilon > 0$, and for all nonnegative integer vectors $\mathbf{n}$ of length k [24]. In this case, the probability of more than one arrival in the interval of length ε is $o(\varepsilon)$ and the process $(X(t), \mathbf{N}(t))$ is a Markov process that is homogeneous in time and in the second component [25]. That is, for all $0 \le \mathbf{k} \le \mathbf{n}$, $i, j \in \mathcal{X}$ and $s, t \ge 0$ we have:

$$\mathbb{P}\{X(s+t) = j, \mathbf{N}(s+t) = \mathbf{n} | X(s) = i, \mathbf{N}(s) = \mathbf{k}\} = p_{i,j}(\mathbf{n} - \mathbf{k}, t).$$

The phase process $X(t)$ is a time-homogeneous Markov chain with a matrix of transition probabilities:

$$\mathbf{P}(t) = \sum_{\mathbf{n} \ge 0} \mathbf{P}(\mathbf{n}, t),$$

where $\mathbf{P}(\mathbf{n}, t) = \left[p_{i,j}(\mathbf{n}, t)\right]$ [25]. It follows from (1) that matrices $\mathbf{A}_v = [A_v(i,j)]$, $i, j \in \mathcal{X}$, $v = 0, 1, \ldots, k$, which characterize MMAP, have the following properties [26,27]:

1. Matrix $\mathbf{A}_0$ has non-negative off-diagonal elements.
2. Matrices $\mathbf{A}_v$, $v = 1, 2, \ldots, k$, are non-negative.

3. The matrix

$$\mathbf{A} = \sum_{v=0}^{k} \mathbf{A}_v \tag{2}$$

is the generator of the phase process.

Transition probability matrices $\mathbf{P}(\mathbf{n}, t)$ can be calculated by the recursion [27]

$$\mathbf{P}(0,t) = e^{\mathbf{A}_0 t},\ \mathbf{P}(\mathbf{n},t) = \sum_{\substack{v=1 \\ n_v>0}}^{k} \int_0^t \mathbf{P}(0,t-x)\mathbf{A}_v\mathbf{P}(\mathbf{n}-\mathbf{e}_v,x)dx,\ \mathbf{n} \neq 0.$$

If matrices $\mathbf{P}(\mathbf{n}, t)$ are known, the joint probability distribution of the number of arrivals in disjoint intervals can be calculated as:

$$\mathbb{P}\{\mathbf{N}(\sum_{j=0}^{r} t_j) - \mathbf{N}(\sum_{j=0}^{r-1} t_j) = \mathbf{k}_r, r = 1,2,\ldots,m\} = \mathbf{q}\mathbf{P}(t_0)\mathbf{P}(\mathbf{k}_1,t_1)\mathbf{P}(\mathbf{k}_2,t_2)\ldots\mathbf{P}(\mathbf{k}_m,t_m)\mathbf{u},$$
$$\mathbf{k}_1,\mathbf{k}_2,\ldots,\mathbf{k}_m \geq 0, t_0,t_1,\ldots,t_m > 0, m = 1,2,\ldots$$

where $\mathbf{q} = [q_i]$ is a row vector of the initial probability distribution of the phase process, $q_i = \mathbb{P}\{X(0) = i\}$.

Consider the MAP of class v customers. It is characterized by transition probability matrices

$$\mathbf{P}_{v,m}(t) = \sum_{\substack{\mathbf{n} \geq 0 \\ n_v = m}} \mathbf{P}(\mathbf{n},t),\ k = 0,1,\ldots,$$

which satisfy the following recursion:

$$\mathbf{P}_{v,0}(t) = e^{(\mathbf{A}-\mathbf{A}_v)t},\ \mathbf{P}_{v,n}(t) = \int_0^t \mathbf{P}_{v,0}(t-x)\mathbf{A}_v\mathbf{P}_{v,n-1}(x)dx,\ n > 0. \tag{3}$$

The joint probability distribution of the number of class v arrivals in disjoint intervals can be calculated as:

$$\mathbb{P}\{N_v(\sum_{j=0}^{r} t_j) - N_v(\sum_{j=0}^{r-1} t_j) = k_r, r = 1,2,\ldots,m\} = \mathbf{q}\mathbf{P}(t_0)\mathbf{P}_{v,k_1}(t_1)\mathbf{P}_{v,k_2}(t_2)\ldots\mathbf{P}_{v,k_m}(t_m))\mathbf{u},$$
$$k_1,k_2,\ldots,k_m \geq 0, t_0,t_1,\ldots,t_m > 0, m = 1,2,\ldots. \tag{4}$$

If $\mathbf{q}$ is the stationary vector of $\mathbf{A}$ and satisfies $\mathbf{q}\mathbf{A}_v = a_v\mathbf{q}$, it follows from (3) that

$$\mathbf{q}\mathbf{P}_{v,n}(t) = e^{-a_v t}\frac{(a_v t)^n}{n!}\mathbf{q},\ n \geq 0.$$

It follows from (4) that, in this case, the arrival process of class v customers is Poisson with rate $a_v = \mathbf{q}\mathbf{A}_v\mathbf{u}$. Similarly, if matrix $\mathbf{A}_v$ satisfies $\mathbf{A}_v\mathbf{u} = a_v\mathbf{u}$, it follows from (3) that

$$\mathbf{P}_{v,n}(t)\mathbf{u} = e^{-a_v t}\frac{(a_v t)^n}{n!}\mathbf{u},\ n \geq 0,$$

and the arrival process of class v customers is Poisson with rate $a_v = \mathbf{q}\mathbf{A}_v\mathbf{u}$ (see also [28]). We next show that the property $\mathbf{q}\mathbf{A}_v = a_v\mathbf{q}$ is important for the existence of product-form distribution, in contrast to the property $\mathbf{A}_v\mathbf{u} = a_v\mathbf{u}$, although in both cases the arrival processes are Poisson.

3. Interacting Markovian Queueing Systems

We consider a Markov network with two nodes having state spaces $\mathcal{X}$ and $\mathcal{Y}$. The first node generates MMAP defined by non-zero matrices $\mathbf{\Lambda}_v = [\Lambda_v(i,j)]$, $i, j \in \mathcal{X}$, $v = 0, 1, \ldots, n$, and the second node generates MMAP defined by non-zero matrices $\mathbf{M}_w = [M_w(k,r)]$, $k, r \in \mathcal{Y}$, $w = 0, 1, \ldots, m$. If the nodes are operated in isolation, the first node could be represented by a homogeneous Markov process with a generator:

$$\mathbf{\Lambda} = \sum_{v=0}^{n} \mathbf{\Lambda}_v. \tag{5}$$

Also, the second could be represented by a homogeneous Markov process with a generator:

$$\mathbf{M} = \sum_{w=0}^{m} \mathbf{M}_w. \tag{6}$$

When a class v customer arrives from the first node to the second, the state of the second node changes according to a stochastic matrix $\mathbf{Q}_v = [Q_v(k,r)]$, $k, r \in \mathcal{Y}$, $v = 1, 2, \ldots, n$. When a class w customer arrives from the second node to the first, the state of the first node changes according to a stochastic matrix $\mathbf{P}_w = [P_w(i,j)]$, $i, j \in \mathcal{X}$, $w = 1, 2, \ldots, m$. The behavior of the system can be represented by a homogeneous Markov process $Z(t) = (X(t), Y(t))$, with the finite state space $\mathcal{Z} = \mathcal{X} \times \mathcal{Y}$ and the generator

$$\mathbf{\Theta} = \mathbf{\Lambda}_0 \otimes \mathbf{I} + \sum_{v=1}^{n} \mathbf{\Lambda}_v \otimes \mathbf{Q}_v + \sum_{w=1}^{m} \mathbf{P}_w \otimes \mathbf{M}_w + \mathbf{I} \otimes \mathbf{M}_0, \tag{7}$$

where $\otimes$ denotes the Kronecker product.

Further, we assume that the generator $\mathbf{\Theta}$ is irreducible. Therefore, the stationary distribution $\pi(i,k)$, $i \in \mathcal{X}$, $k \in \mathcal{Y}$, of the process $Z(t)$ is the unique solution to the system of the following steady-state equations:

$$\begin{gathered}\sum_{i \in \mathcal{X}} \pi(i,r)\Lambda_0(i,j) + \sum_{v=1}^{n} \sum_{i \in \mathcal{X}} \sum_{k \in \mathcal{Y}} \pi(i,k)\Lambda_v(i,j)Q_v(k,r) + \\ + \sum_{w=1}^{m} \sum_{i \in \mathcal{X}} \sum_{k \in \mathcal{Y}} \pi(i,k)P_w(i,j)M_w(k,r) + \sum_{k \in \mathcal{Y}} \pi(j,k)M_0(k,r) = 0, \ j \in \mathcal{X}, r \in \mathcal{Y}\end{gathered} \tag{8}$$

which satisfy the normalizing condition:

$$\sum_{i \in \mathcal{X}} \sum_{k \in \mathcal{Y}} \pi(i,k) = 1. \tag{9}$$

We next derive the conditions under which the stationary distribution has the product form:

$$\pi(i,k) = p(i)q(k), r \in \mathcal{X}, k \in \mathcal{Y} \tag{10}$$

First, however, we need the following auxiliary result.

Theorem 1. *The generators*

$$\mathbf{L} = \mathbf{\Lambda} + \sum_{w=1}^{m} \mu_w(\mathbf{P}_w - \mathbf{I}), \tag{11}$$

$$\mathbf{M} = \mathbf{M} + \sum_{v=1}^{n} \lambda_v(\mathbf{Q}_v - \mathbf{I}) \tag{12}$$

are irreducible for any $\lambda_v > 0$, $v = 1, 2, \ldots, n$, *and* $\mu_w > 0$, $w = 1, 2, \ldots, m$.

Note that matrix **L** is the generator of a Markov process representing the behavior of the first node with a multi-class Poisson arrival process with rates μ_w, $w = 1, 2, \ldots, m$. Matrix **M** is the generator of a Markov process representing the behavior of the second node with a multi-class Poisson arrival process with rates λ_v, $v = 1, 2, \ldots, n$. Therefore, Theorem 1 states that if the MMAP arriving to each network node is replaced by a multi-class Poisson arrival process, then the generators of the Markov processes representing each node in isolation are irreducible.

Theorem 2. *For the stationary distribution of the process $Z(t)$ to have the product form (10), it is necessary and sufficient that vectors $\mathbf{p} = [p(i)]$ and $\mathbf{q} = [q(k)]$ satisfy the following equations:*

$$\mathbf{p}(\mathbf{\Lambda} + \sum_{w=1}^{m} \mu_w(\mathbf{P}_w - \mathbf{I})) = 0, \ \mathbf{pu} = 1, \tag{13}$$

$$\mathbf{q}(\mathbf{M} + \sum_{v=1}^{n} \lambda_v(\mathbf{Q}_v - \mathbf{I})) = 0, \ \mathbf{qu} = 1, \tag{14}$$

$$\sum_{v=1}^{n} \mathbf{p}(\mathbf{\Lambda}_v - \lambda_v \mathbf{I}) \otimes \mathbf{q}(\mathbf{Q}_v - \mathbf{I}) + \sum_{w=1}^{m} \mathbf{p}(\mathbf{P}_w - \mathbf{I}) \otimes \mathbf{q}(\mathbf{M}_w - \mu_w \mathbf{I}) = 0, \tag{15}$$

$$\lambda_v = \mathbf{p}\mathbf{\Lambda}_v\mathbf{u}, \ v = 1, 2, \ldots, n, \ \mu_w = \mathbf{q}\mathbf{M}_w\mathbf{u}, \ w = 1, 2, \ldots, m. \tag{16}$$

Hence, components **p** and **q** of the product form (10) can be found as the stationary distributions of the nodes with Poisson arrival processes. However, this is not an easy task because the systems of Equations (13) and (14) must be solved together with the conditions (16), and therefore the problem of finding vectors **p** and **q** that satisfy the conditions of the theorem is nonlinear. In the next section, we consider particular cases for which this problem can be simplified. The proofs of Theorems 1 and 2 are provided in Appendix A.

Corollary 1. *Let $n = 1$, $m = 0$, vector $\mathbf{p}$ be the solution of equations $\mathbf{p}\mathbf{\Lambda} = 0$, $\mathbf{pu} = 1$, and vector $\mathbf{q}$ be the solution of equations $\mathbf{q}(\mathbf{M} + \lambda_1(\mathbf{Q}_1 - \mathbf{I})) = 0$, $\mathbf{qu} = 1$, where $\lambda_1 = \mathbf{p}\mathbf{\Lambda}_1\mathbf{u}$. Then, for the product form of the stationary distribution of the process $Z(t)$, it is necessary and sufficient that either $\mathbf{p}\mathbf{\Lambda}_1 = \lambda_1\mathbf{p}$ or $\mathbf{q}\mathbf{Q}_1 = \mathbf{q}$.*

Proof. Indeed, according to Theorem 2, for the product-form stationary distribution $\boldsymbol{\pi} = (\pi(i,k))$, it is required that for any $j \in \mathcal{X}, r \in \mathcal{Y}$

$$\left(\sum_{i \in \mathcal{X}} p(i)\Lambda_1(i,j) - \lambda_1 p(j)\right)\left(\sum_{k \in \mathcal{Y}} q(k)Q_1(k,r) - q(r)\right) = 0,$$

which is equivalent to either $\mathbf{p}\mathbf{\Lambda}_1 = \lambda_1\mathbf{p}$ or $\mathbf{q}\mathbf{Q}_1 = \mathbf{q}$.

It was shown in Section 2 that if the condition $\mathbf{p}\mathbf{\Lambda}_1 = \lambda_1\mathbf{p}$ is fulfilled, then, in the stationary mode, the MAP generated by the first node is Poisson with rate λ_1 (see also [9,23]). Because of the PASTA property (Poisson Arrivals See Time Averages) of the Poisson process, the stationary distribution of the second node is equal to the stationary distribution of the Markov chain embedded at times before the customer arrivals. However, then vector $\mathbf{q}\mathbf{Q}_1$ is the stationary distribution of the Markov chain embedded at times after the customer arrivals. Thus, the condition $\mathbf{q}\mathbf{Q}_1 = \mathbf{q}$ implies that, for the second node with Poisson arrivals, the Markov chains embedded before and after customer arrivals have the same stationary distributions. □

Corollary 2. *Let $n = 1$, $m = 1$, vector $\mathbf{p}$ be the solution of equations $\mathbf{p}(\mathbf{\Lambda} + \mu_1(\mathbf{P}_1 - \mathbf{I})) = 0$, $\mathbf{pu} = 1$, where $\mu_1 = \mathbf{q}\mathbf{M}_1\mathbf{u}$, and vector $\mathbf{q}$ be the solution of equations $\mathbf{q}(\mathbf{M} + \lambda_1(\mathbf{Q}_1 - \mathbf{I})) = 0$, $\mathbf{qu} = 1$, where $\lambda_1 = \mathbf{p}\mathbf{\Lambda}_1\mathbf{u}$.*

Then for the product-form stationary distribution of the process $Z(t)$, it is necessary and sufficient that at least one of the following conditions is satisfied:

$$\mathbf{p}\mathbf{\Lambda}_1 = \lambda_1 \mathbf{p} \text{ and } \mathbf{p}\mathbf{P}_1 = \mathbf{p}, \tag{17}$$

$$\mathbf{p}\mathbf{\Lambda}_1 = \lambda_1 \mathbf{p} \text{ and } \mathbf{q}\mathbf{M}_1 = \mu_1 \mathbf{q}, \tag{18}$$

$$\mathbf{q}\mathbf{Q}_1 = \mathbf{q} \text{ and } \mathbf{p}\mathbf{P}_1 = \mathbf{p}, \tag{19}$$

$$\mathbf{q}\mathbf{Q}_1 = \mathbf{q} \text{ and } \mathbf{q}\mathbf{M}_1 = \mu_1 \mathbf{q}. \tag{20}$$

There exists a constant $\psi \neq 0$ such that

$$\sum_{i \in \mathcal{X}} p(i)\Lambda_1(i,j) - \lambda_1 p(j) = \psi \left(\sum_{i \in \mathcal{X}} p(i)P_1(i,j) - p(j)\right) \tag{21}$$

for all $j \in \mathcal{X}$, *with* $\sum\limits_{i \in \mathcal{X}} p(i)P_1(i,j) \neq p(j)$, *and*

$$\sum_{k \in \mathcal{Y}} q(k)M_1(k,r) - \mu_1 q(r) = \psi \left(q(r) - \sum_{k \in \mathcal{Y}} q(k)Q_1(k,r)\right) \tag{22}$$

for all $r \in \mathcal{Y}$ *with* $\sum\limits_{k \in \mathcal{Y}} q(k)Q_1(k,r) \neq q(r)$.

Proof. According to Theorem 2, for the product-form stationary distribution $\boldsymbol{\pi} = (\pi(i,k))$, it is required that for any $j \in \S, r \in \mathcal{Y}$ the following holds:

$$\begin{gathered}\left(\sum_{i \in \mathcal{X}} p(i)\Lambda_1(i,j) - \lambda_1 p(j)\right)\left(\sum_{k \in \mathcal{Y}} q(k)Q_1(k,r) - q(r)\right) + \\ + \left(\sum_{i \in \mathcal{X}} p(i)P_1(i,j) - p(j)\right)\left(\sum_{k \in \mathcal{Y}} q(k)M_1(k,r) - \mu_1 q(r)\right) = 0.\end{gathered} \tag{23}$$

The sufficiency of each of the five conditions given above for the product-form stationary distribution $\pi(i,k)$ is obvious. Let us prove that at least one of them is necessary for the stationary distribution to be of product form.

Suppose that the stationary distribution $\pi(i,k)$ has the product form. From (23), it follows that if one of the vectors $\mathbf{p}\mathbf{\Lambda}_1 - \lambda_1\mathbf{p}$, $\mathbf{q}\mathbf{M}_1 - \mu_1\mathbf{q}$, $\mathbf{p}\mathbf{P}_1 - \mathbf{p}$, $\mathbf{q}\mathbf{Q}_1 - \mathbf{q}$ is a zero vector then there is another zero vector in this group such that one of the conditions (17)–(20) is fulfilled. It remains to consider the case when all these vectors are non-zero. Let $\mathcal{X}^*$ be a set of $j \in \mathcal{X}$ such that $p(j) \neq \sum\limits_{i \in \mathcal{X}} p(i)P_1(i,j)$, and let $\mathcal{Y}^*$ be a set of $r \in \mathcal{Y}$ such that $q(r) \neq \sum\limits_{k \in \mathcal{Y}} q(k)Q_1(k,r)$. Then, the following holds for all $j \in \mathcal{X}^*$ and $r \in \mathcal{Y}^*$:

$$\frac{\sum\limits_{i \in \mathcal{X}} p(i)\Lambda_1(i,j) - \lambda_1 p(j)}{\sum\limits_{i \in \mathcal{X}} p(i)P_1(i,j) - p(j)} = \frac{\sum\limits_{k \in \mathcal{Y}} q(k)M_1(k,r) - \mu_1 q(r)}{q(r) - \sum\limits_{k \in \mathcal{Y}} q(k)Q_1(k,r)}. \tag{24}$$

Since the left-hand side of this equation does not depend on r, and the right-hand side does not depend on j, then for $j \in \mathcal{X}^*$ and $r \in \mathcal{Y}^*$ they are both equal to some constant ψ and therefore (21) and (22) are satisfied. Since vectors $\mathbf{q}\mathbf{M}_1 - \mu_1\mathbf{q}$ and $\mathbf{q}\mathbf{Q}_1 - \mathbf{q}$ are non-zero, it follows from (24) that for each $j \in \mathcal{X}$ the relationships $\sum\limits_{i \in \mathcal{X}} p(i)\Lambda_1(i,j) = \lambda_1 p(j)$ and $\sum\limits_{i \in \mathcal{X}} p(i)P_1(i,j) = p(j)$ are both true or both false. Then, $\sum\limits_{i \in \mathcal{X}} p(i)\Lambda_1(i,j) \neq \lambda_1 p(j)$ for all $j \in \mathcal{X}^*$ in (21), and therefore $\psi \neq 0$.□

The proof of the following Corollary 3 is similar to the proof of Corollary 2.

Corollary 3. *Let $n = 2$, $m = 0$, vector $\mathbf{p}$ be the solution of the linear system $\mathbf{p\Lambda} = 0$, $\mathbf{pu} = 1$, and vector $\mathbf{q}$ be the solution of the linear system $\mathbf{q}(\mathbf{M} + \lambda_1(\mathbf{Q}_1 - \mathbf{I}) + \lambda_2(\mathbf{Q}_2 - \mathbf{I})) = 0$, $\mathbf{qu} = 1$, where $\lambda_1 = \mathbf{p\Lambda}_1\mathbf{u}$, $\lambda_2 = \mathbf{p\Lambda}_2\mathbf{u}$. Then for the product-form stationary distribution of the process $Z(t)$, it is necessary and sufficient that at least one of the following conditions is satisfied:*

$$\mathbf{p\Lambda}_1 = \lambda_1\mathbf{p} \text{ and } \mathbf{p\Lambda}_2 = \lambda_2\mathbf{p}, \tag{25}$$

$$\mathbf{p\Lambda}_1 = \lambda_1\mathbf{p} \text{ and } \mathbf{qQ}_2 = \mathbf{q}, \tag{26}$$

$$\mathbf{qQ}_1 = \mathbf{q} \text{ and } \mathbf{p\Lambda}_2 = \lambda_2\mathbf{p}, \tag{27}$$

$$\mathbf{qQ}_1 = \mathbf{q} \text{ and } \mathbf{qQ}_2 = \mathbf{q}. \tag{28}$$

There exists a constant $\varphi \neq 0$ such that

$$\sum_{i\in\mathcal{X}} p(i)\Lambda_1(i,j) - \lambda_1 p(j) = \varphi\left(\sum_{i\in\mathcal{X}} p(i)\Lambda_2(i,j) - \lambda_2 p(j)\right) \tag{29}$$

for all $j \in \mathcal{X}$ with $\sum_{i\in\mathcal{X}} p(i)\Lambda_2(i,j) \neq \lambda_2 p(j)$, and

$$\sum_{k\in\mathcal{Y}} q(k)Q_2(k,r) - q(r) = \varphi\left(q(r) - \sum_{k\in\mathcal{Y}} q(k)Q_1(k,r)\right) \tag{30}$$

for all $r \in \mathcal{Y}$ with $\sum_{k\in\mathcal{Y}} q(k)Q_1(k,r) \neq q(r)$.

4. Examples

To demonstrate the applicability of the results obtained, let us consider several examples. Examples 1 and 2 illustrate the product-form conditions for $n = 1$ and $m = 0$.

Example 1. *Consider a system with the first node consisting of a bunker with waiting space for one customer and one server (Figure 1). There are always customers in the bunker. They arrive to the first node's server, bypassing the queue, only when the server and the waiting space are empty. In addition to the customers from the bunker, there are customers from an external source forming a Poisson process with rate γ. An external customer waits if, at the time of the customer's arrival, the server is busy. If the waiting space is occupied, the arriving customer is lost. All customers served by the first node arrive to the second node, consisting of one server, without waiting space. The service times are exponentially distributed with parameters α and β for the first and second servers, respectively. Because the first server is always busy, the departure process of the first node is Poisson with rate α. The state of the first node can be represented by the number of customers in the queue, and the state of the second node by the number of customers in service.*

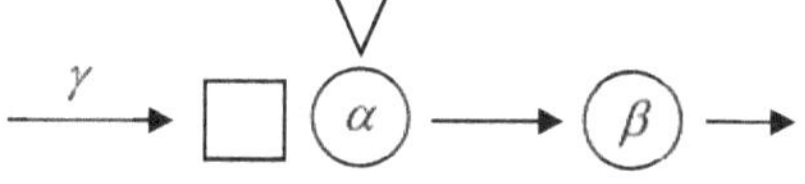

Figure 1. A non-product-form system with a Poisson flow of customers arriving at the second server.

We here show that the stationary probability distribution of the network is not of product form. This system is characterized by matrices

$$\mathbf{\Lambda} = \begin{bmatrix} -\gamma & \gamma \\ \alpha & -\alpha \end{bmatrix}, \mathbf{\Lambda}_1 = \begin{bmatrix} \alpha & 0 \\ \alpha & 0 \end{bmatrix}, \mathbf{Q}_1 = \begin{bmatrix} 0 & 1 \\ 0 & 1 \end{bmatrix}, \mathbf{M} = \begin{bmatrix} 0 & 0 \\ \beta & -\beta \end{bmatrix},$$

and the solutions of the linear systems (13) and (14) are given by

$$p(0) = \frac{\alpha}{\alpha+\gamma}, \; p(1) = \frac{\gamma}{\alpha+\gamma}, \; q(0) = \frac{\beta}{\lambda_1+\beta}, \; q(1) = \frac{\alpha}{\lambda_1+\beta}.$$

where $\lambda_1 = \mathbf{p}\mathbf{\Lambda}_1\mathbf{u} = \alpha$. The steady-state equations for the stationary distribution of the process $Z(t)$ are as follows:

$$(\alpha+\gamma)\pi(0,0) = \beta\pi(0,1), \; (\beta+\gamma)\pi(0,1) = \alpha\, r(0,0) + \alpha\,\pi(1,0) + \alpha\,\pi(1,1),$$
$$\alpha\,\pi(1,0) = \gamma\,\pi(0,0) + \beta\,\pi(1,1), \; (\alpha+\beta)\pi(1,1) = \gamma\,\pi(0,1).$$

It is easy to verify that the normalized solution of these equations is given by

$$\pi(0,0) = \frac{\alpha\beta}{\alpha\beta+(\alpha+\gamma)^2}, \; \pi(0,1) = \frac{\alpha(\alpha+\gamma)}{\alpha\beta+(\alpha+\gamma)^2},$$
$$\pi(1,0) = \frac{\gamma\beta(\alpha+\gamma)}{(\alpha+\beta)(\alpha\beta+(\alpha+\gamma)^2)}, \; \pi(1,1) = \frac{\gamma\alpha(\alpha+\gamma)}{(\alpha+\beta)(\alpha\beta+(\alpha+\gamma)^2)}.$$

It is clear that $\pi(i,k) \neq p(i)q(k)$. Thus, although the second node has a Poisson arrival process, this is not enough for the stationary distribution $\pi(i,k)$ to have the product form. The product-form condition of Corollary 1 does not hold since neither $\mathbf{p}\mathbf{\Lambda}_1 = \lambda_1\mathbf{p}$ nor $\mathbf{q}\mathbf{Q}_1 = \mathbf{q}$.

Example 2. *Let the second node have $w > 0$ waiting spaces and no servers. An arriving customer takes one of the free waiting spaces, if there are any. If all waiting spaces are occupied at the instant of arrival, then the arriving customer and all those in the queue are considered served and leave the system.*

Such a system with a Poisson arrival process can be represented by a Markov process with the state set $\mathcal{Y} = \{0,1,\ldots,w\}$. This process has a uniform stationary distribution $q(k) = 1/(w+1), k = 0,1,\ldots w$, and obviously satisfies the condition $\mathbf{q}\mathbf{Q}_1 = \mathbf{q}$. According to Corollary 1, the stationary distribution $\pi(i,k)$ has the product form.

Example 3. *Now, to illustrate Corollary 2, consider a system with positive and negative customers. Each node consists of one server without waiting space (Figure 2).*

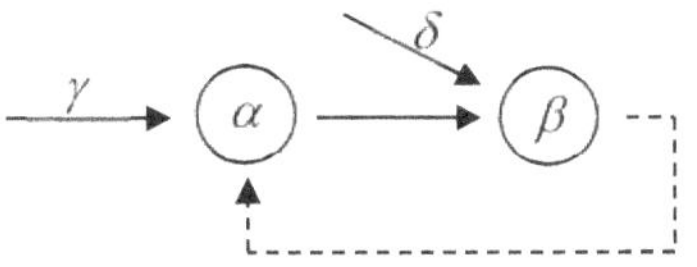

Figure 2. A system with positive and negative customers.

The service times at the first and second servers are exponentially distributed with parameters α and β, respectively. External Poisson processes of positive customers arrive to the servers with rates γ and δ, respectively. A customer departing from the first server arrives to the second server as a positive customer. Arriving positive customers are lost if the servers are busy. A customer leaving the second server comes to the first server as a negative customer and—if the first server is busy—deletes the customer in service.

The matrices that specify the interaction of the nodes are as follows:

$$\mathbf{\Lambda} = \begin{bmatrix} -\gamma & \gamma \\ \alpha & -\alpha \end{bmatrix}, \mathbf{\Lambda}_1 = \begin{bmatrix} 0 & 0 \\ \alpha & 0 \end{bmatrix}, \mathbf{Q}_1 = \begin{bmatrix} 0 & 1 \\ 0 & 1 \end{bmatrix}, \mathbf{M} = \begin{bmatrix} -\delta & \delta \\ \beta & -\beta \end{bmatrix}, \mathbf{M}_1 = \begin{bmatrix} 0 & 0 \\ \beta & 0 \end{bmatrix}, \mathbf{P}_1 = \begin{bmatrix} 1 & 0 \\ 1 & 0 \end{bmatrix},$$

and the solutions of the system of Equations (13) and (14) are given by the formulae

$$p(0) = \frac{\alpha+\mu_1}{\alpha+\gamma+\mu_1},\ p(1) = \frac{\gamma}{\alpha+\gamma+\mu_1},\ q(0) = \frac{\beta}{\lambda_1+\delta+\beta},\ q(1) = \frac{\lambda_1+\delta}{\lambda_1+\delta+\beta}, \tag{31}$$

where

$$\lambda_1 = \alpha p(1),\ \mu_1 = \beta q(1). \tag{32}$$

To check the conditions of Corollary 2, we first calculate the vectors that appear in them:

$$\begin{aligned} \mathbf{p}\mathbf{\Lambda}_1 - \lambda_1\mathbf{p} &= \alpha p^2(1)(1,-1),\ \mathbf{p}\mathbf{P}_1 - \mathbf{p} = p(1)(1,-1), \\ \mathbf{q}\mathbf{M}_1 - \mu_1\mathbf{q} &= \beta q^2(1)(1,-1),\ \mathbf{q}\mathbf{Q}_1 - \mathbf{q} = q(0)(-1,1). \end{aligned}$$

It is clear that the first four conditions of Corollary 2 are not satisfied for any positive parameters $\alpha, \beta, \gamma, \delta$.

In the particular case of $\alpha = \beta = 2, \delta = 1, \gamma = 3$, the system of Equations (31) and (32) has the unique solution $\lambda_1 = 1, \mu_1 = 1$ and the parameter ψ from the fifth condition of Corollary 2 is $\psi = 1$. Thus, in this case, the stationary distribution $\pi(i,k)$ has the product form.

5. Conclusions

In this paper, we have studied a Markov network consisting of two nodes of arbitrary structure, where each node generates a MMAP of customers arriving to the other node. We have derived the necessary and sufficient conditions for the product-form stationary distribution of the network. Simple criteria to check whether a Markov network with one and two customer classes is of product form have been developed. Unlike the existing product-form criteria for Markov networks, the criteria obtained in this article are readily applicable to establish the existence of a product-form stationary distribution for specific Markovian queueing systems. The extension of the developed theory to networks with more than two classes of customers is the subject of our future research.

Author Contributions: Conceptualization, V.A.N. and K.E.S.; methodology and validation, V.A.N.; review and editing, original draft preparation, examples, Y.V.G.; supervision and project administration, K.E.S.

Funding: The publication has been supported by the Ministry of Education and Science of the Russian Federation (project No. 2.3397.2017/4.6).

Acknowledgments: Authors are grateful to anonymous reviewers whose comments have greatly improved this manuscript.

Conflicts of Interest: The authors declare no conflict of interest.

Appendix A

In order to prove Theorem 1, we use a criterion based on the existence of a non-negative vector with positive and zero components, which we call semi-positive.

Lemma A1. *A matrix* $\mathbf{B}$ *with non-negative off-diagonal elements is reducible if and only if there is a semi-positive vector* $\mathbf{x}$ *such that* $\mathbf{B}\mathbf{x}(i) = 0$ *for all indices* i *for which* $x(i) = 0$.

Proof. If matrix $\mathbf{B}$ of order n is reducible, then there exists a partition $\mathcal{Y},\ \mathcal{Z}$ of the set $\mathcal{X} = \{1, 2, \ldots, n\}$ such that $B(k,j) = 0$ for all $k \in \mathcal{Y},\ j \in \mathcal{Z}$. We define a semi-positive vector as follows: $x(j) = 0$ for $j \in \mathcal{Y}, x(j) = 1$ for $j \in \mathcal{Z}$. If $x(k) = 0$, then $k \in \mathcal{Y}$ and we have

$$\mathbf{B}\mathbf{x}(k) = \sum_{j\in\mathcal{Y}} B(k,j)x(j) + \sum_{j\in\mathcal{Z}} B(k,j)x(j) = 0,$$

since $x(j) = 0$ for $j \in \mathcal{Y}$ and $B(k,j) = 0$ for $j \in \mathcal{Z}$.

Let $\mathbf{x}$ be a semi-positive vector and $\mathbf{Bx}(k) = 0$ for all k for which $x(k) = 0$. Let $\mathcal{Y} = \{i|x(i) = 0\}$ and $\mathcal{Z} = \{i|x(i) > 0\}$. Then the sets $\mathcal{Y}$ and $\mathcal{Z}$ form a partition of the set $\mathcal{X}$, and for every $i \in \mathcal{Y}$ we have

$$0 = \mathbf{Bx}(i) = \sum_{j \in \mathcal{Y}} B(i,j)x(j) + \sum_{j \in \mathcal{Z}} B(i,j)x(j) = \sum_{j \in \mathcal{Z}} B(i,j)x(j).$$

Since all terms of the latter sum are non-negative and $x(j) > 0$ for $j \in \mathcal{Z}$, it follows that $B(i,j) = 0$ for all $j \in \mathcal{Y}$, $j \in \mathcal{Z}$. Thus, matrix $\mathbf{B}$ is reducible. □

Proof of Theorem 1. Suppose that matrix $\mathbf{L}$ is reducible. Then, according to the lemma, there exists a semi-positive vector $\mathbf{x}$ such that $\mathbf{Lx}(i) = 0$ for all indices i for which $x(i) = 0$. Since the off-diagonal elements of matrices $\mathbf{\Lambda}$ and $\mathbf{P}_w$ are non-negative, it follows that $\mathbf{\Lambda x}(i) = 0$ and $\mathbf{P}_w\mathbf{x}(i) = 0$ for all $w = 1, 2, \dots, m$ and all i for which $x(i) = 0$.

Vector $\mathbf{y} = \mathbf{x} \otimes \mathbf{u}$ is semi-positive and the following relationships hold:

$$\begin{aligned} \mathbf{\Theta y} &= \mathbf{\Lambda}_0\mathbf{x} \otimes \mathbf{u} + \sum_{v=1}^{n} \mathbf{\Lambda}_v\mathbf{x} \otimes \mathbf{Q}_v\mathbf{u} + \sum_{w=1}^{m} \mathbf{P}_w\mathbf{x} \otimes \mathbf{M}_w\mathbf{u} + \mathbf{x} \otimes \mathbf{M}_0\mathbf{u} = \\ &= \mathbf{\Lambda}_0\mathbf{x} \otimes \mathbf{u} + \sum_{v=1}^{n} \mathbf{\Lambda}_v\mathbf{x} \otimes \mathbf{u} + \sum_{w=1}^{m} \mathbf{P}_w\mathbf{x} \otimes \mathbf{M}_w\mathbf{u} + \mathbf{x} \otimes (\mathbf{M} - \sum_{w=1}^{m} \mathbf{M}_w)\mathbf{u} \\ &= \mathbf{\Lambda x} \otimes \mathbf{u} + \sum_{w=1}^{m} (\mathbf{P}_w - \mathbf{I})\mathbf{x} \otimes \mathbf{M}_w\mathbf{u}. \end{aligned} \tag{A1}$$

Since $y(i,k) = 0$ if and only if $x(i) = 0$, it follows that

$$\begin{aligned} (\mathbf{\Theta y})(i,k) &= (\mathbf{\Lambda x} \otimes \mathbf{u} + \sum_{w=1}^{m} (\mathbf{P}_w - \mathbf{I})\mathbf{x} \otimes \mathbf{M}_w\mathbf{u})(i,k) = \\ &= (\mathbf{\Lambda x})(i) + \sum_{w=1}^{m} ((\mathbf{P}_w - \mathbf{I})\mathbf{x})(i))(\mathbf{M}_w\mathbf{u})(k) = 0 \end{aligned}$$

for all indices (i,k) for which $y(i,k) = 0$. Hence, matrix $\mathbf{\Theta}$ is reducible. The obtained contradiction proves the irreducibility of matrix $\mathbf{L}$. The irreducibility of matrix $\mathbf{M}$ can be proved in a similar fashion. □

Proof of Theorem 2. Note that if a stationary distribution $\boldsymbol{\pi} = (\pi(i,k)), i \in \mathcal{X}, k \in \mathcal{Y}$ has the product form $\boldsymbol{\pi} = \mathbf{p} \otimes \mathbf{q}$, the system of steady-state equations $\boldsymbol{\pi}\mathbf{\Theta} = 0$ can be rewritten as

$$\mathbf{p\Lambda}_0 \otimes \mathbf{q} + \sum_{v=1}^{n} \mathbf{p\Lambda}_v \otimes \mathbf{qQ}_v + \sum_{w=1}^{m} \mathbf{pP}_w \otimes \mathbf{qM}_w + \mathbf{p} \otimes \mathbf{qM}_0 = 0. \tag{A2}$$

First, we show that if the vectors $\mathbf{p} = (p(i)), i \in \mathcal{X}$, and $\mathbf{q} = (q(k)), k \in \mathcal{Y}$, satisfy Equations (13) and (14), then the left-hand sides of (15) and (A2) coincide. We rewrite (13) and (14) as follows:

$$\mathbf{p\Lambda}_0 = \mu\,\mathbf{p} - \sum_{w=1}^{m} \mu_w\mathbf{pP}_w - \sum_{v=1}^{n} \mathbf{p\Lambda}_v, \tag{A3}$$

$$\mathbf{qM}_0 = \lambda\mathbf{q} - \sum_{v=1}^{n} \lambda_v\mathbf{qQ}_v - \sum_{w=1}^{n} \mathbf{qM}_w. \tag{A4}$$

Using these expressions, the following relationships can be derived:

$$\begin{aligned}
&\mathbf{p}\boldsymbol{\Lambda}_0 \otimes \mathbf{q} + \sum_{v=1}^{n} \mathbf{p}\boldsymbol{\Lambda}_v \otimes \mathbf{q}\mathbf{Q}_v + \sum_{w=1}^{m} \mathbf{p}\mathbf{P}_w \otimes \mathbf{q}\mathbf{M}_w + \mathbf{p} \otimes \mathbf{q}\mathbf{M}_0 = \\
&= (\mu\, \mathbf{p} - \sum_{w=1}^{m} \mu_w \mathbf{p}\mathbf{P}_w - \sum_{v=1}^{n} \mathbf{p}\boldsymbol{\Lambda}_v) \otimes \mathbf{q} + \sum_{v=1}^{n} \mathbf{p}\boldsymbol{\Lambda}_v \otimes \mathbf{q}\mathbf{Q}_v + \\
&+ \sum_{w=1}^{m} \mathbf{p}\mathbf{P}_w \otimes \mathbf{q}\mathbf{M}_w + \mathbf{p} \otimes (\lambda \mathbf{q} - \sum_{v=1}^{n} \lambda_v \mathbf{q}\mathbf{Q}_v - \sum_{w=1}^{n} \mathbf{q}\mathbf{M}_w) = \\
&= \sum_{v=1}^{n} \mathbf{p}(\boldsymbol{\Lambda}_v - \lambda_v \mathbf{I}) \otimes \mathbf{q}\mathbf{Q}_v + \sum_{w=1}^{m} \mathbf{p}\mathbf{P}_w \otimes \mathbf{q}(\mathbf{M}_w - \mu_w \mathbf{I}) - \\
&- \sum_{v=1}^{n} \mathbf{p}(\boldsymbol{\Lambda}_v - \lambda_v \mathbf{I}) \otimes \mathbf{q} - \sum_{w=1}^{m} \mathbf{p} \otimes \mathbf{q}(\mathbf{M}_w - \mu_w \mathbf{I}) = \\
&= \sum_{v=1}^{n} \mathbf{p}(\boldsymbol{\Lambda}_v - \lambda_v \mathbf{I}) \otimes \mathbf{q}(\mathbf{Q}_v - \mathbf{I}) + \sum_{w=1}^{m} \mathbf{p}(\mathbf{P}_w - \mathbf{I}) \otimes \mathbf{q}(\mathbf{M}_w - \mu_w \mathbf{I}).
\end{aligned} \tag{A5}$$

Now we prove the necessity of conditions (13)–(15). Let the vector of the stationary probabilities be represented as $\boldsymbol{\pi} = \mathbf{p} \otimes \mathbf{q}$. By post-multiplying (15) by matrix $\mathbf{I} \otimes \mathbf{u}$, we obtain (13), and by post-multiplying (15) by matrix $\mathbf{u} \otimes \mathbf{I}$, we obtain (14). As it was proved previously, the systems of Equations (14) and (A2) are equivalent. Therefore, conditions (13)–(15) are necessary for the product-form stationary distribution. Let us now prove the sufficiency of these conditions.

If the stochastic vectors $\mathbf{p} = (p(i)), i \in \mathcal{X}$, and $\mathbf{q} = (q(k)), k \in \mathcal{Y}$, satisfy conditions (13)–(15), then due to (A5), vector $\boldsymbol{\pi} = \mathbf{p} \otimes \mathbf{q}$ is a solution of the system of steady-state Equation (A2). Since the generator $\boldsymbol{\Theta}$ of process $Z(t)$ is irreducible, the normalized solution of this system is unique and therefore vector $\boldsymbol{\pi} = \mathbf{p} \otimes \mathbf{q}$ is the stationary distribution of the process. □

References

1. Jackson, J.R. Jobshop-like queueing systems. *Manag. Sci.* **1963**, *10*, 131–142. [CrossRef]
2. Gordon, W.J.; Newell, G.F. Closed queueing systems with exponential servers. *Oper. Res.* **1967**, *15*, 254–265. [CrossRef]
3. Kobayashi, H. System design and performance analysis using analytic models. In *Current Trends in Programming Methodology, Vol. III: Software Modeling*; Chandy, K.M., Yeh, R.T., Eds.; Prentice-Hall: Englewood Cliffs, NJ, USA, 1978.
4. Towsley, D.F. Queueing network models with state-dependent routing. *J. ACM* **1980**, *27*, 323–337. [CrossRef]
5. Baskett, F.; Chandy, K.M.; Muntz, R.R.; Palacios, G. Open, closed, and mixed networks of queues with different classes of customers. *J. ACM* **1985**, *22*, 248–260. [CrossRef]
6. Disney, R.L.; König, D. Queueing networks: A survey of their random processes. *SIAM Rev.* **1985**, *27*, 335–403. [CrossRef]
7. Nelson, R.D. The Mathematics of Product-form Queuing Networks. *ACM Comput. Surv.* **1993**, *25*, 339–369. [CrossRef]
8. Balsamo, S. Product-form queueing networks. In *Performance Evaluation: Origins and Directions*; Haring, G., Lindemann, C., Reiser, M., Eds.; Springer: Berlin, Germany, 2000; pp. 377–401.
9. Kelly, F.P. *Reversibility and Stochastic Networks*; Wiley: New York, NY, USA, 1979.
10. Ross, S. *Stochastic Processes*; Wiley: New York, NY, USA, 1983.
11. Whittle, P. *Systems in Stochastic Equilibrium*; Wiley: New York, NY, USA, 1986.
12. Walrand, J. *An Introduction to Queueing Networks*; Prentice-Hall: Englewood Cliffs, NJ, USA, 1988.
13. Gelenbe, E. Random neural networks with negative and positive signals and product-form solution. *Neural Comput.* **1989**, *1*, 502–510. [CrossRef]
14. Gelenbe, E. Product-form queuing-networks with negative and positive customers. *J. Appl. Probab.* **1991**, *28*, 656–663. [CrossRef]
15. Gelenbe, E. G-networks with signals and batch removal. *Probab. Eng. Inf. Sci.* **1993**, *7*, 335–342. [CrossRef]
16. Gelenbe, E.; Fourneau, J.M. G-networks with resets. *Perform. Eval.* **2002**, *49*, 179–191. [CrossRef]
17. Fourneau, J.M.; Gelenbe, E.; Suros, R. G-networks with multiple classes of negative and positive customers. *Theor. Comput. Sci.* **1996**, *155*, 141–156. [CrossRef]

18. Marin, A. Product-form in G-networks. *Probab. Eng. Inf. Sci.* **2016**, *30*, 345–360. [CrossRef]
19. Gelenbe, E. G-Networks and their Applications. *Probab. Eng. Inf. Sci.* **2017**, *31*, 381–575.
20. Chao, X.; Miyazawa, M.; Serfozo, R.F.; Takada, H. Markov network processes with product-form stationary distributions. *Queueing Syst.* **1998**, *28*, 377–401. [CrossRef]
21. Naumov, V.A. On the independent behavior of nodes of a complex system. In *Proceedings 3d Seminar Meeting on Queueing Theory*; Gnedenko, B.V., Gromak, Y., Chepurin, E.V., Eds.; Moscow State University: Moscow, Russian, 1976; pp. 169–177.
22. Chao, X.; Miyazawa, M. Queueing networks with instantaneous movements: A unified approach by quasi-reversibility. *Adv. Appl. Probab.* **2000**, *32*, 284–313. [CrossRef]
23. Chao, X. Networks with customers, signals, and product-form solution. In *Queueing Networks: A Fundamental Approach*; Boucherie, R.J., Dijk, N.M., Eds.; Springer: Cham, Switzerland, 2011; pp. 217–267.
24. He, Q.-M. Queues with marked customers. *Adv. Appl. Probab.* **1996**, *28*, 567–587. [CrossRef]
25. Ezhov, I.I.; Skorokhod, A.V. Markov processes with homogeneous second component. I. *Theory Probab. Its Appl.* **1969**, *14*, 1–13. [CrossRef]
26. Basharin, G.P.; Naumov, V.A. Simple matrix description of peaked and smooth traffic and its applications. In *Fundamentals of Teletraffic Theory. Proceedings of the 3rd International Seminar on Teletraffic Theory*; VINITI: Moscow, Russia, 1984; pp. 38–44.
27. Basharin, G.P.; Naumov, V.A.; Samouylov, K.E. On Markovian modelling of arrival processes. *Stat. Pap.* **2018**, *59*, 1533–1540. [CrossRef]
28. Bean, N.G.; Green, D.A. When is a MAP poisson? *Math. Comput. Model.* **2000**, *31*, 31–46. [CrossRef]

mathematics

Article

On Truncation of the Matrix-Geometric Stationary Distributions

Valeriy A. Naumov [1], Yuliya V. Gaidamaka [2,3,*] and Konstantin E. Samouylov [2,3]

1 Service Innovation Research Institute, 00120 Helsinki, Finland
2 Department of Applied Informatics and Probability, Peoples' Friendship University of Russia (RUDN University), Miklukho-Maklaya St. 6, Moscow 117198, Russia
3 Institute of Informatics Problems, Federal Research Center "Computer Science and Control" of the Russian Academy of Sciences, Vavilov St. 44-2, Moscow 119333, Russia
* Correspondence: gaydamaka-yuv@rudn.ru

Received: 30 July 2019; Accepted: 27 August 2019; Published: 1 September 2019

Abstract: In this paper, we study queueing systems with an infinite and finite number of waiting places that can be modeled by a Quasi-Birth-and-Death process. We derive the conditions under which the stationary distribution for a loss system is a truncation of the stationary distribution of the Quasi-Birth-and-Death process and obtain the stationary distributions of both processes. We apply the obtained results to the analysis of a semi-open network in which a customer from an external queue replaces a customer leaving the system at the node from which the latter departed.

Keywords: Quasi-Birth-and-Death process; matrix-geometric solution; truncated distribution

1. Introduction

Consider a queueing system with an infinite number of waiting places which may be modeled by a Quasi-Birth-and-Death (QBD) process $X(t)$ on a state space $\cup_{l=0}^{\infty} X_l$, where $X_0 = \{(i,0),\ 1 \le i \le n+m\}$ and $X_l = \{(i,l),\ 1 \le i \le m\}, l \ge 1$. Here l represents the number of customers in the queue and i represents the internal state of the system. The generator of the process $X(t)$ has the block tri-diagonal form

$$\mathbf{Q} = \begin{bmatrix} \mathbf{N}_0 & \mathbf{\Lambda}_0 & 0 & 0 & 0 & \cdots \\ \mathbf{M}_1 & \mathbf{N}_1 & \mathbf{\Lambda} & 0 & 0 & \cdots \\ 0 & \mathbf{M} & \mathbf{N} & \mathbf{\Lambda} & 0 & \cdots \\ 0 & 0 & \mathbf{M} & \mathbf{N} & \mathbf{\Lambda} & \cdots \\ \cdots & \cdots & \cdots & \cdots & \cdots & \cdots \end{bmatrix} \tag{1}$$

where $\mathbf{N}_0$ is a square matrix of order n, the matrices $\mathbf{\Lambda}$, $\mathbf{N}$, $\mathbf{M}$ and $\mathbf{N}_1$ are square of order m, and the matrices $\mathbf{\Lambda}_0$ and $\mathbf{M}_1$ are rectangular of appropriate dimensions. We denote by $\mathbf{x}$ the stationary probability vector associated with $\mathbf{Q}$ and we partition it as $\mathbf{x} = (\mathbf{p}, \mathbf{q}_0, \mathbf{q}_1, \ldots)$, where $\mathbf{p}$ is an n-vector, and $\mathbf{q}_l$, $l \ge 0$, are m-vectors. We also define the generator $\mathbf{A} = \mathbf{\Lambda} + \mathbf{N} + \mathbf{M}$.

According to [1], the positive recurrent Markov chain $X(t)$ has a modified matrix-geometric stationary probability vector,

$$\mathbf{q}_l = \mathbf{q}_1 \mathbf{R}^l,\ l \ge 0, \tag{2}$$

where $\mathbf{R}$, which is called the rate matrix, has the spectral radius $\rho < 1$ and is the minimal nonnegative solution of the matrix equation

$$\mathbf{R}^2\mathbf{M} + \mathbf{R}\mathbf{N} + \mathbf{\Lambda} = 0. \tag{3}$$

Once $\mathbf{R}$ is known, vectors $\mathbf{p}$ and $\mathbf{q}_0$ are obtained by solving the linear system

$$\mathbf{p}\mathbf{N}_0 + \mathbf{q}_0\mathbf{M}_1 = 0 \tag{4}$$

$$\mathbf{p}\boldsymbol{\Lambda}_0 + \mathbf{q}_0(\mathbf{N}_1 + \mathbf{R}\mathbf{M}) = 0 \tag{5}$$

with the normalizing condition

$$\mathbf{p}\mathbf{u} + \mathbf{q}_0(\mathbf{I} - \mathbf{R})^{-1}\mathbf{u} = 1, \tag{6}$$

where $\mathbf{u}$ is the column vector of all ones of appropriate size.

The rate matrix $\mathbf{R}$ is usually computed via iterative algorithms [1–4] among which the fastest is the logarithmic reduction algorithm [2]. Available methods for computation of the rate matrix are effective only for relatively small values of m. Exceptions are processes with off-diagonal blocks of a special type. If matrices $\boldsymbol{\Lambda}$ or $\mathbf{M}$ have rank one [5,6], or matrix $\mathbf{A}$ is triangular [7,8], it is possible to find an explicit solution of Equation (3).

Now, consider a queueing system with $L \geq 0$ waiting places, that can be modeled by a Markov chain $X_L(t)$ on a state space $\cup_{l=0}^{L} X_l$ with a generator of the block tri-diagonal form

$$\mathbf{Q}_L = \begin{bmatrix} \mathbf{N}_0 & \boldsymbol{\Lambda}_0 & 0 & 0 & \cdots & 0 \\ \mathbf{M}_1 & \mathbf{N}_1 & \boldsymbol{\Lambda} & 0 & \cdots & 0 \\ 0 & \mathbf{M} & \mathbf{N} & \ddots & \ddots & \vdots \\ 0 & 0 & \ddots & \ddots & \boldsymbol{\Lambda} & 0 \\ \vdots & \vdots & \ddots & \mathbf{M} & \mathbf{N} & \boldsymbol{\Lambda} \\ 0 & 0 & \cdots & 0 & \mathbf{M} & \mathbf{N} + \boldsymbol{\Lambda} \end{bmatrix}. \tag{7}$$

The change in the lower diagonal block is due to the fact that if an arriving customer finds all waiting places occupied it is lost, but the internal state of the system changes in the same manner as if the customer were not lost.

We partition the stationary probability vector $\mathbf{x}_L$ associated with $\mathbf{Q}_L$ as $\mathbf{x}_L = (\mathbf{p}_L, \mathbf{q}_{L,0}, \mathbf{q}_{L,1}, \ldots, \mathbf{q}_{L,L})$, where $\mathbf{p}_L$ is an n-vector, and $\mathbf{q}_{L,l}$, $0 \leq l \leq L$, are m-vectors. Solving finite QBD processes is more complicated than solving infinite processes [9–11]. If for some number s matrices $\boldsymbol{\Lambda}$, $\mathbf{N}$, $\mathbf{M}$ satisfy $\det(s^2\mathbf{M} + s\mathbf{N} + \boldsymbol{\Lambda}) \neq 0$ then the stationary probability vector $\mathbf{x}_L$ can be expressed using two matrix-geometric terms as [12]

$$\mathbf{q}_{L,l} = \mathbf{a}_L\mathbf{R}^l + \mathbf{b}_L\mathbf{S}^{L-l},\ 0 \leq l \leq L,$$

which require computation of the minimal nonnegative solutions of two matrix equations

$$\mathbf{R}^2\mathbf{M} + \mathbf{R}\mathbf{N} + \boldsymbol{\Lambda} = 0,\ \mathbf{M} + \mathbf{S}\mathbf{N} + \mathbf{S}^2\boldsymbol{\Lambda} = 0.$$

The vectors $\mathbf{a}_L$, $\mathbf{b}_L$ and $\mathbf{p}_L$ can be found by solving the linear system

$$\mathbf{p}_L\mathbf{N}_0 + (\mathbf{a}_L + \mathbf{b}_L\mathbf{S}^L)\mathbf{M}_1 = 0 \tag{8}$$

$$\mathbf{p}_L\boldsymbol{\Lambda}_0 + (\mathbf{a}_L + \mathbf{b}_L\mathbf{S}^L)\mathbf{N}_1 + (\mathbf{a}_L\mathbf{R} + \mathbf{b}_L\mathbf{S}^{L-1})\mathbf{M} = 0 \tag{9}$$

$$(\mathbf{a}_L\mathbf{R}^{L-1} + \mathbf{b}_L\mathbf{S})\boldsymbol{\Lambda} + (\mathbf{a}_L\mathbf{R}^L + \mathbf{b}_L)(\mathbf{N} + \boldsymbol{\Lambda}) = 0 \tag{10}$$

and normalizing vector $\mathbf{x}_L$. The condition $\det(s^2\mathbf{M} + s\mathbf{N} + \boldsymbol{\Lambda}) \neq 0$ holds, for example, if the generator $\mathbf{A} = \boldsymbol{\Lambda} + \mathbf{N} + \mathbf{M}$ is irreducible and $\boldsymbol{\pi}\boldsymbol{\Lambda}\mathbf{u} \neq \boldsymbol{\pi}\mathbf{M}\mathbf{u}$, where $\boldsymbol{\pi}$ is the stationary probability vector of $\mathbf{A}$ [13].

The stationary probability vector $\mathbf{x} = (\mathbf{p}, \mathbf{q}_0, \mathbf{q}_1, \ldots)$ of the generator (Equation (1)) satisfies the equation

$$\mathbf{q}_{l+1}\mathbf{M} + \mathbf{q}_l\mathbf{N} + \mathbf{q}_{l-1}\boldsymbol{\Lambda} = 0$$

for all $l \geq 0$. In the finite case the stationary probability vector $\mathbf{x}_L = (\mathbf{p}_L, \mathbf{q}_{L,0}, \mathbf{q}_{L,1}, \ldots, \mathbf{q}_{L,L})$ also satisfies the equation

$$\mathbf{q}_{L,l+1}\mathbf{M} + \mathbf{q}_{L,l}\mathbf{N} + \mathbf{q}_{L,l-1}\boldsymbol{\Lambda} = 0$$

for all $0 \leq l < L$, but the boundary equation is different

$$\mathbf{q}_{L,L}(\mathbf{N} + \mathbf{\Lambda}) + \mathbf{q}_{L,L-1}\mathbf{\Lambda} = 0.$$

It is clear that if the rate matrix $\mathbf{R}$ satisfies $\mathbf{R}^{L+1}\mathbf{M} = \mathbf{R}^{L}\mathbf{\Lambda}$ then the stationary distribution of $X_L(t)$ can be obtained by the truncation of the matrix-geometric stationary distribution (Equation (2)) of $X(t)$ at the level of $l = L$. In turn, this is enough to fulfill the condition $\mathbf{R}^{l+1}\mathbf{M} = \mathbf{R}^{l}\mathbf{\Lambda}$ for some $l \leq L$. In this paper we study the properties of QBD processes with rate matrix $\mathbf{R}$ satisfying $\mathbf{RM} = \mathbf{\Lambda}$. If it is so, then for any L the stationary distribution of the process $X_L(t)$ can be obtained by truncation of the stationary distribution of $X(t)$.

In the next section we show when the rate matrix $\mathbf{R}$ satisfies $\mathbf{RM} = \mathbf{\Lambda}$. In Section 3, we analyze the stationary distributions of infinite and finite QBD processes. In Section 4, we use the obtained results for solving a semi-open network in which customers from an external queue may only arrive to the node from which a departed customer left the system. We assume the process $X(t)$ to be positive recurrent.

The following notation conventions are used throughout the article. Bold lowercase letters denote vectors, and bold capitalized letters denote matrices, $\delta(i,j) = 1$ if $i = j$ and $\delta(i,j) = 0$ otherwise; $\mathbf{I} = [\delta(i,j)]$ is the identity matrix; the ith component of vector $\mathbf{e}_i$ is equal to one and the others are zero.

2. Rate Matrices Satisfying RM = Λ

Here we derive the explicit solution of matrix Equation (3) obtained for a rate matrix satisfying $\mathbf{RM} = \mathbf{\Lambda}$.

Theorem 1. *Let $\mathbf{A}$ be an irreducible generator with stationary probability vector $\boldsymbol{\pi}$, and let $\mathbf{R}$ be the minimal nonnegative solution of Equation (3). Then the matrix $\mathbf{R}$ satisfies the relation $\mathbf{RM} = \mathbf{\Lambda}$ if and only if*

$$\mathbf{\Lambda} = \frac{1}{\boldsymbol{\pi\mu}}\boldsymbol{\lambda\pi}\mathbf{M}, \tag{11}$$

where $\boldsymbol{\lambda} = \mathbf{\Lambda u}$ and $\boldsymbol{\mu} = \mathbf{Mu}$. Moreover, the matrix $\mathbf{R}$ and its spectral radius η are given by

$$\mathbf{R} = \frac{1}{\boldsymbol{\pi\mu}}\boldsymbol{\lambda\pi}, \tag{12}$$

$$\eta = \frac{\boldsymbol{\pi\lambda}}{\boldsymbol{\pi\mu}}. \tag{13}$$

Proof. Let matrix $\mathbf{R}$ satisfy the relation $\mathbf{RM} = \mathbf{\Lambda}$. Then, Equation (3) can be rewritten as

$$\mathbf{RA} = \mathbf{R\Lambda} + \mathbf{RN} + \mathbf{\Lambda} = \mathbf{R}^2\mathbf{M} + \mathbf{RN} + \mathbf{\Lambda} = 0. \tag{14}$$

Since the generator $\mathbf{A}$ is irreducible, each column of the matrix $\mathbf{R}$ is proportional to the vector $\boldsymbol{\pi}$. Therefore $\mathbf{R}$ has the form $\mathbf{R} = \mathbf{w}\boldsymbol{\pi}$, where $\mathbf{w}$ is a row vector satisfying

$$\mathbf{w}\boldsymbol{\pi}\mathbf{M} = \mathbf{\Lambda}. \tag{15}$$

By postmultiplying Equation (15) by vector $\mathbf{u}$ we obtain $\mathbf{w}\boldsymbol{\pi\mu} = \boldsymbol{\lambda}$, which implies Equations (11) and (12).

Let matrix $\mathbf{\Lambda}$ be given by Equation (11). Since the process $X(t)$ is positive recurrent, the matrix $\mathbf{N} + \mathbf{RM}$ is nonsingular [1] and we have

$$\mathbf{R} = -\mathbf{\Lambda}(\mathbf{N} + \mathbf{RM})^{-1}.$$

This relation implies that the rate matrix is of the form $\mathbf{R} = \boldsymbol{\lambda}\boldsymbol{\xi}$, where $\boldsymbol{\xi}$ is the unique solution of the equation

$$\boldsymbol{\xi} = -\frac{1}{\boldsymbol{\pi}\boldsymbol{\mu}}\boldsymbol{\pi}\mathbf{M}(\mathbf{N} + \boldsymbol{\xi}\boldsymbol{\lambda}\mathbf{M})^{-1}. \quad (16)$$

It is easy to check that the vector

$$\boldsymbol{\xi} = \frac{1}{\boldsymbol{\pi}\boldsymbol{\mu}}\boldsymbol{\pi}$$

satisfies Equation (16). Therefore, the rate matrix and its maximum eigenvalue are given by Equations (12) and (13). This completes the proof of the theorem. □

Corollary 1. *Let matrices* $\boldsymbol{\Lambda}$, $\mathbf{M}$, $\mathbf{N}$ *have the block-diagonal form*

$$\boldsymbol{\Lambda} = \begin{bmatrix} \boldsymbol{\Lambda}_1 & 0 & \cdots & 0 \\ 0 & \boldsymbol{\Lambda}_2 & \ddots & \vdots \\ \vdots & \ddots & \ddots & 0 \\ 0 & \cdots & 0 & \boldsymbol{\Lambda}_k \end{bmatrix}, \mathbf{N} = \begin{bmatrix} \mathbf{N}_1 & 0 & \cdots & 0 \\ 0 & \mathbf{N}_2 & \ddots & \vdots \\ \vdots & \ddots & \ddots & 0 \\ 0 & \cdots & 0 & \mathbf{N}_k \end{bmatrix}, \mathbf{M} = \begin{bmatrix} \mathbf{M}_1 & 0 & \cdots & 0 \\ 0 & \mathbf{M}_2 & \ddots & \vdots \\ \vdots & \ddots & \ddots & 0 \\ 0 & \cdots & 0 & \mathbf{M}_k \end{bmatrix}, \quad (17)$$

and suppose that $\mathbf{A}_i = \boldsymbol{\Lambda}_i + \mathbf{N}_i + \mathbf{M}_i$ *is an irreducible generator with a stationary probability vector* $\boldsymbol{\pi}_i$. *Then the minimal nonnegative solution*

$$\mathbf{R} = \begin{bmatrix} \mathbf{R}_1 & 0 & \cdots & 0 \\ 0 & \mathbf{R}_2 & \ddots & \vdots \\ \vdots & \ddots & \ddots & 0 \\ 0 & \cdots & 0 & \mathbf{R}_k \end{bmatrix} \quad (18)$$

of Equation (3) satisfies the relation $\mathbf{RM} = \boldsymbol{\Lambda}$, *if and only if*

$$\boldsymbol{\Lambda}_i = \frac{1}{\boldsymbol{\pi}_i\boldsymbol{\mu}_i}\boldsymbol{\lambda}_i\boldsymbol{\pi}_i\mathbf{M}_i,\ 1 \le i \le k, \quad (19)$$

where $\boldsymbol{\lambda}_i = \boldsymbol{\Lambda}_i\mathbf{u}$ *and* $\boldsymbol{\mu}_i = \mathbf{M}_i\mathbf{u}$. *The matrices* $\mathbf{R}_i$ *and their spectral radii* η_i *are given by*

$$\mathbf{R}_i = \frac{1}{\boldsymbol{\pi}_i\boldsymbol{\mu}_i}\boldsymbol{\lambda}_i\boldsymbol{\pi}_i, \quad (20)$$

$$\eta_i = \frac{\boldsymbol{\pi}_i\boldsymbol{\lambda}_i}{\boldsymbol{\pi}_i\boldsymbol{\mu}_i}. \quad (21)$$

This Corollary 1 follows directly from Theorem 1 applied to each diagonal block of the matrices (17) and (18).

3. Truncation of the Stationary Distribution

Consider a queueing system with $L \ge 0$ waiting places that can be modeled by a Markov chain $X_L(t)$ with a generator of the form (7) with the block-diagonal matrices $\boldsymbol{\Lambda}$, $\mathbf{M}$, $\mathbf{N}$ as in Equation (17). We partition the stationary probability vector $\mathbf{x}_L$ corresponding to $\mathbf{Q}_L$ as $\mathbf{x}_L = (\mathbf{p}, \mathbf{q}_0, \ldots, \mathbf{q}_L)$, where $\mathbf{p}$ is an n-vector, and $\mathbf{q}_l$ is partitioned as $\mathbf{q}_l = (\mathbf{q}_{l,1}, \mathbf{q}_{l,2}, \ldots, \mathbf{q}_{l,k})$, with m_i-vectors $\mathbf{q}_{l,i}$, $1 \le i \le k, 0 \le l \le L$.

Theorem 2. *Let a generator* $\mathbf{Q}_L$ *be irreducible, matrices* $\boldsymbol{\Lambda}$, $\mathbf{M}$, $\mathbf{N}$ *have the block-diagonal form (Equation (17)),* $\mathbf{A}_i = \boldsymbol{\Lambda}_i + \mathbf{N}_i + \mathbf{M}_i$ *be an irreducible generator with the stationary probability vector* $\boldsymbol{\pi}_i$, *the matrices* $\boldsymbol{\Lambda}_i$,$\mathbf{R}_i$ *be given by Equations (19) and (20) respectively, and vectors* $\mathbf{a}$ *and* $\mathbf{b} = (\mathbf{b}_1, \mathbf{b}_2, \ldots, \mathbf{b}_k)$ *satisfy the linear system*

$$\mathbf{a}\mathbf{N}_0 + \mathbf{b}\mathbf{M}_1 = 0, \quad (22)$$

$$\mathbf{a}\boldsymbol{\Lambda}_0 + \mathbf{b}(\mathbf{N}_1 + \boldsymbol{\Lambda}) = 0, \tag{23}$$

$$\mathbf{au} + \mathbf{bu} = 1. \tag{24}$$

Then the stationary probability vector of $\mathbf{Q}_L$ *is given by*

$$\mathbf{p} = c_L \mathbf{a},\ \mathbf{q}_{0,i} = c_L \mathbf{b}_i,\ \mathbf{q}_{l,i} = c_L \rho_i^{l-1} \frac{\mathbf{b}_i \boldsymbol{\lambda}_i}{\boldsymbol{\pi}_i \boldsymbol{\mu}_i} \boldsymbol{\pi}_i,\ 1 \le l \le L,\ 1 \le i \le k, \tag{25}$$

where c_L *is the normalization constant,*

$$c_L = (1 + \sum_{i=1}^{k} s_{L,i})^{-1},\ s_{L,i} = \begin{cases} \frac{\mathbf{b}_i \boldsymbol{\lambda}_i (1-\rho_i^L)}{\boldsymbol{\pi}_i \boldsymbol{\mu}_i (1-\rho_i)}, & \rho_i \ne 1, \\ L \frac{\mathbf{b}_i \boldsymbol{\lambda}_i}{\boldsymbol{\pi}_i \boldsymbol{\mu}_i}, & \rho_i = 1. \end{cases} \tag{26}$$

Proof. Since the matrices $\boldsymbol{\Lambda}_i$ and $\mathbf{R}_i$ are given by Equations (19) and (20), under the Corollary 1 of Theorem 1, matrix (18) is the solution of Equation (3) and satisfies $\mathbf{RM} = \boldsymbol{\Lambda}$. It implies that vectors $\mathbf{p}$ and $\mathbf{q}_l = (\mathbf{q}_{l,1}, \mathbf{q}_{l,2}, \dots, \mathbf{q}_{l,k})$ given by Equation (25) satisfy the following relations:

$$\mathbf{p}\mathbf{N}_0 + \mathbf{q}_0 \mathbf{M}_1 = c_L(\mathbf{a}\mathbf{N}_0 + \mathbf{b}\mathbf{M}_1) = 0,$$

$$\mathbf{p}\boldsymbol{\Lambda}_0 + \mathbf{q}_0\mathbf{N}_1 + \mathbf{q}_1\mathbf{M} = c_L(\mathbf{a}\boldsymbol{\Lambda}_0 + \mathbf{b}\mathbf{N}_1 + \mathbf{bRM}) = c_L(\mathbf{a}\boldsymbol{\Lambda}_0 + \mathbf{b}\mathbf{N}_1 + \mathbf{b}\boldsymbol{\Lambda}) = 0,$$

$$\mathbf{q}_{l+1}\mathbf{M} + \mathbf{q}_l\mathbf{N} + \mathbf{q}_{l-1}\boldsymbol{\Lambda} = \mathbf{q}_{l-1}(\mathbf{R}^2\mathbf{M} + \mathbf{RN} + \boldsymbol{\Lambda}) = 0,\ 0 < l < L,$$

$$\mathbf{q}_L(\boldsymbol{\Lambda} + \mathbf{N}) + \mathbf{q}_{L-1}\boldsymbol{\Lambda} = \mathbf{q}_{L-1}(\mathbf{R}\boldsymbol{\Lambda} + \mathbf{RN} + \boldsymbol{\Lambda}) = \mathbf{q}_{L-1}(\mathbf{R}^2\mathbf{M} + \mathbf{RN} + \boldsymbol{\Lambda}) = 0.$$

It follows that $\mathbf{x}_L = (\mathbf{p}, \mathbf{q}_0, \dots, \mathbf{q}_L)$ is the stationary probability vector of the generator $\mathbf{Q}_L$. Power $\mathbf{R}_i^l$ of the matrix $\mathbf{R}_i$ is given by

$$\mathbf{R}_i^l = \eta_i^{l-1} \frac{\boldsymbol{\lambda}_i \boldsymbol{\pi}_i}{\boldsymbol{\pi}_i \boldsymbol{\mu}_i}.$$

Using this formula and equality (Equation (24)) we obtain the normalization constant (Equation (26)). This completes the proof of the theorem. □

Note that the vector $(\mathbf{a}, \mathbf{b})$ in Equations (22)–(24) gives the stationary distribution of the Markov chain $X_0(t)$ which models a queueing system without waiting places, and the stationary distribution $\mathbf{x}_L$ of the Markov chain $X_L(t)$ depends on the number of waiting places L only via the normalizing constant c_L. Similar results are valid for the infinite QBD process $X(t)$ with generator $\mathbf{Q}$ of the form (1).

Theorem 3. *Let generator* $\mathbf{Q}$ *be irreducible, matrices* $\boldsymbol{\Lambda}$*,* $\mathbf{M}$*,* $\mathbf{N}$ *have the block-diagonal form (17),* $\mathbf{A}_i = \boldsymbol{\Lambda}_i + \mathbf{N}_i + \mathbf{M}_i$ *be an irreducible generator with stationary probability vector* $\boldsymbol{\pi}_i$*, matrices* $\boldsymbol{\Lambda}_i$*,* $\mathbf{R}_i$ *be given by (19) and (20) respectively, and vectors* $\mathbf{a}$ *and* $\mathbf{b} = (\mathbf{b}_1, \mathbf{b}_2, \dots, \mathbf{b}_k)$ *satisfy the linear system (22)–(24). Then process* $X(t)$ *is positive recurrent if and only if* $\boldsymbol{\pi}_i\boldsymbol{\lambda}_i < \boldsymbol{\pi}_i\boldsymbol{\mu}_i$ *for all* $1 \le i \le k$*. In this case the stationary probability vector of* $\mathbf{Q}$ *is given by*

$$\mathbf{p} = c\mathbf{a},\ \mathbf{q}_{0,i} = c\mathbf{b}_i,\ \mathbf{q}_{l,i} = c\eta_i^{l-1} \frac{\mathbf{b}_i \boldsymbol{\lambda}_i}{\boldsymbol{\pi}_i \boldsymbol{\mu}_i} \boldsymbol{\pi}_i,\ l \ge 1, 1 \le i \le k, \tag{27}$$

where c *is the normalization constant,*

$$c = \left(1 + \sum_{i=1}^{k} \frac{\mathbf{b}_i \boldsymbol{\lambda}_i}{\boldsymbol{\pi}_i \boldsymbol{\mu}_i - \boldsymbol{\pi}_i \boldsymbol{\lambda}_i}\right)^{-1}. \tag{28}$$

Proof. The proof of Equations (27) and (28) is similar to the proof of Equations (25) and (26). The spectral radius of the rate matrix **R** is the maximum of the spectral radii if its diagonal blocks are given by Equation (21). Therefore, the process $X(t)$ is positive recurrent if and only if $\pi_i\lambda_i < \pi_i\mu_i$ for all $1 \leq i \leq k$. □

The following result is a direct consequence of Equation (25) for the stationary distribution of the processes $X_L(t)$ and of Equation (27) for the stationary distribution of the process $X(t)$.

Corollary 2. *If the conditions of Theorem 2 hold and $\pi_i\lambda_i < \pi_i\mu_i$ for all $1 \leq i \leq k$ then the stationary distributions of the processes $X_L(t)$ can be obtained by truncation of the stationary distribution of the process $X(t)$.*

4. Semi-Open Networks with Replacement

Since the first papers [14,15] on semi-open networks appeared, the use of these models has become increasingly widespread [16–18]. However, the exact analysis even of the simplest network, with two exponential single-server nodes, Poisson arrival process, and an infinite queue, remains a difficult task [15]. The semi-open networks with external queue are mainly analyzed by approximate methods and the matrix-geometric method [19–23].

Consider a semi-open network consisting of an internal network with v nodes and an external queue. The number of customers in the internal network cannot exceed N. If an arriving customer finds the internal network busy it waits in the external queue or is lost if the number of customers in the queue has reached its maximum value L. Such a network can be considered as a queuing system with N resource units. If all N resource units are occupied, the arriving customer waits in the external queue until a resource unit becomes available. Upon completion of its service in the internal network, the customer releases the occupied resource unit.

We consider semi-open networks in which the customers waiting in the external queue select a node for entering the internal network according to the following rule, which we call the replacement. If a customer leaves the system upon completing service at node i, then a customer from the external queue will start service also at node i, thus "replacing" the leaving customer. As a result, the number of customers in the external queue will decrease by one, but the number of customers at each node of the internal network remains unchanged. An example of a system in which only the replacement of customers leaving the system is possible is shown in Figure 1. It is a semi-open network with only one node where customers can complete service and through which waiting customers can enter the internal network.

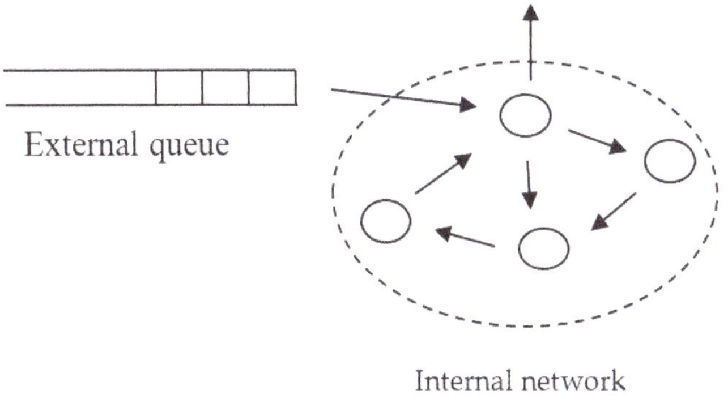

Figure 1. Internal network with a single node in which customers start and finish servicing.

Let the arrival process be Poisson with rate λ. The service rate $\mu_{k_i,i}$ at node i depends on the number k_i of customers at this node: $\mu_{0,i} = 0$ and $\mu_{k_i,i} > 0$ for $1 \leq k_i \leq N$. A customer entering the internal network is sent to node i with probability a_i, $\sum_{i=1}^{n} a_i = 1$. A customer completing service at node i will either require service at node j with a probability $q_{i,j}$ or leave the system with probability $q_{i,0} = 1 - \sum_{j=1}^{n} q_{i,j}$. The visit ratios q_i, $i = 1, 2, \ldots, v$, can be found as a solution of the linear system

$$q_j = a_j + \sum_{i=1}^{n} q_i q_{i,j},\ j = 1, 2, \ldots, v, \tag{29}$$

and the nodes' throughputs are given by $\lambda_i = \lambda q_i$, $i = 1, 2, \ldots, v$. Summing both sides of Equation (1) over j, we obtain a useful relation

$$\sum_{i=1}^{n} q_i q_{i,0} = 1. \tag{30}$$

A semi-open network can be modeled by a Markov chain $\mathbf{Y}_L(t) = (y_0(t), y_1(t), \ldots, y_n(t))$, where $y_0(t)$ represents the number of customers in the system at time t, and $y_j(t)$, $1 \leq j \leq n$, represent the number of customers at node j of the internal network. The process $\mathbf{Y}_L(t)$ has a state space $\cup_{l=0}^{L} Y_l$ with

$$Y_0 = \{(l, \mathbf{k}) \big| \mathbf{k} \in Z_l, 0 \leq l \leq N\},\ Y_l = \{(l + N, \mathbf{k}) \big| \mathbf{k} \in Z_N\},\ 1 \leq l \leq L,$$

where

$$Z_l = \Big\{(k_1, \ldots, k_v) \big| k_j \geq 0,\ j = 1, \ldots, v,\ k_1 + \ldots + k_v = l\Big\},\ 0 \leq l \leq N.$$

We assume that all states of the process $\mathbf{Y}(t)$ communicate and denote its stationary distribution as $\boldsymbol{\theta}_l = (\theta_l(\mathbf{k}))$, $(l, \mathbf{k}) \in Y_l$, $0 \leq l \leq N + L$.

The generator of the process $\mathbf{Y}(t)$ has the block tri-diagonal form (Equation (18)) with the following nonzero transition rates

$$\gamma((l, \mathbf{k}), (l + 1, \mathbf{k} + \mathbf{e}_i)) = \lambda a_i,\ \mathbf{k} \in Z_l,\ 0 \leq l < N,$$

$$\gamma((l, \mathbf{k}), (l + 1, \mathbf{k})) = \lambda,\ \mathbf{k} \in Z_N,\ N \leq l < N + L,$$

$$\gamma((l, \mathbf{k}), (l - 1, \mathbf{k} - \mathbf{e}_i)) = \mu_{k_i} q_{i,0},\ \mathbf{k} \in Z_l,\ 1 \leq l \leq N,$$

$$\gamma((l, \mathbf{k}), (l - 1, \mathbf{k})) = \mu(\mathbf{k}),\ \mathbf{k} \in Z_N,\ N < l \leq N + L,$$

$$\gamma((l, \mathbf{k}), (l, \mathbf{k} + \mathbf{e}_j - \mathbf{e}_i)) = \mu_{k_i} q_{i,j},\ \mathbf{k} \in Z_l,\ 1 \leq l \leq N,$$

where

$$\mu(\mathbf{k}) = \sum_{i=1}^{v} \mu_{k_i,i} q_{i,0}.$$

In Equation (8) the matrices $\boldsymbol{\Lambda}$, $\mathbf{M}$, $\mathbf{N}$ are diagonal with elements $\gamma((l, \mathbf{k}), (l + 1, \mathbf{n})) = \lambda\delta(\mathbf{k}, \mathbf{n})$, $\gamma((l, \mathbf{k}), (l - 1, \mathbf{n})) = \mu(\mathbf{k})\delta(\mathbf{k}, \mathbf{n})$, $\gamma((l, \mathbf{k}), (l, \mathbf{n})) = -(\lambda + \mu(\mathbf{k}))\delta(\mathbf{k}, \mathbf{n})$ respectively. Therefore, the stationary distributions for the finite and infinite semi-open networks with replacement can be found using Theorems 2 and 3.

For a semi-open network without an external queue the closed form solution can be easily derived because there exists an equivalent closed network of capacity N. The equivalent network can be obtained by adding a single-server node 0 with an infinite queue and exponentially distributed service times with parameter λ to the original network. Node 0 plays the role of the external environment of the original network. When a customer arrives to or departs from the original network, it departs from or arrives to node 0 in the equivalent network. The stationary distribution $\mathbf{x}_l = (x_l(\mathbf{k}))$, $\mathbf{k} \in Z_l$,

$l = 0, 1, \ldots, N$ of closed queueing networks with state-dependent service rates have a product form [24] and the solution $\mathbf{a} = (\mathbf{x}_0, \mathbf{x}_1, \ldots, \mathbf{x}_{N-1})$ and $\mathbf{b} = \mathbf{x}_N$ of the linear system (Equations (19)–(21)) is given by

$$x_l(\mathbf{k}) = \frac{1}{G(N)} \prod_{i=1}^{n} \frac{\lambda_i^{k_i}}{\mu_{1,i}\mu_{2,i}\cdots\mu_{k_i,i}}, \mathbf{k} \in Z_l,\ l = 0, 1, \ldots, N, \tag{31}$$

$$G(N) = 1 + \sum_{l=1}^{N} \sum_{\mathbf{k} \in X_l} \prod_{i=1}^{v} \frac{\lambda_i^{k_i}}{\mu_{1,i}\mu_{2,i}\cdots\mu_{k_i,i}}. \tag{32}$$

For closed queueing networks, there are several effective methods for computation of the normalization constant $G(N)$ [25].

By virtue of Theorem 2 for semi-open network with replacement and a finite external queue, the stationary distribution of the process $\mathbf{Y}_L(t)$ is given by

$$\theta_l(\mathbf{k}) = c_L x_l(\mathbf{k}), \mathbf{k} \in Z_l,\ 0 \leq l \leq N, \tag{33}$$

$$\theta_l(\mathbf{k}) = c_L x_N(\mathbf{k}) \rho(\mathbf{k})^{l-N},\ \mathbf{k} \in Z_N,\ N < l \leq N + L, \tag{34}$$

$$c_L = (1 + \sum_{\mathbf{k} \in Z_N} s_L(\mathbf{k}))^{-1},\ s_L(\mathbf{k}) = \begin{cases} x_N(\mathbf{k})\rho(\mathbf{k}) \frac{1-\rho(\mathbf{k})^L}{1-\rho(\mathbf{k})}, & \rho(\mathbf{k}) \neq 1, \\ x_N(\mathbf{k}) L, & \rho(\mathbf{k}) = 1, \end{cases} \tag{35}$$

where

$$\rho(\mathbf{k}) = \frac{\lambda}{\sum_{i=1}^{v} \mu_{k_i,i} q_{i,0}}.$$

It follows from Theorem 3 that a semi-open network with replacement and an infinite external queue is stable if and only if $\rho(\mathbf{k}) < 1$ for all states $\mathbf{k} \in Z_N$. In this case, the stationary distribution of $\mathbf{Y}(t)$ is given by

$$\theta_l(\mathbf{k}) = c_L x_l(\mathbf{k}), \mathbf{k} \in Z_l,\ 0 \leq l \leq N,$$

$$\theta_l(\mathbf{k}) = c_L x_N(\mathbf{k}) \rho(\mathbf{k})^{l-N},\ \mathbf{k} \in Z_N,\ l > N, \tag{36}$$

$$c_L = \left(1 + \sum_{\mathbf{k} \in Z_N} \frac{x_N(\mathbf{k})\rho(\mathbf{k})}{1 - \rho(\mathbf{k})}\right)^{-1}. \tag{37}$$

Similarly, it is possible to analyze semi-open queuing networks partitioned into subnetworks. Each such subnetwork consists of a subset of nodes of the internal network. If a customer leaves the system, then a customer from the external queue will proceed with a certain probability to a node in the same subnetwork in which the customer leaving the system was served. The generator of the related QBD process has the matrices $\mathbf{\Lambda}$ of the form of Equation (17). The results of this paper can be used if the diagonal blocks of $\mathbf{\Lambda}$ satisfy relation (19).

5. Conclusions

We have studied the properties of QBD processes with the rate matrix $\mathbf{R}$ satisfying $\mathbf{RM} = \mathbf{\Lambda}$, which guarantees that the stationary distribution of the finite QBD process is a truncation of the stationary distribution of the infinite QBD process. We have obtained the necessary and sufficient condition for the fulfillment of the equality $\mathbf{RM} = \mathbf{\Lambda}$ and have derived the formulae for the rate matrix and for the superdiagonal blocks of the process generator. These matrices have rank one, and therefore the stationary distribution and the performance parameters of QBD processes can be easily found. The results obtained have been applied to a semi-open network in which customers from the external queue can enter the internal network only through the node from which the departing customer left the system.

Author Contributions: Conceptualization, V.A.N. and K.E.S.; methodology and validation, V.A.N.; review and editing, original draft preparation, examples, Y.V.G.; supervision and project administration, K.E.S.

Funding: The publication has been supported by the Ministry of Education and Science of the Russian Federation (project No. 2.882.2017/4.6).

Conflicts of Interest: The authors declare no conflict of interest.

References

1. Neuts, M.F. *Matrix-Geometric Solutions an Algorithmic Approach*; The Johns Hopkins University Press: Baltimore, MD, USA, 1981.
2. Latouche, G.; Ramaswami, V. A logarithmic reduction algorithm for Quasi-Birth-Death processes. *J. Appl. Probab.* **1993**, *30*, 650–674. [CrossRef]
3. Meini, B. Solving QBD problems: The cyclic reduction algorithm versus the invariant subspace method. *Adv. Perform. Anal.* **1998**, *1*, 215–225.
4. Qiang, Y. On Latouche-Ramaswami's logarithmic reduction algorithm for Quasi-Birth-and-Death processes. *Stoch. Models* **2001**, *18*, 449–467.
5. Neuts, M.F. Explicit steady-state solutions to some elementary queueing models. *Oper. Res.* **1982**, *30*, 480–489. [CrossRef]
6. Latouche, G.; Ramaswami, V. A general class of Markov processes with explicit matrix-geometric solutions. *Oper. Res.* **1986**, *8*, 209–218.
7. Leeuwaarden, J.S.H.; Squillante, M.; Winands, E. Quasi-birth-and-death processes, lattice path counting, and hypergeometric functions. *J. Appl. Probab.* **2009**, *46*, 507–520. [CrossRef]
8. Doroudi, S.; Fralix, B.; Harchol-Balter, M. Clearing analysis on phases: Exact limiting probabilities for skip-free, unidirectional, quasi-birth-death processes. *Stoch. Syst.* **2016**, *6*, 420–458. [CrossRef]
9. Hajek, B. Birth-and-Death Processes on the Integers with Phases and General Boundaries. *J. Appl. Probab.* **1982**, *19*, 488–499. [CrossRef]
10. Naumov, V. Modified Matrix-Geometric Solution for Finite QBD processes. In *Advances in Algorithmic Methods for Stochastic Models*; Latouche, G., Taylor, P., Eds.; Notable Publications: Neshanic Station, NJ, USA, 2000; pp. 257–264.
11. Akar, N.; Sohraby, K. Finite and infinite QBD chains: A simple and unifying algorithmic approach. In Proceedings of the INFOCOM '97, Kobe, Japan, 7–11 April 1997; Volume 3, pp. 1105–1113.
12. Naoumov, V. Matrix-multiplicative approach to quasi-birth-and-death processes analysis. In *Matrix-Anal. Methods Stoch. Models*; Alfa, A.S., Chakravarthy, S.R., Eds.; Marcel Dekker: New York, NY, USA, 1996; pp. 87–106.
13. Krieger, U.R.; Naumov, V. Analysis of a delay-loss system with a superimposed markovian arrival process and state dependent service times. In *Numer. Solut. Markov Chain*; Plateau, B., Stewart, W.J., Silva, M., Eds.; Prensas Universitarias de Zaragoza: Zaragoza, Spain, 1999.
14. Avi-Itzhak, B.; Heyman, D.P. Approximate Queuing Models for Multiprogramming Computer Systems. *Oper. Res.* **1973**, *21*, 1212–1230. [CrossRef]
15. Konheim, A.G.; Reiser, M. Finite Capacity Queuing Systems with Applications in Computer Modeling. *SIAM J. Comput.* **1978**, *7*, 210–229. [CrossRef]
16. Busacott, J.A.; Shanthikumar, J.G. Design of Manufacturing Systems Using Queueing Models. *Queueing Syst.* **1992**, *12*, 135–214. [CrossRef]
17. Cai, X.; Heragu, S.S.; Yang Liu, Y. Modeling Automated Warehouses Using Semi-Open Queueing Networks. In *Handbook of Stochastic Models and Analysis of Manufacturing System Operations*; Smith, J., Tan, B., Eds.; Springer: New York, NY, USA, 2003.
18. Zou, B.P.; de Koster, R.; Xu, X.H. Operating Policies in Robotic Compact Storage and Retrieval Systems. *Transp. Sci.* **2008**, *52*, 788–811. [CrossRef]
19. Jia, J.; Heragu, S.S. Solving Semi-Open Queuing Networks. *Oper. Res.* **2009**, *57*, 391–401. [CrossRef]
20. Roy, D. Semi-Open Queuing Networks: A Review of Stochastic Models, Solution Methods and New Research Areas. *Int. J. Prod. Res.* **2016**, *54*, 1735–1752. [CrossRef]

21. Ekren, B.Y.; Heragu, S.S.; Krishnamurthy, A.; Malmborg, C.J. Matrix-Geometric Solution for Semi-Open Queuing Network Model of Autonomous Vehicle Storage and Retrieval System. *Comput. Ind. Eng.* **2014**, *68*, 78–86. [CrossRef]
22. Dhingraa, V.; Kumawat, G.L.; Roy, D.; de Koster, R. Solving Semi-open Queuing Networks with Time-varying Arrivals: An Application in Container Terminal Landside Operations. *Eur. J. Oper. Res.* **2018**, *267*, 855–876. [CrossRef]
23. Kim, J.; Dudin, A.; Dudin, S.; Kim, C. Analysis of a semi-open queueing network with Markovian arrival process. *Perform. Eval.* **2018**, *120*, 1–19. [CrossRef]
24. Baskett, F.; Chandy, K.M.; Muntz, A.A.; Palacios, F.G. Open, closed, and mixed networks of queues with different classes of customers. *J. ACM* **1975**, *22*, 248–260. [CrossRef]
25. Bolch, G.; Greiner, S.; de Meer, H.; Trivedi, K.S. *Queueing Networks and Markov Chains: Modeling and Performance Evaluation with Computer Science Applications*, 2nd ed.; John Wiley & Sons: Hoboken, NJ, USA, 2006.

mathematics

MDPI

Article

Optimization of Queueing Model with Server Heating and Cooling

Olga Dudina [1,2] and Alexander Dudin [1,2,*]

[1] Department of Applied Mathematics and Computer Science, Belarusian State University, 4, Nezavisimosti Ave., 220030 Minsk, Belarus

[2] Applied Mathematics and Comminications Technology Institute, Peoples' Friendship University of Russia (RUDN University), 6 Miklukho-Maklaya St, Moscow 117198, Russia

* Correspondence: dudin@bsu.by

Received: 29 July 2019; Accepted: 19 August 2019; Published: 21 August 2019

Abstract: The operation of many real-world systems, e.g., servers of data centers, is accompanied by the heating of a server. Correspondingly, certain cooling mechanisms are used. If the server becomes overheated, it interrupts processing of customers and needs to be cooled. A customer is lost when its service is interrupted. To prevent overheating and reduce the customer loss probability, we suggest temporal termination of service of new customers when the temperature of the server reaches the predefined threshold value. Service is resumed after the temperature drops below another threshold value. The problem of optimal choice of the thresholds (with respect to the chosen economical criterion) is numerically solved under quite general assumptions about the parameters of the system (Markovian arrival process, phase-type distribution of service time, and accounting for customers impatience). Numerical examples are presented.

Keywords: processor heating and cooling; markovian arrival process; phase-type service time distribution; impatience

1. Introduction

The goal of operation of many real-world systems is to obtain profit via providing service to some customers. For example, in data centers, the profit is obtained via storing and retrieving the information for users on demand. The operation of such systems is possible only under fulfillment of various limitations. An important problem in organization of the operation of data centers is the effective cooling of servers. High performance servers generate a lot of heat and it is necessary to effectively cool the central processing unit, memory modules, power supplies, graphics processing units and other devices to avoid system overheating and premature failures (see, e.g., [1]).

It is clear that, to maximize the profit, it is necessary to use the power of the available server to a maximum extent. However, this may cause overheating of the server, the loss of a customer who was using the service during the overheating moment and a temporal termination of the service for cooling the server. To prevent overheating of the server during service, it sounds reasonable to stop new services if the temperature of the server reaches some level (threshold). Definitely, this threshold should be less than the critical level but more or less close to this level. Otherwise, a certain part of the server capacity is not utilized and this may lead to the loss of some profit. If service can be postponed or interrupted, it is necessary to specify when service will be resumed. This can be done by means of introducing one more threshold. Service is resumed when the temperature of the server is dropped to this threshold. It is obvious that this threshold should be less than the first threshold. The difference between the thresholds should not be too large. Otherwise, again, the server capacity is under-utilized. However, if the difference is too small, the bans and permissions to start new services can occur too frequently and this can be charged by a decision-maker. Therefore, the problem of optimal choice of the two thresholds is not trivial and challenging.

In this paper, we numerically solve this problem in the following way. Under any fixed pair of thresholds, the behavior of the system is described by a Markov chain. This Markov chain is multi-dimensional because it has to include the components defining the number of customers in the system, the current state of the server (idle, operating, or cooling) and its temperature, underlying processes of customer arrival and service processes. Due to the existence of periods when service is not provided, in our model, we account for the possible impatience of the customers waiting in the queue. Due to considering impatience, this Markov chain does not belong to the class of level-independent quasi-birth-and-death processes and its analysis is non-trivial. We use results from [2,3] for computation of the stationary probabilities of the states of the chain. Having the stationary probabilities computed, we derive formulas for computation of the main performance indicators of the system and the cost criterion for any fixed pair of the thresholds defining behavior of the system. This allows us to numerically solve the problem of choosing the optimal values of the thresholds.

The considered model is very close to the models in which some additional resource is required to provide service to a customer. These models include, in particular, so-called queueing/inventory models (see, e.g., [4]), queueing systems with energy harvesting (see, e.g., [5]), queueing models with paired customers (see [6]), assembly-like queue (see [7]), passenger-taxi or double-ended queues (see [8]), coupled queues (see [9]), etc. In our model, the role of the additional resource is played by the lag between the critical and current value of the server's temperature. The essential difference between our model and the above mentioned models consists of the following. Usually, the additional resource has the influence on the behavior of the queueing system only at the potential service beginning moments. If the resource is available, the known required for service amount of the resource is reserved. Service starts and, then, successfully finishes. Otherwise, if the resource is not available at the potential service beginning moment, service is cancelled or postponed until the required amount of the additional resource becomes available. In our model, we have a more complicated situation: the "additional resource" has a permanent influence on the behavior of the queueing system. Required for a customer service amount of resource is uncertain. It cannot be reserved and, as a consequence, the started service (with available resource at the service beginning moment) can be terminated ahead of the schedule and the customer is lost if the resource becomes unavailable already during service of this customer. Namely, due to uncertainty of the amount of the required resource, it is necessary to ban starting new service when the resource is still available but the number of available units of the resource is less than the threshold value.

The structure of the paper is the following. Section 2 contains the description of the mathematical model of the considered system. The stationary distribution of the multi-dimensional Markov chain describing the number of customers, current operation mode, excess of heating and underlying process of the MAP arrival processes of customers and PH distributed service time is analyzed in Section 3. The generator of this Markov chain is derived. Formulas for computing the key performance measures of the system, including the probabilities of a customer loss (due to the server overheating and due to the customer impatience) are given in Section 4. The results of numerical experiments that illustrate the dependence of the key performance measures of the system on the thresholds are described in Section 5. An optimization problem is considered in brief. Section 6 contains the conclusion of the paper.

2. Description of the Model

We consider a single-server queueing system that has an input buffer of an infinite capacity. Figure 1 illustrates the structure of the system under study.

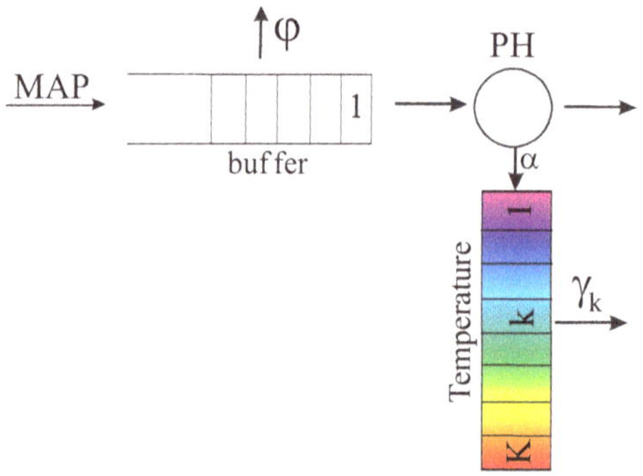

Figure 1. Structure of the system.

Customer arrival is defined as the Markovian Arrival Process (MAP). Arrivals are governed by the underlying Markov chain $\nu_t,\ t \geq 0$, with the finite state space $\{0, 1, ..., W\}$. The residence time of this chain in the state ν has an exponential distribution with the parameter $\lambda_\nu,\ \nu = \overline{0, W}$. Here and in what follows, the notation $\nu = \overline{0, W}$ means that the integer parameter ν takes values from the set $\{0, 1, \ldots, W\}$. When the residence time in the state ν expires, with probability $p_{\nu,\nu'}(0)$, the process ν_t makes a transition to the state ν' without generation of a customer, $\nu' = \overline{0, W}$ $\nu \neq \nu'$, and, with probability $p_{\nu,\nu'}(1)$, the process ν_t makes a transition to the state ν' with a generation of a customer, $\nu, \nu' = \overline{0, W}$. The behavior of this arrival process is completely defined by the matrices D_0 and D_1 consisting of the entries $(D_1)_{\nu,\nu'} = \lambda_\nu p_{\nu,\nu'}(1),\ \nu, \nu' = \overline{0, W},$ and $(D_0)_{\nu,\nu} = -\lambda_\nu, \nu = \overline{0, W},\ (D_0)_{\nu,\nu'} = \lambda_\nu p_{\nu,\nu'}(0),\ \nu, \nu' = \overline{0, W},\ \nu \neq \nu'$. The matrix $D(1) = D_0 + D_1$ is assumed to be irreducible and is the generator of the process $\nu_t, t \geq 0$.

The mean arrival rate λ is computes as $\lambda = \boldsymbol{\theta} D_1 \mathbf{e}$ where $\boldsymbol{\theta}$ is the unique solution to the equations $\boldsymbol{\theta} D(1) = \mathbf{0},\ \boldsymbol{\theta}\mathbf{e} = 1$. Hereinafter, $\mathbf{e}$ denotes a column vector consisting of 1's, and $\mathbf{0}$ is a zero row vector.

For more information about the MAP and its properties, see [10–12].

The service time of a customer has a PH distribution defined by the stochastic row vector $\boldsymbol{\beta}$ and sub-generator S. This time has the following interpretation. Let $m_t,\ t \geq 0$, be the continuous-time Markov process having a finite state space $\{1, \ldots, M, M+1\}$. The states $\{1, \ldots, M\}$ are transient and $M+1$ is the absorbing state. The initial state of this process at the moment of beginning of PH distributed time is randomly selected among the transient states $\{1, \ldots, M\}$ according to the distribution defined by the entries of the row vector $\boldsymbol{\beta} = (\beta_1, \ldots, \beta_M)$. Then, the process m_t makes transitions within the set $\{1, \ldots, M\}$ of the transient states with intensities defined by the entries of the sub-generator S or to the absorbing state. The intensities of the transition to the absorbing state are given by the entries of the column vector $\mathbf{S}_0 = -S\mathbf{e}$. Transition to the absorbing state implies the end of PH distributed time.

The Laplace–Stieltjes transform of the PH distribution is defined as $\boldsymbol{\beta}(sI - S)^{-1}\mathbf{S}_0,\ Re\ s > 0$. The mean service time is equal to $b_1 = \boldsymbol{\beta}(-S)^{-1}\mathbf{e}$. For more detailed information about the PH distribution, see [13]. Its applicability for good approximation of an arbitrary distribution is mentioned, e.g., in [14]. When the server becomes idle, the underlying process $m_t,\ t \geq 0$, does not make any transitions.

The problem of constructing the vector $\boldsymbol{\beta}$ and the matrices D_0 and $D_1,\ S$ based on available statistics regarding the real arrival and service processes is extensively addressed in the existing literature and may be solved following the results from, e.g., the papers [15–17].

During service, the server generates the heat and the temperature of the server is permanently monitored. Without essential loss of generality, we suppose that the temperature of the server is graded in some discrete units, e.g. degrees Celsius. The server can operate when this temperature is in the interval from K' to K''. According to the 2011 version of recommendations of the American Society of Heating, Refrigerating and Air Conditioning Engineers (ASHRAE), for class A1 systems the temperature of the server has to be in the range from 15 C° to 32 C°. To simplify notations, we do not keep track of the absolute temperature of the server, but the excess of the temperature over the lower temperature level. This means that we assume that the (relative) temperature of the server has to be in the range from 0 to K where $K = K'' - K'$. When the temperature of the server reaches the upper level K, service of customers becomes impossible. We say that this server becomes overheated. The server temporarily stops its work, is considered blocked and has to be cooled. A customer using the service when overheating occurs is assumed to be lost. We suggest that the server does not generate the heat when it does not work (is idle or blocked). When the server is working, the rate of the server heating is assumed to be equal to α degrees during unit of time, $\alpha > 0$. In parallel to heating of the server, it is permanently cooled. We assume that the cooling rate is equal to $\gamma_k,\ \gamma_k \geq 0$, when the current temperature of the server is equal to $k,\ k = \overline{0, K}$.

When the server becomes overheated, it stops generation of the heat and only is cooled. We assume that the server remains blocked until its temperature drops to the level (threshold) $K_1,\ K_1 < K$. After that, the server becomes unblocked and can start service. We assume that the customers residing in the buffer are impatient. Each of these customers departs from the buffer without receiving service (is lost) independently of other waiting customers after a "patience time" expires. This time is exponentially distributed with the parameter ϕ.

The overheating of the server implies the loss of the potential profit gained by customers' service. This loss is related to the loss of customers, during service of which the overheating occurs, and the loss of a capacity (throughput) of the server spent on service of such customers. The overheating may require server recovery, not only cooling. Therefore, it is desirable to avoid the overheating. To prevent the overheating occurrence, it is reasonable to stop new services when the temperature of the server becomes pretty high. We assume that the threshold K_2 is fixed such that $K_1 < K_2 \leq K$. The server cannot start new services if its temperature is equal or greater than K_2. However, the ongoing service continues. It cannot be interrupted unless the server becomes overheated, i.e., its temperature becomes equal to K. If this service is successfully finished while the server does not become overheated, the server remains blocked and does not start new services until its temperature drops to K_1.

It is obvious that the values of performance indicators of the system depend on the choice of the pair of thresholds (K_1, K_2) and our first goal is to provide a way for computing the values of these measures for any fixed values of thresholds. To this end, we elaborate the algorithm for computation of the stationary distribution of the system states.

3. Process of System States and Its Analysis

Let the critical temperature K and thresholds K_1 and K_2 be fixed, $0 \leq K_1 < K_2 \leq K$.

It is easy to see that the behavior of the considered system can be described by the following regular irreducible continuous-time Markov chain

$$\xi_t = \{n_t, r_t, k_t, \nu_t, m_t\},\ t \geq 0,$$

where, during the epoch $t,\ t \geq 0$,

- n_t is the number of customers in the buffer, $n_t \geq 0$.
- $r_t,\ r_t = \overline{0, 2}$, is the server state: $r_t = 0$ if the server is idle, $r_t = 1$ if the server is busy, and $r_t = 2$ if the server is blocked.
- k_t is the temperature of the server, $k_t = \overline{0, K}$.
- ν_t is the state of the underlying process of the MAP, $\nu_t = \overline{0, W}$.
- m_t is the state of the underlying process of the PH service process, $m_t = \overline{1, M}$.

The Markov chain ξ_t, $t \geq 0$, has the following state space:

$$\left(\{0,0,k,v\},\, k=\overline{0,K_2-1}\right)\bigcup\left(\{n,1,k,v,m\},\, n\geq 0,\, k=\overline{0,K-1},\ m=\overline{1,M}\right)\bigcup$$

$$\left(\{n,2,k,v\},\, n\geq 0,\, k=\overline{K_1+1,K}\right),\ v=\overline{0,W}.$$

To formally define the Markov chain ξ_t, we need to specify its transition rates within this state space. Since this chain has five components when the server is not idle or blocked and four components when the server is idle or blocked, to avoid operations with multi-dimensional arrays, it is necessary to enumerate the states in some order. We assume the lexicographic ordering. This means that firstly the states of the Markov chain ξ_t are numbered in the increasing order of the component n_t. Within the set of the states having the same value, say n, $n \geq 0$, of this component, the states are numbered in the increasing order of the component r_t. Within the set of the states having the same values, say (n,r), $n \geq 0$, $r = 0,1,2$, of these components, the states are numbered in the increasing order of the component k_t. Within the set of the states having the same values, say (n,r,k), $n \geq 0$, $r = 0,1,2$, $k = \overline{0,K}$, of the three components, the states are numbered in the increasing order of the component v_t, $v_t = \overline{0,W}$. Finally, the states from the sets $(n,1,k,v)$ are numbered in the increasing order of the component m_t, $m_t = \overline{1,M}$.

Let us denote by G the generator of the Markov chain ξ_t. It follows from the introduced enumeration of the components of the chain that G is the matrix consisting of the blocks $G_{n,n'}$, $n,n' \geq 0$, $|n-n'| \leq 1$, defining the intensities of transitions from the states having the value n of the component n_t to the states having the value n' of this component.

Theorem 1. *The infinitesimal generator G of the Markov chain ξ_t, $t \geq 0$, has the following block-tridiagonal structure:*

$$G=\begin{pmatrix} G_{0,0} & G_{0,1} & O & O & O & \dots \\ G_{1,0} & G_{1,1} & G_{1,2} & O & O & \dots \\ O & G_{2,1} & G_{2,2} & G_{2,3} & O & \dots \\ \vdots & \vdots & \vdots & \vdots & \vdots & \ddots \end{pmatrix}.$$

Here, the blocks $(G_{0,0}^{r,r'})_{r,r'=\overline{0,2}}$ of the matrix $G_{0,0}$, whose diagonal entries are negative and define, up to the sign, the intensities of the exit of the Markov chain ξ_t from the corresponding states and the non-diagonal entries define the intensities of transitions that do not imply customers appearance in the empty buffer, have the following form:

$$G_{0,0}^{0,0} = I_{K_2} \otimes D_0 + (E_{K_2}^- - C_{K_2}) \otimes I_{\bar{W}},$$

$$G_{0,0}^{0,1} = I_{K_2,K} \otimes D_1 \otimes \boldsymbol{\beta},$$

$$G_{0,0}^{0,2} = O_{K_2\bar{W},(K-K_1)\bar{W}},$$

$$G_{0,0}^{1,0} = I_{K,K_2} \otimes I_{\bar{W}} \otimes \mathbf{S}_0,$$

$$G_{0,0}^{1,1} = I_K \otimes D_0 \oplus S + (E_K^- - C_K + \alpha(E^+ - I_K)) \otimes I_{\bar{W}M},$$

$$G_{0,0}^{1,2} = \tilde{I}_{K,K-K_1} \otimes I_{\bar{W}} \otimes \mathbf{S}_0 + \alpha\hat{I} \otimes I_{\bar{W}} \otimes \mathbf{e}_M,$$

$$G_{0,0}^{2,0} = \gamma_{K_1+1}\tilde{I}_{K-K_1,K_2} \otimes I_{\bar{W}},$$

$$G_{0,0}^{2,1} = O_{(K-K_1)\bar{W},K\bar{W}M},$$

$$G_{0,0}^{2,2} = I_{K-K_1} \otimes D_0 + (\tilde{E} - \tilde{C}) \otimes I_{\bar{W}}.$$

The matrix $G_{0,1}$, whose entries define the intensities of transitions when a customer arrives to the empty buffer, has the form

$$G_{0,1} = \begin{pmatrix} O_{K_2\bar{W},K\bar{W}M} & O_{K_2\bar{W},(K-K_1)\bar{W}} \\ G_{0,1}^{1,1} & O_{K\bar{W}M,(K-K_1)\bar{W}} \\ O_{(K-K_1)\bar{W},K\bar{W}M} & G_{0,1}^{2,2} \end{pmatrix}$$

where

$$G_{0,1}^{1,1} = I_K \otimes D_1 \otimes I_M,$$

$$G_{0,1}^{2,2} = I_{K-K_1} \otimes D_1.$$

The matrix $G_{1,0}$, whose entries define the intensities of transitions when the single customer staying in the buffer, departs from the buffer (due to the impatience or service beginning), has the form

$$G_{1,0} = \begin{pmatrix} O_{K\bar{W}M,K_2\bar{W}} & G_{1,0}^{1,1} & O_{(K-K_1)\bar{W},K_2\bar{W}} \\ O_{(K-K_1)\bar{W},K_2\bar{W}} & G_{1,0}^{2,1} & G_{1,0}^{2,2} \end{pmatrix},$$

where

$$G_{1,0}^{1,1} = \phi I_{K\bar{W}M} + B \otimes I_{\bar{W}} \otimes \mathbf{S}_0 \boldsymbol{\beta},$$

$$G_{1,0}^{2,1} = \gamma_{K_1+1} \tilde{I}_{K-K_1,K} \otimes I_{\bar{W}} \otimes \boldsymbol{\beta},$$

$$G_{1,0}^{2,2} = \phi I_{(K-K_1)\bar{W}}.$$

The blocks $(G_{n,n}^{r,r'})_{r,r'=\overline{1,2}}$ of the matrix $G_{n,n}$, $n \geq 1$, whose diagonal entries are negative and define, up to the sign, the intensities of the exit of the Markov chain ξ_t from the corresponding states when the number of customers in the buffer is equal to n, $n \geq 1$, and the non-diagonal entries define the intensities of transitions that do not imply the change of the number of customers in the buffer, have the following form:

$$G_{n,n}^{1,1} = I_K \otimes D_0 \oplus S + (E_K^- - C_K + \alpha(E^+ - I_K) - n\phi I_K) \otimes I_{\bar{W}M},$$

$$G_{n,n}^{1,2} = \bar{I}_{K,K-K_1} \otimes I_{\bar{W}} \otimes \mathbf{S}_0 + \alpha \hat{I} \otimes I_{\bar{W}} \otimes \mathbf{e}_M,$$

$$G_{n,n}^{2,1} = O_{(K-K_1)\bar{W},K\bar{W}M},$$

$$G_{n,n}^{2,2} = I_{K-K_1} \otimes D_0 + (\tilde{E} - \tilde{C}) \otimes I_{\bar{W}} - n\phi I_{(K-K_1)\bar{W}}, \; n \geq 1.$$

The blocks $(G_{n,n+1}^{r,r'})_{r,r'=\overline{1,2}}$ of the matrix $G_{n,n+1}$, $n \geq 1$, whose entries define the intensities of increasing the number of customers in the buffer from n to $n+1$, have the following form:

$$G_{n,n+1}^{1,1} = I_K \otimes D_1 \otimes I_M,$$

$$G_{n,n+1}^{1,2} = O_{K\bar{W}M,(K-K_1)\bar{W}},$$

$$G_{n,n+1}^{2,1} = O_{(K-K_1)\bar{W},K\bar{W}M},$$

$$G_{n,n+1}^{2,2} = I_{K-K_1} \otimes D_1, \; n \geq 1.$$

The non-zero blocks $(G_{n,n-1}^{r,r'})_{r,r'=\overline{1,2}}$ of the matrix $G_{n,n-1}$, $n \geq 2$, whose entries define the intensities of decreasing the number of customers in the buffer from n to $n-1$, have the following form:

$$G_{n,n-1}^{1,1} = n\phi I_{K\bar{W}M} + B \otimes I_{\bar{W}} \otimes \mathbf{S}_0 \boldsymbol{\beta},$$

$$G_{n,n-1}^{2,1} = \gamma_{K_1+1} \tilde{I}_{K-K_1,K} \otimes I_{\bar{W}} \otimes \boldsymbol{\beta},$$

$$G^{2,2}_{n,n-1} = n\phi I_{(K-K_1)\bar{W}},\ n \geq 2.$$

Here,

- *I is the identity matrix, and O is a zero matrix of an appropriate dimension.*
- $\bar{W} = W + 1$;
- *$\otimes$ and $\oplus$ are the symbols of the Kronecker product and the sum of matrices, respectively.*
- *E_l^- is a square matrix of size l with all zero entries except the entries $(E_l^-)_{k,k-1} = \gamma_k,\ k = \overline{1, l-1}$.*
- *C_l is a square matrix of size l with all zero entries except the entries $(C_l)_{k,k} = \gamma_k,\ k = \overline{1, l-1}$.*
- *$I_{K_2,K}$ is a matrix of size $K_2 \times K$ with all zero entries except the entries $(I_{K_2,K})_{n,n},\ n = \overline{0, K_2 - 1}$, which are equal to 1.*
- *I_{K,K_2} is a matrix of size $K \times K_2$ with all zero entries except the entries $(I_{K,K_2})_{n,n},\ n = \overline{0, K_2 - 1}$, which are equal to 1.*
- *E^+ is a square matrix of size K with all zero entries except the entries $(E^+)_{k,k+1},\ k = \overline{0, K-2}$, which are equal to 1.*
- *$\bar{I}_{K,K-K_1}$ is a matrix of size $K \times (K - K_1)$ with all zero entries except the entries $(\bar{I}_{K,K-K_1})_{n,n-K_1-1},\ n = \overline{K_2, K-1}$, which are equal to 1.*
- *$\hat{I}$ is a matrix of size $K \times (K - K_1)$ with all zero entries except the entry $(\hat{I})_{K-1,K-K_1-1}$, which is equal to 1.*
- *$\tilde{I}_{K-K_1,K_2}$ is a matrix of size $(K - K_1) \times K_2$ with all zero entries except the entry $(\tilde{I}_{K-K_1,K_2})_{0,K_1}$, which is equal to 1.*
- *$\tilde{E}$ is a square matrix of size $K - K_1$ with all zero entries except the entries $(\tilde{E})_{k,k-1} = \gamma_{K_1+k+1},\ k = \overline{1, K - K_1 - 1}$.*
- *$\tilde{C}$ is a square matrix of size $K - K_1$ with all zero entries except the entries $(\tilde{C})_{k,k} = \gamma_{K_1+k+1},\ k = \overline{0, K - K_1 - 1}$.*
- *B is a square matrix of size K with all zero entries except the entries $(B)_{k,k} = 1,\ k = \overline{0, K_2 - 1}$.*
- *$\tilde{I}_{K-K_1,K}$ is a matrix of size $(K - K_1) \times K$ with all zero entries except the entry $(\tilde{I}_{K-K_1,K})_{0,K_1} = 1$.*

The proof of the theorem is implemented via the careful analysis of various scenarios of the system behavior at the moments of changing the states of the underlying processes of arrivals and service, changing the temperature of the server due to heating and cooling, customers departure due to impatience. The symbols of Kronecker product and sum of matrices are very helpful for description of transition intensities of several independent Markov processes.

It can be easily shown that the Markov chain ξ_t belongs to the class of Asymptotically Quasi–Toeplitz Markov Chains ($AQTMC$) (see [2]).

Theorem 2. *The stationary distribution of the Markov chain ξ_t exists for any values of the system parameters.*

The assertion of the theorem stems from the fact that the customers staying in the buffer are assumed to be impatient ($\phi > 0$). The strict proof of Theorem 2 can be done by using the results from [2]. This proof is straightforward and rather routine, thus it is omitted here.

Let us denote by $\pi(n, r, k)$ the row vector of stationary probabilities of the states of the chain having the value (n, r, k) of the first three components listed in the described above order.

In addition, denote

$$\pi(0,0) = (\pi(0,0,0), \ldots, \pi(0,0,K_2 - 1)),$$

$$\pi(n,1) = (\pi(n,1,0), \ldots, \pi(n,1,K-1)),$$

$$\pi(n,2) = (\pi(n,2,K_1+1), \ldots, \pi(n,2,K)),\ n \geq 0,$$

$$\pi(0) = (\pi(0,0), \pi(0,1), \pi(0,2)),\ \pi(n) = (\pi(n,1), \pi(n,2)),\ n \geq 1.$$

Because the state space of the Markov chain ξ_t is infinite and the generator G of this chain does not have Toeplitz-like structure, the problem of computation of the vectors $\pi(n),\ n \geq 0$, is not easy.

Fortunately, the chains with the generator of such a type were analysed in [2,3] and the algorithms developed in those papers allow computing these vectors.

4. Performance Indicators

Once the vectors $\boldsymbol{\pi}(n),\ n \geq 0$, have been computed, we can calculate various performance indicators of the system.

The mean number N of customers in the buffer is computed by

$$N = \sum_{n=1}^{\infty} n\boldsymbol{\pi}(n)\mathbf{e}.$$

The average temperature T of the server is computed by

$$T = \sum_{k=1}^{K_2-1} k\boldsymbol{\pi}(0,0,k)\mathbf{e} + \sum_{n=0}^{\infty}\sum_{k=1}^{K-1} k\boldsymbol{\pi}(n,1,k)\mathbf{e} + \sum_{n=0}^{\infty}\sum_{k=K_1+1}^{K} k\boldsymbol{\pi}(n,2,k)\mathbf{e}.$$

The variance of the temperature of the server is equal to

$$\sum_{k=1}^{K_2-1} k^2\boldsymbol{\pi}(0,0,k)\mathbf{e} + \sum_{n=0}^{\infty}\sum_{k=1}^{K-1} k^2\boldsymbol{\pi}(n,1,k)\mathbf{e} + \sum_{n=0}^{\infty}\sum_{k=K_1+1}^{K} k^2\boldsymbol{\pi}(n,2,k)\mathbf{e} - T^2.$$

The probability P_{idle} that the server is idle at an arbitrary moment is

$$P_{idle} = \boldsymbol{\pi}(0,0)\mathbf{e}.$$

The probability P_{imm} that the server is idle at the moment of an arbitrary customer arrival (and this customer immediately starts service) is

$$P_{imm} = \frac{1}{\lambda}\boldsymbol{\pi}(0,0)(I_{K_2} \otimes D_1)\mathbf{e}.$$

The probability P_{busy} that the server is busy at an arbitrary moment is

$$P_{busy} = \sum_{n=0}^{\infty} \boldsymbol{\pi}(n,1)\mathbf{e}.$$

The average number N_{system} of customers in the system is computed by $N_{system} = N + P_{busy}$.
The probability P_{block} that the server is blocked at an arbitrary moment is

$$P_{block} = \sum_{n=0}^{\infty} \boldsymbol{\pi}(n,2)\mathbf{e}.$$

The probability P_{imp} of an arbitrary customer loss due to impatience is

$$P_{imp} = \frac{\phi N}{\lambda}.$$

The intensity λ_{out} of the flow of served customers is

$$\lambda_{out} = \sum_{n=0}^{\infty} \boldsymbol{\pi}(n,1)(\mathbf{e}_{K\bar{W}} \otimes \mathbf{S}_0).$$

The probability $P_{overheating}$ of an arbitrary customer loss due to the server overheating is

$$P_{overheating} = \alpha\lambda^{-1}\sum_{n=0}^{\infty}\pi(n,1,K-1)\mathbf{e}.$$

The intensity $l_{vacation}$ of the transition after the service completion to the vacation regime (the server is overheated or is preventively switched-off for cooling) is

$$l_{vacation} = \sum_{n=0}^{\infty}\pi(n,1)(\tilde{I}_{K,K-K_1}\otimes I_{\bar{W}}\otimes \mathbf{S}_0)\mathbf{e}.$$

The probability P_{loss} of an arbitrary customer loss is computed as

$$P_{loss} = P_{imp} + P_{overheating} = 1 - \frac{\lambda_{out}}{\lambda}.$$

5. Numerical Example

Let the MAP arrival flow be defined by the matrices

$$D_0 = \begin{pmatrix} -0.3379101412 & 0 \\ 0 & -0.0109675577 \end{pmatrix},\ D_1 = \begin{pmatrix} 0.3356635104 & 0.0022466308 \\ 0.0061087109 & 0.0048588468 \end{pmatrix}.$$

The mean arrival rate is $\lambda = 5$, the coefficient of correlation of two successive intervals between arrivals $c_{cor} = 0.2$, and the squared coefficient of variation of these intervals $c_{var} = 12.4$.

Let the maximum value of the temperature be $K = 50$, and the rate of the heat generation be $\alpha = 1$. The rate of the server cooling when its temperature is equal to k is $\gamma_k = \frac{0.3k}{K}$, $k = \overline{1, K-1}$, $\gamma_K = 0.01$.

The rate of a customer departure from the buffer due to impatience is $\phi = 0.0015$.

The service time of a customer has the PH distribution with the irreducible representation (β, S) where $\beta = (0.9, 0.1)$, $S = \begin{pmatrix} -6 & 0 \\ 0 & -0.2 \end{pmatrix}$. The average service time is equal to 0.65 and the squared coefficient of variation of the service time is equal to 10.95.

Let us vary the threshold K_1 over the interval $[0, K)$ and the parameter K_2 over the interval $[K_1 + 1, K]$.

Figures 2–4 illustrate the dependence of the average number N of customers in the buffer, the average intensity λ_{out} of the flow of served customers and the average temperature T of the server on the values of K_1 and K_2.

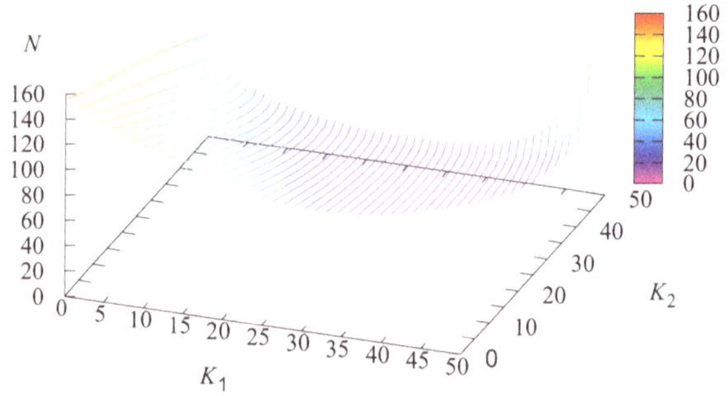

Figure 2. Dependence of the average number N of customers in the buffer on the values of K_1 and K_2.

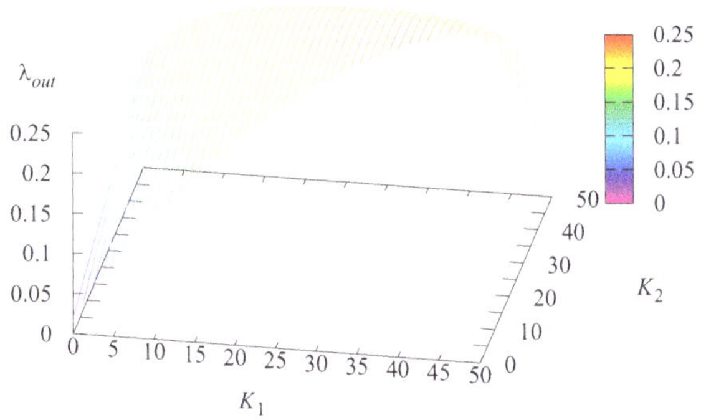

Figure 3. Dependence of the average average intensity λ_{out} of the flow of served customers on the values of K_1 and K_2.

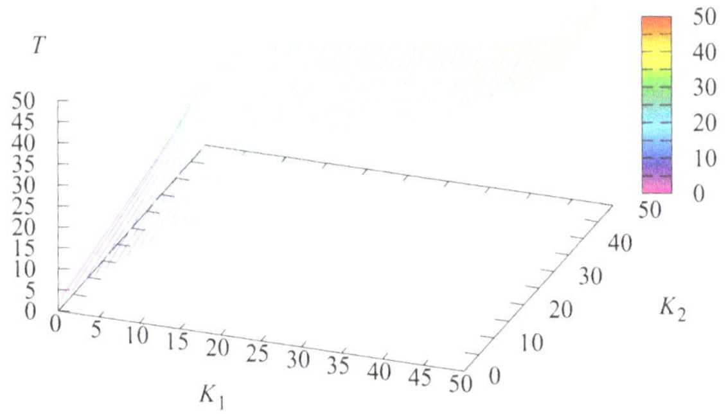

Figure 4. Dependence of the average temperature T of the server on the values of K_1 and K_2.

It can be observed in Figure 2 that the average number N of customers in the buffer increases when the threshold K_2 grows. This is easily explained by the fact that the probability of overheating occurrence increases when the threshold K_2 becomes close to the critical temperature. It is assumed that the rate of the server cooling is small after overheating occurrence, the blocking period of the server becomes large and a lot of customers stay in the buffer. It should be noted that, for any fixed value of K_2 there exists a value of K_1, which minimizes the average number N. The average intensity λ_{out} of the flow of served customers is small when both thresholds K_1 and K_2 are small (low temperature of the server is guaranteed at expense of managing too long blocking periods) and when both thresholds K_1 and K_2 are large (the server is quite often overheated, which causes the corresponding loss of customers). This intensity λ_{out} is much higher for intermediate values of the thresholds K_1 and K_2. Figure 4 well matches to the just given explanation of the surface in Figure 3.

Figures 5–7 illustrate the dependence of the probability P_{idle} that the server is idle, the probability P_{busy} that the server is busy, and the probability P_{block} that the server is blocked at an arbitrary moment on the values of K_1 and K_2.

The probability P_{idle} that the server is idle and the probability P_{busy} that the server is busy also are maximal for intermediate values of thresholds K_1 and K_2. As expected, the probability P_{block} that the server is blocked decreases when the threshold K_1 grows.

Figures 8–10 illustrate the dependence of the probability P_{imp} that an arbitrary customer is lost due to impatience, the loss probability $P_{overheating}$ that an arbitrary customer is lost due to server overheating, and the probability P_{loss} that an arbitrary customer is lost on the values of K_1 and K_2.

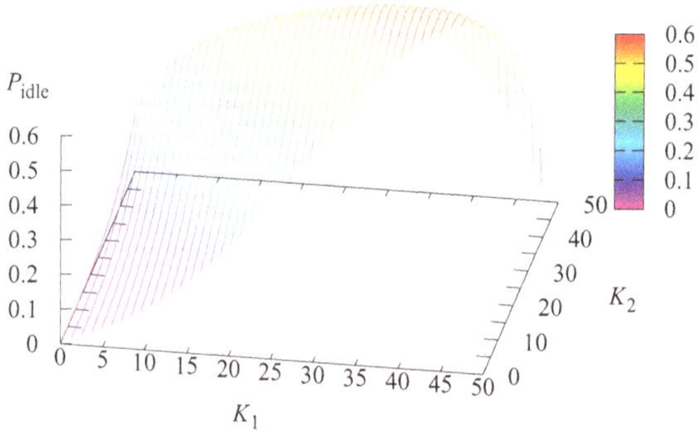

Figure 5. Dependence of the probability P_{idle} that the server is idle on the values of K_1 and K_2.

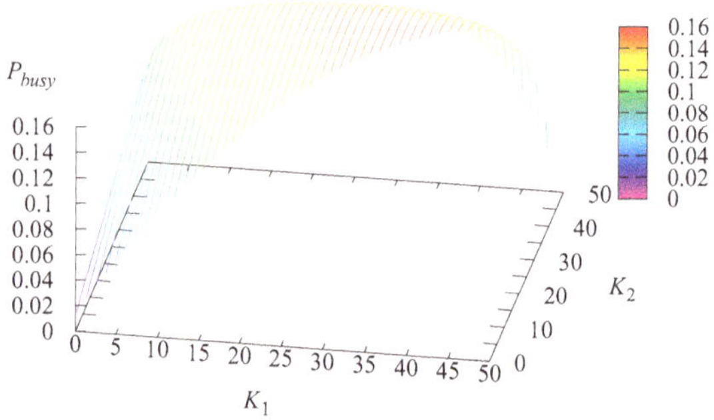

Figure 6. Dependence of the probability P_{busy} that the server is busy on the values of K_1 and K_2.

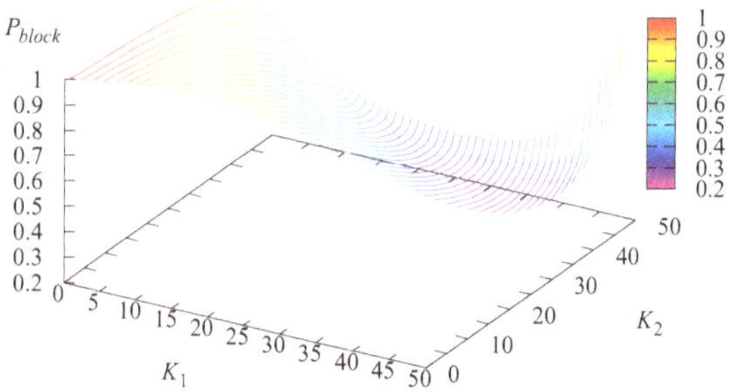

Figure 7. Dependence of the probability P_{block} that the server is blocked on the values of K_1 and K_2.

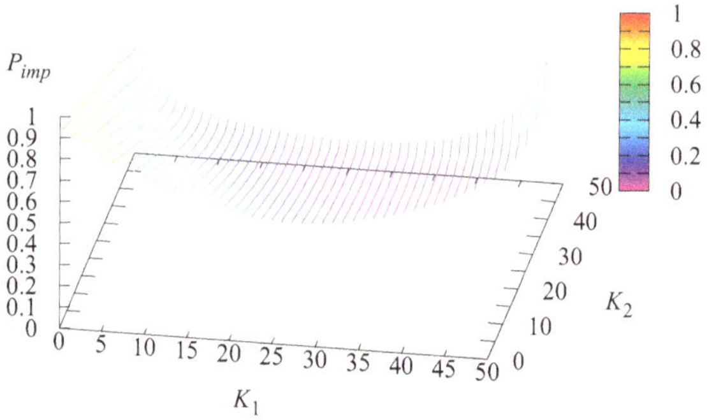

Figure 8. Dependence of the probability P_{imp} that an arbitrary customer is lost due to impatience on the values of K_1 and K_2.

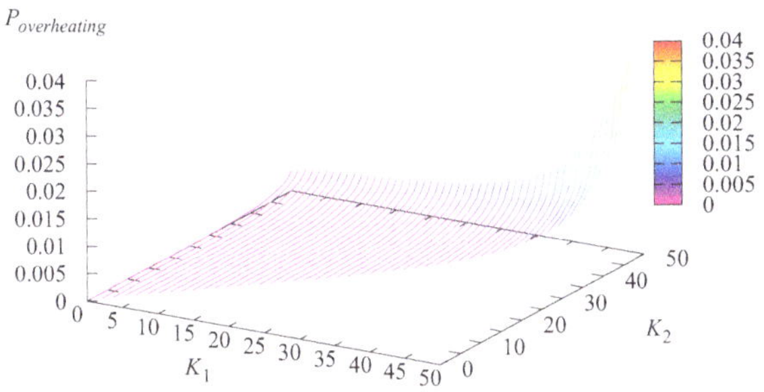

Figure 9. Dependence of the probability $P_{overheating}$ that an arbitrary customer is lost due to server overheating on the values of K_1 and K_2.

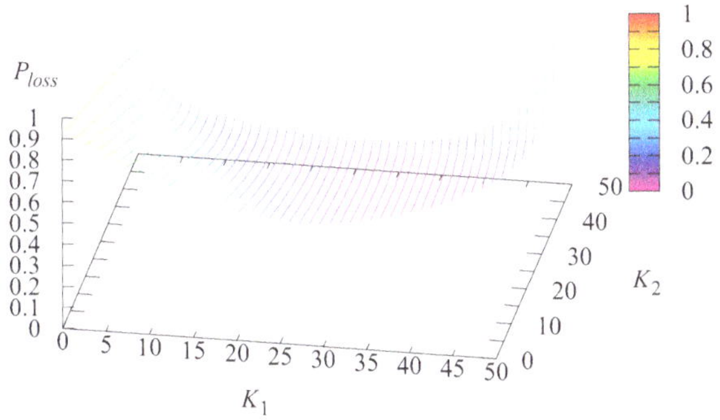

Figure 10. Dependence of the probability P_{loss} of an arbitrary customer loss on the values of K_1 and K_2.

Figure 9 evidently shows that the loss probability $P_{overheating}$ that an arbitrary customer is lost due to server overheating sharply increases when the thresholds K_1 and K_2 grow. Therefore, the proposed mechanism for preventing overheating is highly effective. It can be observed in Figures 8–10 that there exists a pair of the thresholds that minimizes the loss probability P_{loss}. The minimal value of the probability P_{loss} in this example is equal to 0.065255 and is achieved under the following values of the thresholds: $K_1 = 36$ and $K_2 = 37$.

Customers loss in the considered system occurs due to overheating of the server during ongoing service and due to impatience of customers. The charges paid for these types of losses may be different. The charge paid for the loss due to overheating can be much higher because the loss due to impatience is just the loss of *potential* profit, while the loss due to overheating means the real loss of a customer, violation of service level agreement and possible expenditures to return the overheated server to the operable mode. Therefore, various other optimization problems can be formulated.

In this paper, we consider the following economical criterion of the quality of the system operation:

$$E = a\lambda P_{imp} + (b + c)\lambda P_{overheating} + cI_{vacation}.$$

This economical criterion indicates the charge paid by the system per unit of time, where a is the charge paid by the system for each customer loss due to impatience, b is the charge paid by the system for customer loss due to overheating, and c is the charge paid by the system for managing operation of the system via each transition to the blocking regime.

Let us fix the following values of the cost coefficients: $a = 1,\ b = 10,\ c = 2$.

Figure 11 illustrates the dependence of the economical criterion E on the values of K_1 and K_2.

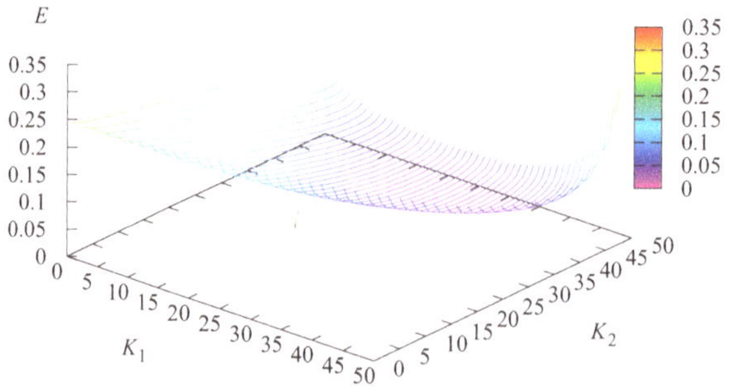

Figure 11. Dependence of the economical criterion E on the values of K_1 and K_2.

The minimal value of the economical criterion E here is equal to $E^* = 0.0308509$ and is achieved when $K_1 = 29$ and $K_2 = 42$. Note that, for the same system but without control, when no prevention of overheating is assumed (i.e., $K_2 = 50$, indicating the server is switched-off only when it becomes overheated, and $K_1 = 49$), the value of the economical criterion is more than ten times higher: $E(49, 50) = 0.309677$.

6. Conclusions

In this paper, a novel in the literature queueing model is considered. This model considers the possible heating of a server during the service process that causes the necessity of its permanent cooling. Such a model can be applied, e.g., for optimization of operation of servers of data centers that generate a lot of heat during their operation and the proper cooling mechanisms have to be used to avoid a collapse of the server. We offer the discipline for control by the system operation aiming to prevent premature overheating of a server and the loss of customers. This discipline is defined by two thresholds. One threshold is used to define the temperature of a server that when exceeded causes the stop of new services and to block the server when its temperature becomes close to the critical temperature. One more threshold is used to define the temperature when the blocking of the server can be finished and service can be resumed. The system is analyzed under quite general assumptions about the arrival and service processes. The generator of the multi-dimensional Markov chain, which describes the behavior of the system under any values of thresholds, is derived. This allows computing the stationary distribution of the states of the Markov chain and the key performance indicators of the system. Usefulness of the proposed strategy of preventive control is demonstrated via numerical experiments.

The obtained results can be used for managerial goals. In fact, the results can be used for the choice of the proper equipment for service provisioning (accounting for the different speeds of operation and heat generation by the different servers), its cooling and optimal management by periodical switching-off the server via the optimal choice of the thresholds.

As directions for future research, systems with the Batch Markov Arrival Process, phase-type distribution of heating and cooling times, several possible modes of the server operation (with various service and heating rates), etc., can be considered.

Author Contributions: Conceptualization O.D. and A.D.; methodology A.D. and O.D.; software O.D.; validation O.D.; formal analysis O.D. and A.D.; investigation A.D.; writing—original draft preparation A.D.; writing—review and editing A.D.; supervision A.D.; and project administration O.D. and A.D.

Funding: The publication was prepared with the support of the "RUDN University Program 5-100".

Conflicts of Interest: The authors declare no conflict of interest.

References

1. Available online: https://searchdatacenter.techtarget.com/answer/Whats-the-highest-server-temperature-you-can-handle (accessed on 25 July 2019).
2. Klimenok, V.I.; Dudin, A.N. Multi-dimensional asymptotically quasi-Toeplitz Markov chains and their application in queueing theory. *Queueing Syst.* **2006**, *54*, 245–259. [CrossRef]
3. Dudin, S.; Dudina O. Retrial multi-server queueing system with PHF service time distribution as a model of a channel with unreliable transmission of information. *Appl. Math. Model.* **2019**, *65*, 676–695. [CrossRef]
4. Krishnamoorthy, A.; Manikandan, R.; Lakshmy, B. A revisit to queueing-inventory system with positive service time. *Ann. Oper. Res.* **2015**, *233*, 221–236. [CrossRef]
5. Kim, C.; Dudin, S.; Dudin, A.; Samouylov, K. Multi-threshold control by a single-server queuing model with a service rate depending on the amount of harvested energy. *Perform. Eval.* **2018**, *127–128*, 1–20. [CrossRef]
6. Latouche, G. Queues with paired customers. *J. Appl. Probab.* **1981**, *18*, 684–696. [CrossRef]
7. Harrison, J.M. Assembly-like queues. *J. Appl. Probab.* **1973**, *10*, 354–367. [CrossRef]
8. Kendall, D.G. Some problems in the theory of queues. *J. R. Stat. Soc. Ser. Methodol.* **1951**, *13*, 151–173. [CrossRef]
9. Evdokimova, E.; De Turck, K.; Fiems, D. Coupled queues with customer impatience. *Perform. Eval.* **2018**, *118*, 33–47. [CrossRef]
10. Chakravarthy, S.R. The batch Markovian arrival process: A review and future work. In *Advances in Probability Theory and Stochastic Processes*; Krishnamoorthy, A., Raju, N., Ramaswami, V., Eds.; Notable Publications Inc.: Branchburg, NJ, USA, 2001; pp. 21–29.
11. Lucantoni, D. New results on the single server queue with a batch Markovian arrival process. *Commun. Stat. Stoch. Models* **1991**, *7*, 1–46. [CrossRef]
12. Vishnevski, V.M.; Dudin, A.N. Queueing systems with correlated arrival flows and their applications to modeling telecommunication networks. *Autom. Remote Control* **2017**, *78*, 1361—1403. [CrossRef]
13. Neuts, M. *Matrix-Geometric Solutions in Stochastic Models: An Algorithmic Approach*; John Hopkins University Press: Baltimore, MD, USA, 1981.
14. Asmussen, S. *Applied Probability and Queues*; Springer: New York, NY, USA, 2003.
15. Buchholz, P., Kemper, P., Kriege, J. Multi-class Markovian arrival processes and their parameter fitting. *Perform. Eval.* **2010**, *67*, 1092–1106. [CrossRef]
16. Buchholz, P., Kriege, J. Fitting correlated arrival and service times and related queueing performance. *Queueing Syst.* **2017**, *85*, 337–359. [CrossRef]
17. Okamura, H., Dohi, T. Mapfit: An R-Based Tool for *PH*/*MAP* Parameter Estimation. *Lect. Notes Comput. Sci.* **2015**, *9259*, 105–112.

mathematics

MDPI

Article

Monte Carlo Methods and the Koksma-Hlawka Inequality

Sergey Ermakov * and Svetlana Leora

The Faculty of Mathematics and Mechanics, St. Petersburg State University, 199034 St. Petersburg, Russia
* Correspondence: sergej.ermakov@gmail.com

Received: 30 June 2019; Accepted: 7 August 2019; Published: 9 August 2019

Abstract: The solution of a wide class of applied problems can be represented as an integral over the trajectories of a random process. The process is usually modeled with the Monte Carlo method and the integral is estimated as the average value of a certain function on the trajectories of this process. Solving this problem with acceptable accuracy usually requires modeling a very large number of trajectories; therefore development of methods to improve the accuracy of such algorithms is extremely important. The paper discusses Monte Carlo method modifications that use some classical results of the theory of cubature formulas (quasi-random methods). A new approach to the derivation of the well known Koksma-Hlawka inequality is pointed out. It is shown that for high ($s > 5$) dimensions of the integral, the asymptotic decrease of the error comparable to the asymptotic behavior of the Monte Carlo method, can be achieved only for a very large number of nodes N. It is shown that a special criterion can serve as a correct characteristic of the error decrease (average order of the error decrease). Using this criterion, it is possible to analyze the error for reasonable values of N and to compare various quasi-random sequences. Several numerical examples are given. Obtained results make it possible to formulate recommendations on the correct use of the quasi-random numbers when calculating integrals over the trajectories of random processes.

Keywords: Monte Carlo method; quasi-Monte Carlo method; Koksma-Hlawka inequality; quasi-random sequences; stochastic processes

---

1. Introduction

Let $\xi(t,\omega)$ be a random process, $t_1 < t_2 < \ldots < t_n$ are given time moments, and $F_n(t_1,x_1,\ldots,t_n,x_n) = P(\xi(t_1,\omega) \leq x_1;\ldots;\xi(t_n,\omega) \leq x_n)$ are finite-dimensional distributions of the process. The Monte Carlo method of modeling a process $\xi(t,\omega)$ usually consists of modeling its finite-dimensional distributions. Extensive literature is devoted to the algorithms of such modeling (for example, [1] and some cited sources). The methods developed in these papers allow us to express realizations of a random vector $\Xi = P(\xi(t_1,\omega);\ldots;\xi(t_n,\omega))$ through a certain set of randomizations of a random variable $(\alpha_1,\ldots,\alpha_M)$ uniformly distributed on $[0,1]$, where $M \geq n$ and, generally speaking, these can be random.

Ultimately, the problem solution Φ is represented as the expectation of a certain function Ψ of a random vector Ξ for selected finite values n and M. That means $\Phi = E\Psi(\Xi)$, and the computational process consists in multiple (N times) calculations of the independent realizations Ξ^j of the vector Ξ and in estimation of Φ using the arithmetic mean $E\Psi(\Xi) \approx 1/N\sum_{j=1}^{N}\Psi(\Xi^j)$. Let us notice that the value $M = M_j$ may depend on the number of the realization j. The maximum value $\overline{M} = \max_j M_j$ is called a constructive dimension of the algorithm. Finally, recall that Ξ^j is expressed in terms of realizations of uniformly distributed numbers, i.e., Φ, the integral over the $\overline{M}$-dimensional unit hypercube, has the following form

$$\Phi = E\Psi(\Xi(\alpha_1,\ldots,\alpha_{\overline{M}})) = \int_0^1 d\alpha_1 \ldots \int_0^1 d\alpha_{\overline{M}}.$$

Further, it is convenient to change the designations and talk about the calculation of the integral $J = \int_{D_s} f(X)dX$ using the s-dimensional unit hypercube, $X = (x_1,\ldots,x_s)$, $D_s = \{X : 0 \le x_i \le 1;\ i = 1,\ldots,s\}$. The problem of calculating this integral using the Monte Carlo method is well known. If it is estimated using the sum $J \approx 1/N \sum_{j=1}^{N} f(\alpha_1^j,\ldots,\alpha_s^j)$, where α_i^j are independent realizations of a random variable uniformly distributed on $[0,1]$, and f is a quadratically integrable function, then for the error it is possible to construct a confidence interval of width $O(N^{-1/2})$. At the same time, a number of articles [2–4] based on the theory of numbers considerations, pointed out sequences of s-dimensional vectors $Y_1,\ldots,Y_N$, for which the error $J - 1/N \sum_{j=1}^{N} f(Y_j)$ decreased as $ln^s N/N$, for functions that have the first partial derivative with respect to each variable. This result is obviously almost $\sqrt{N}$ times better than the Monte Carlo method. The sequences $Y_1,\ldots,Y_N$, possessing the property mentioned above, are called quasi-random . Extensive literature is devoted to their properties and applications (see, for example, [5] and the bibliography available there).

As a rule, the authors consider that quasi-random sequences are significantly better than the pseudo-random sequences used in the Monte Carlo method. The legitimacy of such comparisons is studied in detail below.

2. Koksma-Hlawka Inequality and Random Quadrature Formulas

One of the well-known approaches to constructing the sum

$$K_N[f] = \sum_{j=1}^{N} A_j f(X_j), \tag{1}$$

where A_j are constants, $X_j = (x_1^{(j)},\ldots,x_s^{(j)}) \in D \subset R^s$, which is used in the integral $\int_D f(X)dX$ calculation, is as follows. It is assumed that $f \subset \mathcal{F}$ belongs to a linear normed space of functions, the error of the integration formula

$$R_N[f] = \int_D f(X)dX - K_N[f] \tag{2}$$

is considered as a functional in this space and parameters of the formula (1) are A_j and X_j, $j = 1,\ldots,N$ are chosen so as to minimize the norm of the functional [6].

This task is usually very difficult. It is enough to note that it is possible to obtain the explicit expression of the above-mentioned functional norm only in a few particular cases. One of these cases is $A_j = 1/N$, $j = 1,\ldots,N$, D is a unit hypercube $D_s = \{X : 0 \le x_l \le 1, l = 1,\ldots,s\}$ and $\mathcal{F}$ is a space of functions of bounded variation in the Hardy-Krause sense. In this case one can use the Koksma-Hlawka inequality [7,8]

$$|R_N[f]| \le V(f) \cdot D^*(X_1,\ldots,X_N), \tag{3}$$

where $V(f)$ is the variation mentioned above, and $D^*(X_1,\ldots,X_N)$ is the error norm, which is called a discrepancy in the non-Russian literature or a deviation in the Russian one. Sometimes it is also called a star discrepancy. In the number-theoretical sense, this quantity characterizes the uniformity of the sequences distribution and equals to

$$\sup_X |S(X) - \frac{1}{N} A(X_1,\ldots,X_N)|, \tag{4}$$

where $S(X)$ is the volume of the multidimensional box $\Delta(X) = \{Y : y_l \leq x_l, l = 1, \ldots, s\}$, and $A(X_1, \ldots, X_N)$ is a number of points of sequences belonging to $\Delta(X)$.

Similar explanations regarding the value $V(f)$ can be found in Appendix A for this article. For our purposes, it suffices to note that for $N \to \infty$, one knows the upper bound for the asymptotic behavior of $D^*(X_1, \ldots, X_N)$ [9], namely

$$D^*(X_1, \ldots, X_N) \leq O(\frac{ln^s N}{N}). \tag{5}$$

The authors of [2–4] indicate of the algorithms for constructing sequences $X_1, \ldots, X_N$ for which equality is achieved in (5). These sequences are called quasi-random due to the formal similarity of algorithms of quasi-random numerical integration with the Monte Carlo method of calculating multiple integrals. As we have already said, the authors of numerous articles devoted to the study and use of quasi-random numbers usually note that the Monte-Carlo method has an asymptotic error decrease of the same order as $O(N^{-1/2})$, while from (5) we can conclude that quasi-random methods provide an error decrease as $O(N^{-1})$, more precisely $O(N^{-1+\varepsilon})$ for any arbitrarily small ε.

The proof of Inequality (3) given in the literature is quite large and complex. As we have already noted, it can easily be obtained by means of functional analysis. For $s = 2$ we showed it in the application. The general case simply requires more complex definitions. It can be noted that using the theory of cubature formulas [1], one can obtain analogs of Inequality (3) for Sobolev functions and many other classes of functions and specify sequences for these classes that have faster order of the error decrease. However, at the same time, computational algorithms are usually significantly complicated.

Other important applications of classical computational mathematics arise while estimating the error of quasi-random methods. The construction of the confidence interval in the Monte Carlo method automatically gives an estimate of the error, but for the quasi-Monte Carlo in its pure form there is only the Inequality (3), which is of little use when solving a specific problem. This difficulty can be overcome by randomizing quasi-random points.

Suppose $Y_1, \ldots, Y_N$ are quasi-random vectors of dimension s, and $\overrightarrow{\beta_1}, \ldots, \overrightarrow{\beta_m}$ are realizations of vectors of the same dimension, uniformly distributed in D_s. The sum $S_{m,N}(f) = 1/m \sum_{l=1}^{m} 1/N \sum_{k=1}^{N} f(\{Y_k + \overrightarrow{\beta_l}\})$ is an unbiased estimate of the integral

$$ES_{m,N}(f) = \int_{D_s} f(X)dX. \tag{6}$$

The curly brackets denote the operation of taking the fractional part performed on the components of the vector. The error for this randomized sum can be approximated using the central limit theorem. The following statement is trivial

$$J \approx \frac{1}{N} \sum_{k=1}^{N} f(\{Y_k + \overrightarrow{\beta}\}), \tag{7}$$

where $\overrightarrow{\beta}$ is a vector uniformly distributed in D_s, is a random quadrature formula with one free node. The theory of such formulas is given in detail in [1]. With the help of this theory, one can establish for which class of functions the formula is exact. A number of papers consider other methods for randomizing quasi-random numbers, known as scrambling methods.

3. Numerical Error Estimation Experiment: Monte Carlo and Quasi-Monte Carlo

In this paper we show that references to Inequality (3) when evaluating a computational algorithm are not completely correct, at least for $s > 5$, and we suggest some correct approaches to determining the asymptotical error decrease of the quasi-random methods. It can be immediately noted that in the case of the Monte Carlo (MC) method, we are talking about the width of the confidence interval and the asymptotic behavior in the central limit theorem is already seen for small N ($N > 5$, for example). The situation is different for the expression (5). The multiplier $ln^s N$ plays a significant role already for

$s \geq 4$. The asymptotics of order $O(N^{-1+\varepsilon})$ can be reached when $N \to \infty$, but as follows from Table 1, for $s = 5$ the rate of error decrease of the two methods becomes approximately equal with N^{12}, for $s = 10$, with N^{40}, for $s = 15$ with $N = 10^{65}$.

Table 1. Asymptotical error decrease.

s		5			10			15	
N	10^6	10^{12}	10^{14}	10^6	10^{18}	10^{40}	10^6	10^{30}	10^{65}
$\frac{ln^s N}{N}$	0.5	10^{-5}	10^{-8}	10^5	10^{-2}	10^{-21}	10^{11}	10^{-3}	10^{-33}
$\frac{1}{\sqrt{N}}$	10^{-3}	10^{-6}	10^{-8}	10^{-3}	10^{-9}	10^{-20}	10^{-3}	10^{-15}	10^{-33}

The data given in Table 1 clearly confirm the unsuitability of Inequality (5) to show the advantages of the quasi-Monte Carlo (QMC) method for calculating integrals of large multiplicity and, with increasing multiplicity of the integral, this situation worsens. Many authors (for example, [10]) confirm that the real error decrease for different values of N for $s > 5$ does not obey inequality (5), but behaves like $N^{-1+\varepsilon}$ with $0 \leq \varepsilon \leq 1/2$. Thus, one should speak about the quality of a particular quasi-random sequence only for reasonable values of N, when the asymptotic behavior indicated by equality (5) hasn't been fulfilled yet and one should introduce a reasonable quality criterion only from empirical considerations. First, we discuss the behavior of the integration error in some numerical examples. Let $f(X)$ be defined and integrable in the s-dimensional unit cube D_s and the integral $\int_{D_s} f(X)dX$ is calculated using the cubature formula (1).

Consider the integrals:

$$I_1 = \int_{D_s} 2^s e^s/(e-1)^s \prod_{i=1}^{s} x_i e^{-x_i^2} dX, I_2 = \int_{D_s} (8/\pi)^s \left(\prod_{i=1}^{s} x_i(1-x_i)\right)^{1/2} dX. \quad (8)$$

$$I_3 = \int_{D_s} \prod_{i=1}^{s} \frac{|4x_i - 2| + i}{i+1} dX, I_4 = \int_{D_s} \sqrt{1/|\sum_{i=1}^{s}(x_i - 1/2)^2 - 1/4|} dX. \quad (9)$$

The exact value of the integrals I_1, I_2, I_3 are known, they are equal to 1 for all s. Table 2 shows the absolute error of the QMC method, calculated for $N = 10^6$ and for the quasi-random Sobol (ErrS) and Halton (ErrH) sequences.

Table 2. The absolute error of the quasi-Monte Carlo (QMC) method.

I_i		I_1			I_2			I_3	
s	10	15	20	10	15	20	10	15	20
ErrS	10^{-4}	10^{-4}	10^{-3}	10^{-6}	10^{-4}	10^{-4}	10^{-7}	10^{-6}	10^{-6}
ErrH	10^{-4}	10^{-3}	10^{-3}	10^{-6}	10^{-4}	10^{-4}	10^{-7}	10^{-7}	10^{-7}
$\frac{ln^s N}{N}$	10^5	10^{11}	10^{16}	10^5	10^{11}	10^{16}	10^5	10^{11}	10^{16}

The obtained calculations lead to the following conclusions. The error values differ by no more than one order of magnitude for the considered sequences. The well-known estimate $ln^s N/N$ cannot serve as a reasonable estimate of the error in this case (see the last line in the Table 2). As already noted, the practical application of quasi-Monte Carlo for large s is limited. The calculations of integrals by the QMC method, the exact value of which is unknown, do not allow us to compare the accuracy of the various quasi-sequences used among themselves.

4. The Criterion of Decreasing Residual

We show that the use of the randomized quasi-Monte Carlo (RQMC) procedure allows us to effectively estimate the error of the quasi-Monte Carlo method. Consider a new criterion, that will allow us to judge about the error decrease depending on N, and will provide an opportunity to compare the quality of various quasi-random integration methods.

As a quality criterion, we will use the average order of the error decrease.

Consider the error change interval $[N_1, N_2]$ for some randomized quasi-random quadrature formula. The average value for a given number of realizations N, $N \in [N_1, N_2]$ is denoted α and we will approximate $R_f(N)$ using the least squares method with the function $y = a + b \cdot N^{-\alpha}$. The value of α is called the average order of the error decrease. In the case when the exact value of the integral is unknown we use the error estimate obtained by randomization.

Let us estimate the average order of the error decrease of the numerical integration for integrals I_4. The exact value of these integrals is not known. The calculations were carried out for $n = 10^5$ (the number of nodes) and $M = 10$ (the number of randomizations) for quasi-random sequences of the Sobol and Halton method. Moreover, the total number of nodes is $N = 10^6$. The value of random error $R_f(N)$ is approximated on the following intervals: $\Delta r_1 = [99,700, 99,800], \Delta r_2 = [99,800, 99,900], \Delta r_3 = [99,900, 100,000]$. The results of the calculations are given in Tables 3 and 4, where δ is the average order of the error decrease value on a given interval.

Table 3. The average decreasing order of the error $\alpha = \alpha(\Delta)$ for the integral I_4. Randomized quasi-Monte Carlo (RQMC).

	$s = 15$				$s = 20$			
Method	Δr_1	Δr_2	Δr_3	δ	Δr_1	Δr_2	Δr_3	δ
RQMC, Sobol	0.84	0.85	0.87	0.85	0.95	1.0	1.0	0.98
RQMC, Halton	0.70	0.64	0.61	0.65	0.85	0.92	0.98	0.92

Table 4. The average decreasing order of the error $\alpha = \alpha(\Delta)$ for the integral I_1.

	$s = 15$				$s = 20$			
Method	Δr_1	Δr_2	Δr_3	δ	Δr_1	Δr_2	Δr_3	δ
RQMC, Sobol	0.59	0.61	0.61	0.61	0.65	0.63	0.52	0.47
RQMC, Halton	0.61	0.61	0.60	0.61	0.53	0.52	0.52	0.60

Conducted calculations show the relative stability of the estimate of α for changing N_1 and N_2.

5. Conclusions

Error analysis of numerical methods plays a major role in the choice of algorithms for solving a problem. The main goals of this article are to propose a new quality criterion for algorithms for calculating multidimensional integrals; point out the incorrect use of the Koksma-Hlawka inequality when comparing the asymptotic error behavior of the Monte Carlo and the quasi-Monte Carlo methods for calculating integrals; and propose a new quality criterion for calculation algorithm integrals with large multiplicity.

This will allow, in particular, to choose the ratio between the number of random and quasi-random components of the nodes used in quadrature formulas, when their number (constructive dimension) is very large. The results obtained are confirmed by numerical examples. It makes possible to judge the comparative quality of various quasi-random sequences in some cases.

The results obtained, can be useful in solving other problems (for example, optimization problems). However, this requires separate studies that are beyond the scope of this work.

Author Contributions: Conceptualization, S.E.; methodology, S.E. and S.L.; software, S.L.; validation, S.E. and S.L.; formal analysis, S.E.; investigation, S.E. and S.L.; writing—original draft preparation, S.E.; writing—review and editing, S.E. and S.L.; supervision, S.E.; project administration, S.E.

Funding: This research is supported by the Russian Foundation for Basic Research, under Grant number 17-01-00267-a.

Conflicts of Interest: The authors declare no conflict of interest.

Appendix A

For the reader's convenience a brief proof of the inequality Koksma-Hlawki in the two-dimensional case is given.

Let $f(x,y)$ have integrable derivatives $\partial f(x,1)/\partial x$, $\partial f(1,y)/\partial y$ and $\partial^2 f(x,y)/\partial x \partial y$ in unit square $D_2 = \{0 \le x \le 1; 0 \le y \le 1\}$. Then we have

$$f(x,y) = f(1,1) + \int_x^1 f'_x(u,1)du + \int_y^1 f'_y(1,v)dv - \int_x^1 \int_y^1 f''_{xy}(u,v)dudv.$$

Elementary transformations allow to obtain from

$$R_N[f] = \int_{D_2} f(x,y)dxdy - \frac{1}{N}\sum_{j=1}^{N} f(x_j,y_j)$$

the expression

$$\begin{aligned} R_N[f] = &\int_0^1 f'_x(u,1)[-u + \frac{1}{N}\sum \Theta(x_j - u)]du \\ &+ \int_0^1 f'_y(1,v)[-v + \frac{1}{N}\sum \Theta(y_j - v)]dv \\ &+ \int_0^1 \int_0^1 f''_{xy}(u,v)[uv - \frac{1}{N}\sum \Theta(x_j - u)\Theta(y_j - v)]dudv, \end{aligned} \tag{A1}$$

where

$$\Theta(z) = \begin{cases} 1 & z < 0 \\ 0 & z > 0 \\ 1/2 & z = 0. \end{cases}$$

If we denote

$$V(f) = \int_0^1 |f'_x(u,1)|du + \int_0^1 |f'_y(1,v)|dv + \int_0^1 \int_0^1 |f''_{xy}(u,v)|dudv,$$

that we have

$$|R_N[f]| \le V(f) \cdot max(K_1(x_1,\ldots,x_N), K_1(y_1,\ldots,y_N), K_2(x_1,y_1;\ldots;x_N,y_N),$$

where

$$K_1(x_j,\ldots,x_N) = \sup_u |u - \frac{1}{N}\sum \Theta(x_j - u)|,$$

$$K_1(y_j,\ldots,y_N) = \sup_v |v - \frac{1}{N}\sum \Theta(y_j - u)|,$$

$$K_2 = \sup_{uv} |uv - \frac{1}{N}\sum \Theta(x_j - u)\Theta(y_j - v)|.$$

However, as you can see,

$$K_1(x_1,\ldots,x_N) \le K_2(x_1,y_1;\ldots;x_N,y_N), K_1(y_1,\ldots,y_N) \le K_2(x_1,y_1;\ldots;x_N,y_N)$$

and $|R_N[f]| \leq V(f) \cdot K_2(x_1, y_1; \ldots; x_N, y_N)$, $K_2(x_1, y_1; \ldots; x_N, y_N)$ and there is discrepance of a sequence of points.

In general, $D^*(X_1, \ldots, X_N)$ and $V(f)$ are defined as follows

$$D^*[f] = \sup_{x_1, \cdots, x_s} |\prod_{l=1}^{s} x_l - \frac{1}{N} A(X_1, \ldots, X_N)]|,$$

here $\prod_{l=1}^{s} x_l$ is the volume of the hyperparallelepiped $\Gamma = \{Y : 0 \leq y_l \leq x_l, l = 1, \ldots, s\}$, and $A(X_1, \ldots, X_N)$ is the number of sequence points $X_1, \cdots, X_N$, belonging to this parallelepiped.

References

1. Ermakov, S.M. *Monte Carlo Method and Related Questions*; Fizmatlit: Moscow, Russia, 1975. (In Russian)
2. Halton, J.H. On the Efficiency of Certain Quasi-random Sequence of Points Inevaluating Multi-dimensional Integrals. *Numer. Math.* **1960**, 1988, 84–90. [CrossRef]
3. Sobol, I.M. *Multidimensional Quadrature Formulas and Haar Functions*; Nauka: Moscow, Russia, 1969. (In Russian)
4. Korobov, N.M. *Number-Theoretic Methods in Approximate Analysis*; MZNMO: Moscow, Russia, 2004. (In Russian)
5. Lemieux, C. *Monte Carlo and Quasi-Monte Carlo Sampling*; Springer: New York, NY, USA, 2009.
6. Sobolev, S.L. *Introduction to the Theory of Cubature Formulas*; Nauka: Moscow, Russia, 1974. (In Russian)
7. Koksma, J.F. Een algemeene stelling inuit de theorie der gelijkmatige verdeelingmodulo 1. *Math. (Zutphen B)* **1942**, *11*, 7–11.
8. Hlawka, E. Discrepancy and Riemann Integration. In *Studies in Pure Mathematics*; Academic Press: London, UK, 1971; pp. 21–129.
9. Kuipers, L; Niederreiter, H. *Uniform Distribution of Sequences*; John Wiley and Sons: New York, NY, USA, 1974.
10. Levitan, Y.L.; Markovich, N.I.; Rozin, S.R.; Sobol, I.M. On quasi-random sequences for numerical calculations. *Zh. Vychisl. Math Mat. Phys.* **1988**, *28*, 755–759.

mathematics

Article

Exact Time-Dependent Queue-Length Solution to a Discrete-Time *Geo/D/1* Queue

Jung Woo Baek [1] and **Yun Han Bae** [2,*]

[1] Department of Industrial Engineering, Chosun University, Gwangju 61452, Korea
[2] Department of Mathematics Education, Sangmyung University, Seoul 03016, Korea
* Correspondence: yhbae@smu.ac.kr

Received: 18 July 2019; Accepted: 6 August 2019; Published: 7 August 2019

Abstract: Time-dependent solutions to queuing models are beneficial for evaluating the performance of real-world systems such as communication, transportation, and service systems. However, restricted results have been reported due to mathematical complexity. In this study, we present a time-dependent queue-length formula for a discrete-time $Geo/D/1$ queue starting with a positive number of initial customers. We derive the time-dependent formula in closed form.

Keywords: time-dependent queue-length probability; discrete-time $Geo/D/1$ queue; closed-form solution

1. Introduction

Queuing models have been widely used for performance evaluation of practical systems such as communication, transportation, and service systems. For the most part, the previous studies have addressed stationary solutions of queuing models for the optimal control of long-term system performance. However, time-dependent solutions to queuing systems are more meaningful for practical systems; a good example was presented in Griffiths et al. [1], where the authors applied the transient solution of an $M/E_k/1$ queue to model and analyze a 24-hour traffic profile on the Severn Bridge in England.

The pioneering results on the transient results were presented by Luchak [2] and Saaty [3]. Using a modified Bessel function, they presented the transient queue-length distribution of an $M/M/1$ queue starting with a positive number of initial customers. Later, computationally efficient solutions were presented in many other studies [4–8]. For $M/E_k/1$ queues, several results have been presented. Griffiths et al. [1,9] and Leonenko [10] derived transient solutions for $M/E_k/1$ queues starting with a positive number of initial customers, and Baek et al. [11] extended their results to be applied to analysis in a single busy period. Furthermore, Kapodistria et al. [12] presented the time-dependent solutions to a linear birth/immigration-death process with binomial catastrophes. They presented general results that include a background random environment.

The first study on queues with deterministic service times was conducted by Garcia et al. [13]. Using the matrix-analytic method, they derived the transient queue-length distribution of an $M/D/1/N$ queue in a computationally efficient form. Later, the queue-length formula was extended to an $M/D/c$ queue by Franx [14]. Recently, extensive results on the transient solution of the $M/D/1$ queue were presented by Baek et al. [15]. They derived not only the queue-length probability but also the waiting time distribution in closed form.

The aforementioned studies focused only on the queues in a continuous-time domain. Time-dependent solutions to queues in a discrete-time domain are even scarcer. Parthasarathy and Sudhesh [16] were the first to derive the transient queue-length solution to a discrete-time $Geo/Geo/1$ queue. Employing a generating function and continued fractions, they obtained a computationally efficient solution. Kim [17] studied the same model and presented the solution in a formal form.

However, we failed to find any result related to the time-dependent solution to the discrete-time queue with a constant service time. From this perspective, we present a time-dependent queue-length solution to a $Geo/D/1$ queue. To the best of our knowledge, our proposed results have not been noted in existing literature.

2. Main Results

In this section, we present the main results. Consider a discrete-time $Geo/D/1$ queuing system with an infinite buffer in which the timeline is divided into intervals of equal length called time slots. Customers arrive into the system according to the Bernoulli process at a rate of p, and the service times are assumed to be a constant of D. We assume that arrivals and service completion can occur only at the slot boundaries. More specifically, we consider the late-arrival model [18]. Therefore, we assume that customer arrivals can occur only in $(n-, n)$ and services can be completed only in $(n, n+)$.

Let N_0 be the number of new initial customers at time slot 0. We define $N(n)$ as the number of customers at the end of the time slot n and with the following probability:

$$Q_k^{(j)}(n) = Pr[N(n) = k | N_0 = j], \quad (k \geq 0,\ j \geq 0),\ (n > 0).$$

Because $Q_0^{(j)}(n)$ plays a major role in deriving $Q_k^{(j)}(n)$, we first derive the probability $Q_0^{(j)}(n)$ in the following theorem.

Theorem 1. *We have*

$$Q_0^{(j)}(n) = \sum_{r=0}^{\lfloor \frac{n}{D} \rfloor - j} \binom{n}{r} p^r (1-p)^{n-r} \left(1 - \frac{r}{n} D\right). \tag{1}$$

Proof. Let us define $A(n)$ as the number of arrivals during the time interval $(0, n]$ and $\tau_{i(r)}(n)$ as the i-th arrival epoch under the condition that $A(n) = r$. We note that the service time is a constant D and the system starts with j initial customers. Therefore, conditioning on $A(n)$, we have

$$\begin{aligned} Q_0^{(j)}(n) = \sum_{r=0}^{\infty} & Pr\Big(\tau_{1(r)}(n) < \tau_{2(r)}(n) < \cdots < \tau_{r(r)}(n), \\ & \tau_{1(r)}(n) \leq n - rD, \tau_{2(r)}(n) \leq n-(r-1)D, \ldots, \tau_{r(r)}(n) \leq n - D | A(n) = r\Big) \\ & \times Pr(A(n) = r) U(n \geq (j+r)D) \\ = \sum_{r=0}^{\lfloor \frac{n}{D} \rfloor - j} & Pr\Big(\tau_{1(r)}(n) \leq n - rD, \tau_{2(r)}(n) \leq n-(r-1)D, \ldots, \tau_{r(r)}(n) \leq n - D | A(n) = r\Big) Pr(A(n) = r), \end{aligned} \tag{2}$$

where $U(A)$ is an indicator function which takes 1, if A is true, or 0, and the second equality holds because it is obvious that $\tau_{1(r)}(n) < \tau_{2(r)}(n) < \cdots < \tau_{r(r)}(n)$.

Equation (2) means that the system can be empty at time slot n only when the i-th customer among r customers arrives before $[n - (r-(i-1))D]$-th time slot $(1 \leq i \leq r)$ under the condition that $A(n) = r$. Since the arrival process is Bernoulli process with rate p, we have $Pr(A(n) = r) = \binom{n}{r} p^r (1-p)^{n-r}$. Next, to complete Equation (2), we need to derive $Pr(\tau_{1(r)}(n) \leq n - rD, \tau_{2(r)}(n) \leq n-(r-1)D, \ldots, \tau_{r(r)}(n) \leq n - D | A(n) = r)$. For $1 \leq n_1 < n_2 < \cdots < n_r \leq n$, it is not difficult to have

$$\begin{aligned} & Pr\Big(\tau_{1(r)}(n) = n_1, \tau_{2(r)}(n) = n_2, \cdots, \tau_{r(r)}(n) = n_r | A(n) = r\Big) \\ = & \frac{Pr(\tau_{1(r)}(n) = n_1, \tau_{2(r)}(n) = n_2, \cdots, \tau_{r(r)}(n) = n_r, A(n) = r)}{Pr(A(n) = r)} = \frac{p^r(1-p)^{n-r}}{\binom{n}{r} p^r (1-p)^{n-r}} = \frac{1}{\binom{n}{r}}. \end{aligned} \tag{3}$$

Therefore, we have

$$
\begin{aligned}
&Pr\left(\tau_{1(r)}(n) \le n - rD, \tau_{2(r)}(n) \le n-(r-1)D, \dots, \tau_{r(r)}(n) \le n - D | A(n) = r\right) \\
&= \sum_{n_1=1}^{n-rD} \sum_{n_2=n_1+1}^{n-(r-1)D} \cdots \sum_{n_r=n_{r-1}+1}^{n-D} Pr\left(\tau_{1(r)}(n) = n_1, \tau_{2(r)}(n) = n_2, \cdots, \tau_{r(r)}(n) = n_r | A(n) = r\right) \\
&= \sum_{n_1=1}^{n-rD} \sum_{n_2=n_1+1}^{n-(r-1)D} \cdots \sum_{n_r=n_{r-1}+1}^{n-D} \frac{1}{\binom{n}{r}}.
\end{aligned} \tag{4}
$$

Using Equation (4) in Equation (2), we obtain

$$
Q_0^{(j)}(n) = \sum_{r=0}^{\lfloor \frac{n}{D} \rfloor - j} \left[\sum_{n_1=1}^{n-rD} \sum_{n_2=n_1+1}^{n-(r-1)D} \cdots \sum_{n_r=n_{r-1}+1}^{n-D} \frac{1}{\binom{n}{r}} \right] \binom{n}{r} p^r (1-p)^{n-r}. \tag{5}
$$

Now, to obtain Equation (1), it is sufficient to prove that

$$
\sum_{n_1=1}^{n-rD} \sum_{n_2=n_1+1}^{n-(r-1)D} \cdots \sum_{n_r=n_{r-1}+1}^{n-D} \frac{1}{\binom{n}{r}} = 1 - \frac{r}{n} D. \tag{6}
$$

We use mathematical induction for the proof of Equation (6). For $r = 1$, we trivially have $\sum_{n_1=1}^{n-D} \frac{1}{\binom{n}{1}} = 1 - \frac{1}{n} D$. We now assume that Equation (6) holds for $r = l$. Then, for $r = l+1$, we have

$$
\begin{aligned}
&\sum_{n_1=1}^{n-(l+1)D} \sum_{n_2=n_1+1}^{n-lD} \cdots \sum_{n_{l+1}=n_l+1}^{n-D} \frac{1}{\binom{n}{l+1}} \\
&= \sum_{n_1=1}^{n-(l+1)D} \left[\sum_{i_1=1}^{n-n_1-lD} \cdots \sum_{i_l=i_{l-1}+1}^{n-n_1D} \frac{1}{\binom{n-n_1}{l}} \right] \frac{\binom{n-n_1}{l}}{\binom{n}{l+1}} = \sum_{n_1=1}^{n-(l+1)D} \left[1 - \frac{lD}{n-n_1} \right] \frac{\binom{n-n_1}{l}}{\binom{n}{l+1}},
\end{aligned} \tag{7}
$$

where we use $i_k = n_{k+1} - n_1$ for $k \ge 1$.

We now have

$$
\begin{aligned}
&\sum_{n_1=1}^{n-(l+1)D} \left[1 - \frac{lD}{n-n_1} \right] \frac{\binom{n-n_1}{l}}{\binom{n}{l+1}} = \sum_{n_1=1}^{n-(l+1)D} \frac{n-n_1-lD}{n-n_1} \cdot \frac{(n-n_1)! \cdot (n-l-1)!}{n! \cdot (n-n_1-l)!} \cdot (l+1) \\
&= \frac{n-1-lD}{n-1} \cdot \frac{(n-1)! \cdot (n-l-1)!}{n! \cdot (n-1-l)!} \cdot (l+1) + \frac{n-2-lD}{n-2} \cdot \frac{(n-2)! \cdot (n-l-1)!}{n! \cdot (n-2-l)!} \cdot (l+1) \\
&\qquad + \cdots + \frac{(l+1)D - lD}{(l+1)D} \cdot \frac{((l+1)D)! \cdot (n-l-1)!}{n! \cdot ((l+1)D - l)!} \cdot (l+1) \\
&= \sum_{m=(l+1)D}^{n-1} \frac{m-lD}{m} \cdot \frac{m! \cdot (n-l-1)!}{n! \cdot (m-l)!} \cdot (l+1) \\
&= \left(1 - \frac{(l+1)D}{n}\right) \cdot \frac{(n-l-1)!}{(n-1)!} \cdot \frac{l+1}{n-(l+1)D} \sum_{m=(l+1)D}^{n-1} \frac{(m-lD) \cdot (m-1)!}{(m-l)!}.
\end{aligned} \tag{8}
$$

In Equation (8), we have

$$\begin{aligned}
&\sum_{m=(l+1)D}^{n-1} \frac{(m-lD)\cdot(m-1)!}{(m-l)!} \\
&= \sum_{m=0}^{n-1}\left[\frac{m!}{(m-l)!} - lD\frac{(m-1)!}{(m-l)!}\right] - \sum_{m=0}^{(l+1)D-1}\left[\frac{m!}{(m-l)!} - lD\frac{(m-1)!}{(m-l)!}\right] \\
&= l!\left[\cdot \sum_{m=0}^{n-1}\binom{m}{l} - D\cdot\sum_{m=0}^{n-1}\binom{m-1}{l-1} - \sum_{m=0}^{(l+1)D-1}\binom{m}{l} + D\cdot\sum_{m=0}^{(l+1)D-1}\binom{m-1}{l-1}\right] \\
&= l!\left[\binom{n}{l+1} - D\cdot\binom{n-1}{l} - \binom{(l+1)D}{l+1} + D\cdot\binom{(l+1)D-1}{l}\right] \\
&= l!\cdot\binom{n}{l+1} - l!\cdot D\cdot\binom{n-1}{l} \\
&= \frac{(n-1)!}{(n-l-1)!}\cdot\frac{n-(l+1)D}{l+1},
\end{aligned} \tag{9}$$

where we use $\sum_{j=0}^{n}\binom{j}{m} = \binom{n+1}{m+1}$ for the third equality.

Using Equations (8) and (9) in Equation (7), we obtain

$$\sum_{n_1=1}^{n-(l+1)D}\sum_{n_2=n_1+1}^{n-lD}\cdots\sum_{n_{l+1}=n_l+1}^{n-D}\frac{1}{\binom{n}{l+1}} = 1 - \frac{l+1}{n}D.$$

Therefore, we prove that Equation (6) holds for $r = 1, 2, 3, \cdots$. □

Next, we derive $Q_k^{(j)}(n)$, $(k \geq 1)$. We have the following theorem.

Theorem 2. *We have*

$$\begin{aligned}
Q_k^{(j)}(n) = &\binom{n}{k+\lfloor\frac{n}{D}\rfloor - j} p^{k+\lfloor\frac{n}{D}\rfloor-j}(1-p)^{n-(k+\lfloor\frac{n}{D}\rfloor-j)} \\
&+ \sum_{r=0}^{\lfloor\frac{n}{D}\rfloor-j-1}\sum_{m=0}^{\lfloor\frac{n}{D}\rfloor-j-1-r}\binom{n-(r+1)D}{m}p^m(1-p)^{n-(r+1)D-m} \\
&\times\left(\frac{(k+r+1)-p-p[(r+1)D]}{(1-p)(k+r+1)}\right)\left(1-\frac{m}{\frac{n}{D}-r-1}\right)\binom{(r+1)D}{k+r}p^{k+r}(1-p)^{(r+1)D-(k+r)}.
\end{aligned} \tag{10}$$

Proof. Let us define A^{last} and τ^{last} as the number of arrivals during the last busy period and the time slot at which the last busy period starts, respectively. We then have

$$\begin{aligned}
Q_k^{(j)}(n) &= Pr[N(n) = k | N_0 = j] \\
&= Pr\left[N(n) = k, 0 \leq \tau^{last} \leq n - \left(\lfloor\frac{n}{D}\rfloor - j\right)D | N_0 = j\right] \\
&\quad + \sum_{r=0}^{\lfloor\frac{n}{D}\rfloor-j-1} Pr[N(n) = k, n-(r+1)D < \tau^{last} \leq n - rD | N_0 = j].
\end{aligned} \tag{11}$$

To complete Equation (11), we first derive the first term in the second equality. We again note that the system starts with j customers, and the service time is a constant D. Therefore, if the last busy period starts before the $n - \left(\lfloor\frac{n}{D}\rfloor - j\right)D$-th time slot, $\lfloor\frac{n}{D}\rfloor - j$ customers among newly arrived customers should be served during the last busy period. Then, to become $N(n) = k$ under this situation, $k + \lfloor \frac{n}{D} \rfloor - j$ customers should arrive in the system, and we have

$$\begin{aligned} & Pr\left[N(n) = k, 0 \le \tau^{last} \le n - \left(\lfloor \frac{n}{D} \rfloor - j\right) D | N_0 = j\right] \\ & = Pr\left[N(n) = k, 0 \le \tau^{last} \le n - \left(\lfloor \frac{n}{D} \rfloor - j\right) D, A^{last} = k + \lfloor \frac{n}{D} \rfloor - j | N_0 = j\right]. \end{aligned} \tag{12}$$

Next, we have

$$\begin{aligned} & Pr\left[N(n) = k, 0 \le \tau^{last} \le n - \left(\lfloor \frac{n}{D} \rfloor - j\right) D, A^{last} = k + \lfloor \frac{n}{D} \rfloor - j | N_0 = j\right] \\ & = Pr\left[N(n) = k, 0 \le \tau^{last} \le n, A^{last} = k + \lfloor \frac{n}{D} \rfloor - j | N_0 = j\right] \\ & \quad - Pr\left[N(n) = k, n - \left(\lfloor \frac{n}{D} \rfloor - j\right) D < \tau^{last} \le n, A^{last} = k + \lfloor \frac{n}{D} \rfloor - j | N_0 = j\right] \\ & = Pr\left[k \le N(n) \le k + \lfloor \frac{n}{D} \rfloor - j, 0 \le \tau^{last} \le n, A^{last} = k + \lfloor \frac{n}{D} \rfloor - j | N_0 = j\right] \\ & \quad - Pr\left[k+1 \le N(n) \le k + \lfloor \frac{n}{D} \rfloor - j, 0 \le \tau^{last} \le n, A^{last} = k + \lfloor \frac{n}{D} \rfloor - j | N_0 = j\right] \\ & \quad - Pr\left[k \le N(n) \le k + \lfloor \frac{n}{D} \rfloor - j, n - \left(\lfloor \frac{n}{D} \rfloor - j\right) D < \tau^{last} \le n, A^{last} = k + \lfloor \frac{n}{D} \rfloor - j | N_0 = j\right] \\ & \quad + Pr\left[k+1 \le N(n) \le k + \lfloor \frac{n}{D} \rfloor - j, n - \left(\lfloor \frac{n}{D} \rfloor - j\right) D < \tau^{last} \le n, A^{last} = k + \lfloor \frac{n}{D} \rfloor - j | N_0 = j\right] \\ & = Pr\left[k \le N(n) \le k + \lfloor \frac{n}{D} \rfloor - j, 0 \le \tau^{last} \le n, A^{last} = k + \lfloor \frac{n}{D} \rfloor - j | N_0 = j\right] \\ & \quad - Pr\left[k \le N(n) \le k + \lfloor \frac{n}{D} \rfloor - j, n - \left(\lfloor \frac{n}{D} \rfloor - j\right) D < \tau^{last} \le n, A^{last} = k + \lfloor \frac{n}{D} \rfloor - j | N_0 = j\right] \\ & \quad - Pr\left[k+1 \le N(n) \le k + \lfloor \frac{n}{D} \rfloor - j, 0 \le \tau^{last} \le n - \left(\lfloor \frac{n}{D} \rfloor - j\right) D, A^{last} = k + \lfloor \frac{n}{D} \rfloor - j | N_0 = j\right]. \end{aligned} \tag{13}$$

In Equation (13), we have

$$Pr\left[k+1 \le N(n) \le k + \lfloor \tfrac{n}{D} \rfloor - j, 0 \le \tau^{last} \le n - (\lfloor \tfrac{n}{D} \rfloor - j) D, A^{last} = k + \lfloor \tfrac{n}{D} \rfloor - j | N_0 = j\right] = 0 \tag{14}$$

because $\lfloor \frac{n}{D} \rfloor - j$ customers among the newly arrived customers should be served during the last busy period, if the busy period starts in the time interval $\left[0, n - (\lfloor \frac{n}{D} \rfloor - j) D\right]$.

Furthermore, we note that

$$Pr[N(n) = k, n - \left(\lfloor \frac{n}{D} \rfloor - j\right) D < \tau^{last} \le n, A^{last} = k + \lfloor \frac{n}{D} \rfloor - j | N_0 = j] = 0. \tag{15}$$

We then have

$$\begin{aligned} & Pr\left[N(n) = k, 0 \le \tau^{last} \le n - \left(\lfloor \frac{n}{D} \rfloor - j\right) D, A^{last} = k + \lfloor \frac{n}{D} \rfloor - j | N_0 = j\right] \\ & = Pr\left[k \le N(n) \le k + \lfloor \frac{n}{D} \rfloor - j, 0 \le \tau^{last} \le n, A^{last} = k + \lfloor \frac{n}{D} \rfloor - j | N_0 = j\right] \\ & \quad - Pr\left[k+1 \le N(n) \le k + \lfloor \frac{n}{D} \rfloor - j, n - \left(\lfloor \frac{n}{D} \rfloor - j\right) D < \tau^{last} \le n, A^{last} = k + \lfloor \frac{n}{D} \rfloor - j | N_0 = j\right] \\ & = \binom{n}{k + \lfloor \frac{n}{D} \rfloor - j} \cdot p^{k + \lfloor \frac{n}{D} \rfloor - j} \cdot (1-p)^{n - (k + \lfloor \frac{n}{D} \rfloor - j)} \\ & \quad - Q_0^{(j)}\left[n - \left(\lfloor \frac{n}{D} \rfloor - j\right) D\right] \cdot \binom{(\lfloor \frac{n}{D} \rfloor - j) D}{k + \lfloor \frac{n}{D} \rfloor - j} \cdot p^{k + \lfloor \frac{n}{D} \rfloor - j} \cdot (1-p)^{(\lfloor \frac{n}{D} \rfloor - j) D - (k + \lfloor \frac{n}{D} \rfloor - j)}. \end{aligned} \tag{16}$$

Next, we need to derive $Pr[N(n) = k, n-(r+1)D < \tau^{last} \leq n-rD|N_0 = j]$, $(0 \leq r \leq \lfloor \frac{n}{D} \rfloor - j - 1)$ in Equation (11). When $r = 0$, it is not difficult to have

$$Pr[N(n) = k, n-D < \tau^{last} \leq n|N_0 = j] = Q_0^{(j)}(n-D)\left[\binom{D}{k} \cdot p^k \cdot (1-p)^{D-k}\right] \tag{17}$$

because no service can be completed during the time interval $(n-D, n]$.

In Equation (17), we assume that $\binom{D}{k} = 0$, if $k > D$. Next, we consider the case with $1 \leq r \leq \lfloor \frac{n}{D} \rfloor - j - 1$. Applying a similar approach used in Equations (12)–(16), we have

$$\begin{aligned}
&Pr[N(n) = k, n-(r+1)D < \tau^{last} \leq n-rD|N_0 = j] \qquad (18)\\
&= Pr\left[N(n) = k, n-(r+1)D < \tau^{last} \leq n-rD, A^{last} = k+r|N_0 = j\right]\\
&= Pr\left[k \leq N(n) \leq k+r, n-(r+1)D < \tau^{last} \leq n, A^{last} = k+r|N_0 = j\right]\\
&\qquad - Pr\left[k+1 \leq N(n) \leq k+r, n-rD < \tau^{last} \leq n, A^{last} = k+r|N_0 = j\right]\\
&= Q_0^{(j)}(n-(r+1)D) \cdot \binom{(r+1)D}{k+r} \cdot p^{k+r} \cdot (1-p)^{(r+1)D-(k+r)}\\
&\qquad - Q_0^{(j)}(n-rD) \cdot \binom{rD}{k+r} \cdot p^{k+r} \cdot (1-p)^{rD-(k+r)}, \quad \left(1 \leq r \leq \lfloor \frac{n}{D} \rfloor - j - 1\right).
\end{aligned}$$

Using Equations (12) and (16)–(18) in Equation (11), we can obtain the simplified form as

$$\begin{aligned}
Q_k^{(j)}(n) = &\binom{n}{k+\lfloor \frac{n}{D} \rfloor - j} \cdot p^{k+\lfloor \frac{n}{D} \rfloor - j} \cdot (1-p)^{n-(k+\lfloor \frac{n}{D} \rfloor - j)} \qquad (19)\\
&+ \sum_{r=0}^{\lfloor \frac{n}{D} \rfloor - j - 1} Q_0^{(j)}(n-(r+1)D) \cdot \binom{(r+1)D}{k+r} \cdot p^{k+r} \cdot (1-p)^{(r+1)D-(k+r)}\\
&- \sum_{r=0}^{\lfloor \frac{n}{D} \rfloor - j - 1} Q_0^{(j)}(n-(r+1)D) \cdot \binom{(r+1)D}{k+r+1} \cdot p^{k+r+1} \cdot (1-p)^{(r+1)D-(k+r+1)}.
\end{aligned}$$

From Equation (1), we have

$$Q_0^{(j)}[(n-(r+1)D)] = \sum_{m=0}^{\lfloor \frac{n}{D} \rfloor - (r+1) - j} \binom{n-(r+1)D}{m} p^m (1-p)^{n-(r+1)D-m} \left(1 - \frac{mD}{n-(r+1)D}\right). \tag{20}$$

Then, using the above equation in Equation (19), we have

$$\begin{aligned}
&Q_k^{(j)}(n) \qquad (21)\\
&= \binom{n}{k+\lfloor \frac{n}{D} \rfloor - j} p^{k+\lfloor \frac{n}{D} \rfloor - j}(1-p)^{n-(k+\lfloor \frac{n}{D} \rfloor - j)}\\
&+ \sum_{r=0}^{\lfloor \frac{n}{D} \rfloor - j - 1} \sum_{m=0}^{\lfloor \frac{n}{D} \rfloor - j - 1 - r} \binom{n-(r+1)D}{m} \binom{(r+1)D}{k+r} \left(1 - \frac{m}{\frac{n}{D} - r - 1}\right) p^{m+k+r}(1-p)^{n-(m+k+r)}\\
&- \sum_{r=0}^{\lfloor \frac{n}{D} \rfloor - j - 1} \sum_{m=0}^{\lfloor \frac{n}{D} \rfloor - j - 1 - r} \binom{n-(r+1)D}{m} \binom{(r+1)D}{k+r+1} \left(1 - \frac{m}{\frac{n}{D} - r - 1}\right) p^{m+k+r+1}(1-p)^{n-(m+k+r+1)}.
\end{aligned}$$

We then have the following simplified form:

$$
\begin{aligned}
Q_k^{(j)}(n) = {} & \binom{n}{k+\lfloor \frac{n}{D} \rfloor - j} p^{k+\lfloor \frac{n}{D} \rfloor - j}(1-p)^{n-(k+\lfloor \frac{n}{D} \rfloor - j)} \\
& + \sum_{r=0}^{\lfloor \frac{n}{D} \rfloor - j - 1} \sum_{m=0}^{\lfloor \frac{n}{D} \rfloor - j - 1 - r} \binom{n-(r+1)D}{m} p^m (1-p)^{n-(r+1)D-m} \\
& \qquad \times \binom{(r+1)D}{k+r} p^{k+r}(1-p)^{(r+1)D-(k+r)} \left(1 - \frac{m}{\frac{n}{D} - r - 1}\right) \\
& - \sum_{r=0}^{\lfloor \frac{n}{D} \rfloor - j - 1} \sum_{m=0}^{\lfloor \frac{n}{D} \rfloor - j - 1 - r} \binom{n-(r+1)D}{m} p^m (1-p)^{n-(r+1)D-m} \\
& \times \binom{(r+1)D}{k+r} p^{k+r}(1-p)^{(r+1)D-(k+r)} \left(1 - \frac{m}{\frac{n}{D} - r - 1}\right) \frac{p[(r+1)D-(k+r)]}{(1-p)(k+r+1)}.
\end{aligned}
\tag{22}
$$

Using algebra, we can now obtain Equation (10) to complete the proof. □

3. Numerical Examples

In this section, we show the numerical results. We use Equations (1) and (10) to compute the transient probabilities $Q_0^{(2)}(n)$, $Q_1^{(2)}(n)$, $Q_2^{(2)}(n)$ and $Q_3^{(2)}(n)$. Figure 1 shows the computation results of the probabilities.

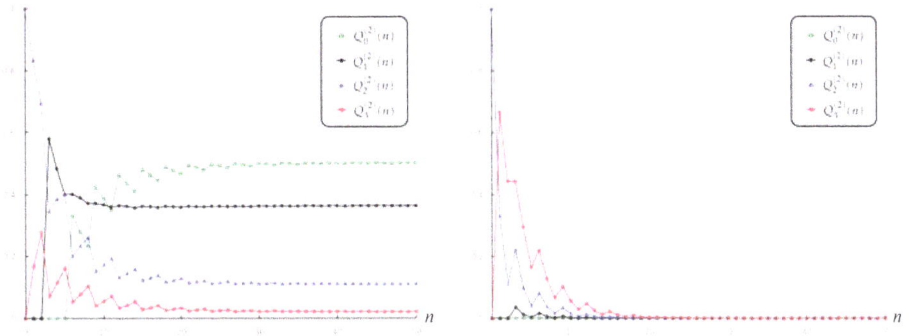

(a) $p = 1/6$, $D = 3$, $(\rho = pD = 1/2)$ (b) $p = 4/6$, $D = 3$, $(\rho = pD = 2)$

Figure 1. Probabilities $Q_0^{(2)}(n)$, $Q_1^{(2)}(n)$, $Q_2^{(2)}(n)$ and $Q_3^{(2)}(n)$.

Figure 1a shows the transient probabilities under the condition that $p = 1/6$ and $D = 3$ (i.e., $\rho = 0.5$). Let P_n be the stationary probability of the $Geo/D/1$ queue. If $\rho < 1$, we can use the result in Gravey et al. [19] to compute the stationary probability and obtain $P_0 = 0.5$, $P_1 = 0.364$, $P_2 = 0.11059$, and $P_3 = 0.02107$. Then, in Figure 1a, we can confirm that each transient queue-length probability converges to the stationary value. Figure 1b shows the transient probabilities under the condition that $p = 4/6$ and $D = 3$ (i.e., $\rho = 2$). When $\rho > 1$, all the probabilities should be 0 as n increases; we can confirm this in Figure 1b.

Author Contributions: Conceptualization: Y.H.B. and J.W.B.; Formal analysis: Y.H.B. and J.W.B.; Writing and original draft preparation: Y.H.B. and J.W.B.

Funding: The work of Y.H.B. was supported by the Basic Science Research Program through the National Research Foundation of Korea (NRF) funded by the Ministry of Education (NRF-2018R1D1A1B07043360). The work of J.W.B. was supported by the Basic Science Research Program through the National Research Foundation of Korea (NRF) funded by the Ministry of Education (NRF-2018R1D1A1B07043146).

Conflicts of Interest: The authors declare no conflict of interest.

References

1. Griffiths, G.D.; Leonenko, G.M.; Williams, J.E. Time-dependent analysis of non-Empty $M/E_k/1$ queue. *Qual. Technol. Quant. Manag.* **2008**, *5*, 309–320. [CrossRef]
2. Luchak, G. The continuous time solution of the equations of the single channel queue with a general class of service-time distributions by the method of generating functions. *J. R. Stat. Soc. Ser. B (Methodol.)* **1958**, *20*, 176–181. [CrossRef]
3. Saaty, L.T. *Elements of Queueing Theory*; Mcgraw Hill: New York, NY, USA, 1961.
4. Al-Seedy, R.O.; El-Sherbiny, A.A.; El-Shehawy, S.A.; Ammar, S.I. Transient solution of the $M/M/c$ queue with balking and reneging. *Comput. Math. Appl.* **2009**, *57*, 1280–1285. [CrossRef]
5. Cheon, G.-S.; Choi, B.D.; Jin, S.-T. An application of Riordan arrays to the transient analysis of $M/M/1$ queues. *Appl. Math. Comput.* **2014**, *237*, 659–671. [CrossRef]
6. Kijima, M. The transient solution to a class of Markovian queues. *Comput. Math. Appl.* **1992**, *24*, 17–24. [CrossRef]
7. Sharma, O.P.; Gupta, U.C. Transient Behaviour of an M/M/1/N queue. *Stoch. Process. Their Appl.* **1982**, *13*, 327–331. [CrossRef]
8. Tarabia, A.M.K. A new formula for the transient behaviour of a non-empty $M/M/1/\infty$ queue. *Appl. Math. Comput.* **2002**, *132*, 1–10. [CrossRef]
9. Griffiths, G.D.; Leonenko, G.M.; Williams, J.E. The transient solution to $M/E_k/1$ queue. *Oper. Res. Lett.* **2006**, *34*, 349–354. [CrossRef]
10. Leonenko, G.M. A new formula for the transient solution of the Erlang queueing Model. *Stat. Probab. Lett.* **2009**, *79*, 400–406. [CrossRef]
11. Baek, J.W.; Lee, H.W.; Moon, S.K. A time-dependent busy period queue length formula for the $M/E_k/1$ queue. *Stat. Probab. Lett.* **2014**, *87*, 98–104. [CrossRef]
12. Kapodistria, S.; Phung-Duc, T.; Resing, J. Linear Birth/Immigration-Death Process with Binomial Catastrophes. *Probab. Eng. Inf. Sci.* **2016**, *30*, 79–111. [CrossRef]
13. Garcia, J.M.; Brun, O.; Gauchard, D. Transient analytical solution of $M/D/1/N$ queues. *J. Appl. Probab.* **2002**, *39*, 853–864. [CrossRef]
14. Franx, G.J. A simple solution for the M/D/c waiting time distribution. *Oper. Res. Lett.* **2001**, *29*, 221–229. [CrossRef]
15. Baek, J.W.; Lee, H.W.; Ahn, S.; Bae, Y.H. Exact timedependent solutions for the M/D/1 queue. *Oper. Res. Lett.* **2016**, *44*, 692–695. [CrossRef]
16. Parthasarathy, P.R.; Sudhesh, R. Exact transient solution of a discrete time queue with state-dependent rates. *Am. J. Math. Manag. Sci.* **2006**, *26*, 253–276. [CrossRef]
17. Kim, J. Transient analysis of the Geo/Goe/1 queue. *J. Chungcheong Math. Soc.* **2008**, *21*, 385–393.
18. Takagi, H. *Queueing Analysis: A Foundation of Performance Evaluation*; North Holland: Amsterdam, The Netherlands, 1993
19. Gravey, A.; Louvion, J.-R.; Boyer, P. On the $Geo/D/1$ and $Geo/D/1/n$ queues. *Perform. Eval.* **1990**, *11*, 117–125. [CrossRef]

mathematics

Article

Analysis of a Semi-Open Queuing Network with a State Dependent Marked Markovian Arrival Process, Customers Retrials and Impatience

Chesoong Kim [1,*], Sergey Dudin [2,3], Alexander Dudin [2,3] and Konstantin Samouylov [3]

1 Department of Industrial Engineering, Sangji University, Wonju, Kangwon 26339, Korea
2 Department of Applied Mathematics and Computer Science, Belarusian State University, 4, Nezavisimosti Ave., 220030 Minsk, Belarus
3 Department of Applied Informatics and Probability Theory, Peoples' Friendship University of Russia (RUDN University), 6 Miklukho-Maklaya St, 117198 Moscow, Russia
* Correspondence: dowoo@sangji.ac.kr

Received: 10 July 2019; Accepted: 5 August 2019; Published: 7 August 2019

Abstract: We consider a queuing network with single-server nodes and heterogeneous customers. The number of customers, which can obtain service simultaneously, is restricted. Customers that cannot be admitted to the network upon arrival make repeated attempts to obtain service. The service time at the nodes is exponentially distributed. After service completion at a node, the serviced customer can transit to another node or leave the network forever. The main features of the model are the mutual dependence of processes of customer arrivals and retrials and the impatience and non-persistence of customers. Dynamics of the network are described by a multidimensional Markov chain with infinite state space, state inhomogeneous behavior and special structure of the infinitesimal generator. The explicit form of the generator is derived. An effective algorithm for computing the stationary distribution of this chain is recommended. The expressions for computation of the key performance measures of the network are given. Numerical results illustrating the importance of the account of the mentioned features of the model are presented. The model can be useful for capacity planning, performance evaluation and optimization of various wireless telecommunication networks, transportation and manufacturing systems.

Keywords: queuing network; retrials; state-dependent marked Markovian arrival process; wireless telecommunication networks

1. Introduction

The theory of queuing networks has a wide range of applications for modeling various real-world systems including telecommunication and logistic networks, health care, public transportation, production and manufacturing systems, see, for example, References [1–4] and so forth.

The queuing networks with homogeneous customers are usually classified (see, e.g., Reference [5]) into three categories: open networks where customers arrive from the outside and depart from the network after receiving service; closed networks where the number of customers circulating in the network is constant; and semi-open networks where customers arrive from the outside and depart from the network but only the finite number of customers can stay inside the network at any time.

In our paper, we deal with a queuing network which belongs to a relatively new category of semi-open queuing networks that recently were applied for the analysis of various real-world systems. For the review of the state of the art and the references see, for example, References [5–11].

Salient features of the considered in our paper model are the following:

- **Account of retrial phenomenon.** We assume that at most N customers can receive service in the network simultaneously. If a primary customer (customer arriving from the outside) arrives when N customers receive service, the customer joins the so-called orbit having an infinite capacity from which he/she retries to obtain access to the network after a random amount of time. A customer from the orbit can enter the network if the number of customers receiving service at the retrial moment is less than N. The theory of retrial *queues* is essentially less developed than the theory of queues with losses or buffers due to the higher complexity of the processes defining the behavior of the system, for references see, for example, References [12,13]. To the best of our knowledge, the results devoted to the exact analysis of the retrial *queuing networks*, which are more general than the tandem queues, are absent in the existing literature except the recent paper, Reference [14], which is briefly cited below. Here and thereafter, we only occasionally cite the papers where the corresponding networks are analyzed by means of approximations rather than exact solutions, see, for example, Reference [15].
- **More complex customers arrival process.** We consider a more general and realistic arrival process than those known in the literature. The overwhelming majority of the existing queuing networks consider the input as a stationary Poisson process. However, in most real-world systems, the input rate is time dependent. It is already well-recognized in the literature that the flows in modern telecommunication systems and networks are bursty and inadequately modelled by a stationary Poisson process. Instead, the so-called Markovian arrival process (MAP), see, for example, References [16,17], is a much better choice for description of real-world arrival processes which exhibit variation of the instantaneous arrival rate and correlation of inter-arrival times. Surveys on queuing *systems* with the MAP can be found in References [16,18]. Concerning the queuing *networks*, there are only a few papers on this topic, see, for example, Reference [19]. This paper deals with an approximation of the queuing networks with the MAP and phase-type distribution of service times. Exact results are known only for a specific kind of queuing networks, namely, the tandem queues, with the MAP, see, for example, in References [20–23]. Note that tandem queues considered in References [22,23] take into account retrials of customers. In this paper, we consider more general than the MAP-marked Markovian arrival process ($MMAP$), see, for example, Reference [24]. This flow is heterogeneous and has several types of customers. Type defines the node of the network at which the customer arrives.
- **More complex process of customers retrials and dependence of arrivals of primary customers and retrials.** Traditionally, it is assumed in the literature that the processes of customers arrivals and retrials are independent. More, it is usually assumed (except the special case when it is suggested that only one customer from the orbit can make retrials) that, under a fixed number of customers in orbit, the inter-retrial times have the exponential distribution with a fixed parameter. We can refer only to Reference [25] where the $BMAP/SM/1$ retrial queue is studied under the assumption that the intensities of the individual retrials are modulated by a continuous-time Markov chain. This Markov chain is independent of the underlying Markov chain of the $BMAP$ arrival process of primary customers. Such independence is not very realistic in real-world systems because when the arrival rate of primary customers fluctuates depending on the time of a day or a night or due to some external factors, it is very likely that the rate of retrials also depends on the time and the same external factors. In our paper, we consider the model with dependent processes of the arrival of primary customers and customers from the orbit.
- **Account of possible impatience of customers during staying in the orbit and waiting times in the nodes as well as non-persistence of customers staying in the orbit.** Impatience of customers, that is, a possibility of abandonment during the waiting time after some period of waiting and non-persistence of customers staying in the orbit, that is, a possibility to renege from the orbit after any unsuccessful retrial, are typical for many real-world systems and networks. Therefore, they should be taken into account during performance evaluation and capacity planning.

Semi-open queuing networks with MAP arrival process are analysed in References [7,10]. However, the retrial phenomenon is not taken into account in those papers. The model considered in Reference [7] is simpler than the one studied in Reference [10] because it is assumed in Reference [7] that the arrival flow is described by the $MMPP$ (Markov Modulated Poisson Process), which is a particular case of the MAP considered in Reference [10] and the network has a linear topology, that is, a type of network topology in which each node is connected one after the other in a sequential chain. An arbitrary topology is supposed in Reference [10]. Exact algorithmic results are obtained in Reference [7] only for the case of a tandem consisting of two stations. In the case of a larger number of stations, approximate results are obtained. The analysis presented in Reference [10] is exact algorithmic for an arbitrary finite number of nodes, network topology and customer routing. In the recent paper, Reference [14], the model from Reference [10] is generalized to the case when there is no input buffer in the network and a customer arriving to the network when N customers receive service moves to the orbit and makes the retrials to obtain service. A classical retrial strategy is applied. This strategy assumes that the total retrial rate is proportional to the number of customers staying in the orbit.

In the model considered in this paper, we significantly extend the results of References [10,14] to the networks with state dependent processes of arrival of primary customers and retrials, account of impatience and non-persistence of customers in the orbit and customers impatience in the buffers of the nodes of the network. Considered mutual dependence of the flows of primary customers and retrials is typical for many real-world systems but is not studied in the existing literature even for simple queuing systems, not to mention queuing networks.

Examples of potential applications of the obtained results to the analysis of real-world systems can be found, for example, in References [7,10].

The rest of the paper has the following structure. The mathematical model of the queuing network is completely described in Section 2. The multidimensional stochastic process describing the dynamics of the orbit and nodes of the network is presented in Section 3. This process is a Markov chain with one denumerable and several finite space components. It belongs to the class of level-dependent Quasi-Birth-and-Death processes. The generator of this Markov chain is presented. The problem of computing the stationary distribution of the chain is discussed in this section. In Section 4, expressions for the main performance indicators of the network are presented. Illustrative numerical examples giving insights into behavior of the network are provided in Section 5. Section 6 contains some concluding remarks.

2. Mathematical Model

A semi-open queuing network consists of L nodes which are single-server queuing systems with finite buffers. The structure of the network is presented on Figure 1. We assume that the capacity of the buffer at the lth node is equal to $N_l - 1,\ 1 \le N_l \le \infty$. The capacity of the network, that is, the maximum number of customers, which can be processed in the network at the same time, is $N,\ 1 \le N < \infty$. We assume that $N \le \min\limits_{l \in \{1,2,\dots,L\}} N_l$. This guarantees that customers admitted to the network are not lost during processing in the network due to a buffer overflow.

The admission of arriving (primary) customers to the network is implemented as follows. If the capacity of the network is not exhausted on customer's arrival, that is, the number of customers in the network is less than N, the customer enters the service. Otherwise, the customer moves to the orbit and retries for service later. The capacity of the orbit is assumed to be infinite. The process of generation of primary customers and retrials of customers from the orbit is defined by the underlying process $\psi_t,\ t \ge 0$, with a finite state space $\{0, 1, \dots, W\}$. The intensities of transitions of the process ψ_t depend on the current number i_t of customers in the orbit, $i_t \ge 0$.

When $i_t = 0$, that is, the orbit is empty, the intensities of transitions are given by the set of square matrices $D_l,\ l = \overline{1, L}$, of size $\bar{W} = W + 1$ and by the non-diagonal entries of the matrix D_0. Here and thereafter, the notation $l = \overline{1, L}$ means that the parameter l admits the values from the set $\{1, 2, \dots, L\}$.

The diagonal entries of the matrix D_0 are negative such as the relation $\sum_{l=0}^{L} D_l\mathbf{e} = \mathbf{0}^T$ holds true where $\mathbf{e} = (1,1,\ldots,1)^T$, $\mathbf{0} = (0,0,\ldots,0)$ and ^{T} is the vector transpose symbol. Transitions, the intensities of which are given by the non-diagonal entries of the matrix D_0, do not cause the generation of primary customers. Transitions, the intensities of which are given by the entries of the matrix D_l, cause the generation of a type-l customer, $l = \overline{1,L}$. If the capacity of the network is not exhausted at the moment of type-l customer generation, this customer enters the lth node of the network. If the server of this node is idle, the customer starts service. Otherwise, the customer enters the buffer of this node. By comparing the presented description of the arrival process when the orbit is idle with the definition given in Reference [24] we can conclude that this process coincides with the $MMAP$.

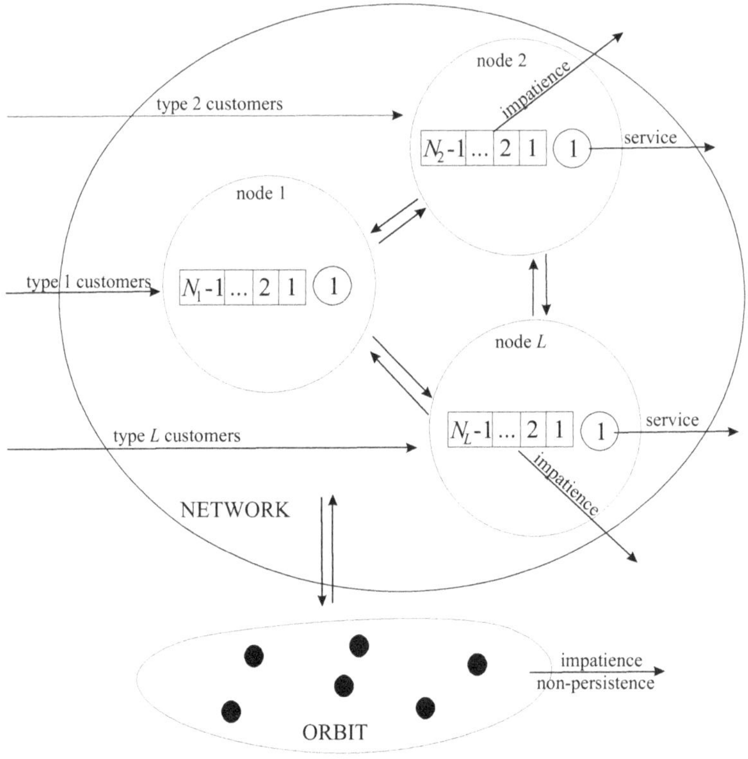

Figure 1. Retrial queuing network under study.

Under any fixed number i, $i > 0$, of customers in the orbit, the transitions of the process ψ_t can cause either the arrival of a primary customer or a retrial from the orbit. These transitions are given by a set of the same matrices D_l, $l = \overline{0,L}$, and the matrix $D^{(i)}$. As above, the intensities of transitions, which are given by the entries of the matrix D_l, cause the generation of a primary type-l, $l = \overline{1,L}$, customer. Transitions, the intensities of which are given by the entries of the matrix $D^{(i)}$, cause the generation of a repeated attempt from the orbit. The non-diagonal entries of the matrix $D_0^{(i)} = D_0 - \text{diag}\{D^{(i)}\mathbf{e}\}$ define the intensities of transitions which do not cause the generation of primary customers or customers from the orbit. Here, diag$\{\mathbf{b}\}$ denotes the diagonal matrix with the diagonal entries given by the entries of the vector $\mathbf{b}$. The matrix $\sum_{l=1}^{L} D_l + D^{(i)} + D_0^{(i)}$ is the generator. Let us denote by $\boldsymbol{\theta}^{(i)}$ the left stochastic eigenvector of this matrix. Conditional on the fact that i customers are staying in the orbit, the arrival rate of type l primary customers is equal to $\lambda_l^{(i)} = \boldsymbol{\theta}^{(i)} D_l \mathbf{e}$, $l = \overline{1,L}$. The conditional arrival rate of customers from the orbit is equal to $\lambda^{(i)} = \boldsymbol{\theta}^{(i)} D^{(i)} \mathbf{e}$, $i > 0$.

We assume that customers staying in the orbit are impatient and non-persistent. Impatience means that after a time interval, whose duration is exponentially distributed with parameter γ, $\gamma > 0$, the customer from the orbit leaves the network without service and is lost. The non-persistence means that after unsuccessful retrial attempts, the customer leaves the network forever with probability h, $0 \leq h \leq 1$, and returns to the orbit with the complementary probability. Analogously, the customers waiting in the buffers of the network are impatient. After the time interval, whose duration is exponentially distributed with the parameter β_l, a customer waiting in the buffer of the l-th node leaves the node and departs from the network without service.

A retrial customer, which is admitted for service, moves to the l-th node with probability p_l, $l = \overline{1,L}$, $\sum\limits_{l=1}^{L} p_l = 1$.

We assume that the service time at the node l has an exponential distribution with parameter μ_l, $0 < \mu_l < \infty$, $l = \overline{1,L}$. When this time expires, the serviced customer can move to another node or leave the network forever. The probability of the transition of the serviced customer from the l-th node to the l'-th node is defined as $q_{l,l'}$, $l' = \overline{1,L}$, $l' \neq l$. The probability of leaving the network after service in the l-th node is $q_{l,0}$, $q_{l,0} = 1 - \sum\limits_{l'=1, l' \neq l}^{L} q_{l,l'}$.

3. Process of System States

It is easy to see that the dynamics of the considered network is described by the continuous-time Markov chain

$$\xi_t = \{i_t, n_t, \psi_t, n_t^{(1)}, \ldots, n_t^{(L)}\}, \; t \geq 0,$$

where, at the moment t, $t \geq 0$,

- i_t is the number of customers in the orbit, $i_t \geq 0$;
- n_t is the total number of customers in the network, $n_t = \overline{0,N}$;
- ψ_t is the state of the underlying process of the arrival process of primary and retrial customers, $\psi_t = \overline{0,W}$;
- $n_t^{(l)}$ is the number of customers in the l-th node, $l = \overline{1,L}$, $n_t^{(l)} = \overline{0,n_t}$, $\sum\limits_{l=1}^{L} n_t^{(l)} = n_t$.

To compute the stationary distribution of the states of the Markov chain ξ_t, $t \geq 0$, we need to derive the generator of this chain. To this end, we use the matrix-analytic methods. Therefore, to simplify derivation of the generator it is useful to deal not with separate states of the chain but with whole groups of the states having the same value of the two first components of the Markov chain. We call such a group having the value (i,n) of these components a macro-state (i,n). There are many possibilities to enumerate the states of the Markov chain that belong to a fixed macro-state. Here, we assume that the states are numerated in the reverse lexicographic order of the components $n_t^{(1)}, \ldots, n_t^{(L)}$ and the direct lexicographic order of the component ψ_t. We call the set of macro-states $\{(i,0), (i,1), \ldots, (i,N)\}$ as the level i, $i \geq 0$.

To simplify derivations, in what follows, we use the following notation:

- I is the identity matrix and O is a zero matrix of an appropriate dimension;
- $\otimes$ and $\oplus$ are the symbols of Kronecker product and sum of matrices, respectively, see Reference [26];
- diag$\{\ldots\}$ is the diagonal matrix with the diagonal entries listed or defined by the entries of the vector in the brackets;
- diag$_+\{\ldots\}$ is the updiagonal matrix with the updiagonal entries listed in the brackets;
- diag$_-\{\ldots\}$ is the subdiagonal matrix with the subdiagonal entries listed in the brackets;

- $\boldsymbol{\beta}$ is the column vector defined as $\boldsymbol{\beta} = (\beta_1, \ldots, \beta_L)^T$;
- $\mathbf{p}$ is the row vector defined as $\mathbf{p} = (p_1, \ldots, p_L)$;
- $\mathbf{a}_l$ is the row vector of size L that has all zero components except the lth component which is equal to 1;
- $\mathbf{q}$ is the column vector defined as $\mathbf{q} = (q_1, q_2, \ldots, q_L)^T = (q_{1,0}\mu_1, q_{2,0}\mu_2, \ldots, q_{L,0}\mu_L)^T$;
- $T_n = \binom{n+L-1}{L-1} = \frac{(n+L-1)!}{n!(L-1)!}$ is the number of states of the process $(n_t^{(1)}, \ldots, n_t^{(L)})$ when $\sum\limits_{l=1}^{L} n_t^{(l)} = n$, $n = \overline{0, N}$.

Because we will operate with macro-states, we need to analyse the vector process $\mathbf{n}_t = \{n_t^{(1)}, \ldots, n_t^{(L)}\}$, $t \geq 0$, that defines the dynamics of the number of customers in the nodes of the network. The following events can change this number:

(1) An admission to the network of a customer from the orbit. Let us denote as $P_n(\mathbf{p})$, $n = \overline{0, N-1}$, the matrix consisting of probabilities of the transitions of the process $\mathbf{n}_t$ during the epoch when n, $n < N$, customers are obtaining service in the network and a customer from the orbit makes a retrial attempt;
(2) An admission to the network of an arriving type-l customer. Let us denote as $P_n(\mathbf{a}_l)$, $n = \overline{0, N-1}$, the matrix consisting of the transitions probabilities of the process $\mathbf{n}_t$ during the epoch when n, $n < N$, customers are obtaining service in the network and a type-l customer arrives to the system;
(3) A customer finishes the service in one of the nodes and transits to some another one. Let us denote as B_n, $n = \overline{1, N}$, the matrix which components define the intensities of the process $\mathbf{n}_t$ transitions in this case, conditioned on the fact that n customers are in the network;
(4) A customer finishes the service in some node and leaves the network. Let us denote as $S_n(\mathbf{q})$, $n = \overline{0, N-1}$, the matrix which components define the intensities of the process $\mathbf{n}_t$ transitions in this case, conditioned on the fact that there are $N - n$ customers in the network;
(5) A customer leaves some node due to impatience. The matrix $Z_n(\boldsymbol{\beta})$, $n = \overline{1, N}$, defines the intensities of the process $\mathbf{n}_t$ transitions in the considered case when there are n customers in the network.

It is worth noting that the matrices P_n, $S_n(\mathbf{q})$, $n = \overline{0, N-1}$, and B_n, $n = \overline{1, N}$, can be computed based on the use of the corresponding results from Reference [10]. Derivation of matrices $Z_n(\boldsymbol{\beta})$ is novel because the impatience of the customers staying in the buffers of the network nodes was not considered in the related literature previously. This derivation can be performed as follows.

Step (1) We compute the matrices $Z_n^{(k)}(\boldsymbol{\beta})$ using the recursive formulas:

$$Z_n^{(0)}(\boldsymbol{\beta}) = (n-1)\beta_L, \; n = \overline{1, N},$$

$$Z_n^{(k)}(\boldsymbol{\beta}) = \begin{pmatrix} (n-1)\beta_{L-k}I & O & \cdots & O \\ Z_1^{(k-1)}(\boldsymbol{\beta}) & (n-2)\beta_{L-k}I & \cdots & O \\ O & Z_2^{(k-1)}(\boldsymbol{\beta}) & \cdots & O \\ \vdots & \vdots & \ddots & \vdots \\ O & O & \cdots & 0\beta_{L-k}I \\ O & O & \cdots & Z_n^{(k-1)}(\boldsymbol{\beta}) \end{pmatrix}, \; k = \overline{1, L-1}, \; n = \overline{1, N}.$$

Step (2) Compute the matrices $Z_n(\boldsymbol{\beta})$ as $Z_n(\boldsymbol{\beta}) = Z_n^{(L-1)}(\boldsymbol{\beta})$, $n = \overline{0, N-1}$.

Also, let us introduce the matrix Γ_n. It is the diagonal matrix the diagonal entries of which define the intensities of exit of the process $\mathbf{n}_t$ from its states when n customers obtain service in the network. The matrices Γ_n are defined as follows:

$$\Gamma_0 = 0, \; \Gamma_n = -\text{diag}\{B_n\mathbf{e} + S_{N-n}(\mathbf{q})\mathbf{e} + Z_n(\boldsymbol{\beta})\mathbf{e}\}, \; n = \overline{1, N}.$$

We denote as $\mathbf{G}$ the infinitesimal generator of the Markov chain $\xi_t,\ t \geq 0$. The matrix $\mathbf{G}$ has an infinite size. It consists of the blocks $\mathbf{G}_{i,j},\ i,j \geq 0$, defining transition intensities from the level i to the level j. In turn, each matrix $\mathbf{G}_{i,j}$ consists of sub-blocks $\mathbf{G}_{i,j}^{(n,n')}$ of transition intensities from the macro-state (i,n) to the macro-state (j,n'). The diagonal entries of the blocks $\mathbf{G}_{i,i}^{(n,n)}$ are negative and are equal, up to the sign, to the rates of the exit of the Markov chain ξ_t from the corresponding states.

Lemma 1. *The infinitesimal generator* $\mathbf{G}$ *of the Markov chain* $\xi_t,\ t \geq 0$, *has a block-tridiagonal structure:*

$$\mathbf{G} = \mathrm{diag}\{\mathbf{G}_{i,i},\ i \geq 0\} + \mathrm{diag}_+\{\mathbf{G}_{i,i+1},\ i \geq 0\} + \mathrm{diag}_-\{\mathbf{G}_{i,i-1},\ i \geq 1\}. \tag{1}$$

The non-zero blocks $\mathbf{G}_{i,j},\ i,j \geq 0$, *have the following form:*

$$\mathbf{G}_{i,i} = \mathrm{diag}\{\mathbf{G}_{i,i}^{(n,n)},\ n = \overline{0,N}\} + \mathrm{diag}_+\{\mathbf{G}_{i,i}^{(n,n+1)},\ n = \overline{0,N-1}\} + \mathrm{diag}_-\{\mathbf{G}_{i,i}^{(n,n-1)},\ n = \overline{1,N}\},\ i \geq 0, \tag{2}$$

$$\mathbf{G}_{i,i-1} = \mathrm{diag}\{\mathbf{G}_{i,i-1}^{(n,n)},\ n = \overline{0,N}\} + \mathrm{diag}_+\{\mathbf{G}_{i,i-1}^{(n,n+1)},\ n = \overline{0,N-1}\},\ i \geq 1, \tag{3}$$

$$\mathbf{G}_{i,i+1} = \mathrm{diag}\{O,\dots,O,\sum_{k=1}^{L} D_k \otimes I_{T^N}\},\ i \geq 0, \tag{4}$$

where

$$\mathbf{G}_{i,i}^{(0,0)} = D_0^{(i)} - i\gamma I_{\bar{W}}, \tag{5}$$

$$\mathbf{G}_{i,i}^{(n,n)} = (D_0^{(i)} - i\gamma I_{\bar{W}}) \oplus (B_n + \Gamma_n),\ n = \overline{1,N-1}, \tag{6}$$

$$\mathbf{G}_{i,i}^{(N,N)} = (D_0^{(i)} + (1-h)D^{(i)} - i\gamma I_{\bar{W}}) \oplus (B_N + \Gamma_N), \tag{7}$$

$$\mathbf{G}_{i,i}^{(n,n+1)} = \sum_{k=1}^{L} D_k \otimes P_n(\mathbf{a}_k),\ n = \overline{0,N-1}, \tag{8}$$

$$\mathbf{G}_{i,i}^{(n,n-1)} = I_{\bar{W}} \otimes (S_{N-n}(\mathbf{q}) + Z_n(\boldsymbol{\beta})),\ n = \overline{1,N}, \tag{9}$$

$$\mathbf{G}_{i,i-1}^{(n,n+1)} = D^{(i)} \otimes P_n(\mathbf{p}),\ n = \overline{0,N-1}, \tag{10}$$

$$\mathbf{G}_{i,i-1}^{(n,n)} = i\gamma I_{\bar{W}T_n},\ n = \overline{0,N-1}, \tag{11}$$

$$\mathbf{G}_{i,i-1}^{(N,N)} = i\gamma I_{\bar{W}T_N} + hD^{(i)} \otimes I_{T_N}. \tag{12}$$

A brief proof of the Lemma is as follows. Block tridiagonal structure (1) of the generator $\mathbf{G}$ is explained by the fact that the probability of two or more customers arrival or departure from the orbit during the interval of the infinitesimal length is negligible. The matrices $\mathbf{G}_{i,i}, \mathbf{G}_{i,i-1}, \mathbf{G}_{i,i+1}$ have block structures (2)–(4) were the blocks define the intensities of transitions that lead to the corresponding change of the number of customers presenting in the network.

The most simple form (4) has the matrix $\mathbf{G}_{i,i+1}$ defining the intensities of customers arriving at the orbit. Because a customer joins the orbit only when the capacity of the server is exhausted (the number of customers in the network is equal to N), only one block of the matrix $\mathbf{G}_{i,i+1}$, namely, $\mathbf{G}_{i,i+1}^{(N,N)}$, is not equal to zero. This matrix block corresponds to the arrival of a primary customer of any type when the number of customers in the network is equal to N. This customer moves to the orbit and the number of customers in the network does not change.

The matrix $\mathbf{G}_{i,i-1}$ defining the intensities of customers departure from the orbit is the block two-diagonal matrix of form (3). It has the non-zero diagonal blocks corresponding to the case when the customer departing from the orbit does not enter the network but is lost. Such a departure occurs due to the impatience of the customers in the orbit when the number of customers in the network is equal to n, $n < N$, or the impatience and non-persistence of the customers staying in the orbit when this number is equal to N. The corresponding intensities of the departure are given by the matrices of form (11) and (12), respectively. The matrix $\mathbf{G}_{i,i-1}$ has also the up-diagonal blocks corresponding to the case when a customer from the orbit makes a retrial when the number of customers in the network is equal to n, $n < N$, and enters the network. After entering the network, this customer joins some of the network nodes. The intensities of occurrence of these events are given by the matrices of form (10).

The matrix $\mathbf{G}_{i,i}$ has block tridiagonal structure (2). The diagonal entries of its diagonal blocks $\mathbf{G}_{i,i}^{(n,n)}$ are negative. Their values with the opposite sign define the rate of the exit of the Markov chain ξ_t from the corresponding states. These diagonal entries are the corresponding entries of the matrices of form (5), if $n = 0$, (6), if $n = \overline{1, N-1}$, and (7) if $n = N$. The non-diagonal entries of the diagonal blocks $\mathbf{G}_{i,i}^{(n,n)}$ are non-negative and define the intensities of the transitions of the Markov chain ξ_t that do not cause the change of the number of customers neither in the orbit and in the network. These entries are given by the corresponding entries of the matrices of form (5), if $n = 0$, (6), if $n = \overline{1, N-1}$, and (7) if $n = N$. The up-diagonal blocks $\mathbf{G}_{i,i}^{(n,n+1)}$ correspond to the arrival of a primary customer of any type when the number of the customers in the network is less than N and its entering the corresponding node of the network. These blocks are defined by formula (8). The sub-diagonal blocks $\mathbf{G}_{i,i}^{(n,n-1)}$ correspond to a customer departure from the network due to the service completion or due to impatience and have the form (9). Lemma is proven.

It can be shown that the Markov chain ξ_t, $t \geq 0$, belongs to the class of Asymptotically Quasi-Toeplitz Markov chains, see Reference [27]. Using the results from Reference [27], it is easily verified that because the customers in orbit are impatient, that is, $\gamma > 0$, the Markov chain ξ_t, $t \geq 0$, is ergodic. Then, the following limits (stationary probabilities) exist for any set of the network parameters:

$$\pi(i, n, \psi, n^{(1)}, \ldots, n^{(L)}) = \lim_{t \to \infty} P\{i_t = i,\ n_t = n, \psi_t = \psi,\ n_t^{(1)} = n^{(1)}, \ldots, n_t^{(L)} = n^{(L)}\},$$

$$i \geq 0,\ n = \overline{0, N},\ \psi = \overline{0, W},\ n^{(l)} = \overline{0, n},\ l = \overline{1, L},\ \sum_{l=1}^{L} n^{(l)} = n.$$

Let us denote by $\boldsymbol{\pi}(i, n)$ the row vectors of the stationary probabilities that belong to the macro-state (i, n) and by $\boldsymbol{\pi}_i$ the row vectors of the stationary probabilities that belong to the level i, $i \geq 0$.

It is well known that the probability vectors $\boldsymbol{\pi}_i$, $i \geq 0$, satisfy the following system of linear algebraic equations:

$$(\boldsymbol{\pi}_0, \boldsymbol{\pi}_1, \ldots)\mathbf{G} = \mathbf{0}, \quad (\boldsymbol{\pi}_0, \boldsymbol{\pi}_1, \ldots)\mathbf{e} = 1. \tag{13}$$

Comparing this system with the corresponding system for the probability vectors $\boldsymbol{\pi}_i$, $i \geq 0$, for the analogous queuing network with the buffer considered in Reference [10], we see the following. The size of the vectors $\boldsymbol{\pi}_i$ (and the size of the corresponding square blocks of the generator) in Reference [10] is equal to $\bar{W}T_i$ if $i = \overline{0, N}$, and $\bar{W}T_N$ if $i > N$. The size of all vectors $\boldsymbol{\pi}_i$, $i \geq 0$, in the considered in our paper model of the queuing network with retrials is equal to $\bar{W} \sum_{n=0}^{N} T_n$. Therefore, the blocks of generator (1) are much larger than the blocks of the generator in Reference [10]. Thus, the algorithm from Reference [28], which was used for numerical work in Reference [10], is not very effective for solving system (13). To solve this system, we recommend to use a more effective numerically stable algorithm recently developed in Reference [29].

4. Performance Measures

The average number N_{orbit} of customers in the orbit is computed by

$$N_{orbit} = \sum_{i=1}^{\infty} i \boldsymbol{\pi}_i \mathbf{e}.$$

The average number $N_{network}$ of customers in the network at an arbitrary moment is computed by

$$N_{network} = \sum_{i=0}^{\infty}\sum_{n=1}^{N} n\boldsymbol{\pi}(i,n)\mathbf{e}.$$

The probability P_{imm} that an arbitrary customer is admitted to the network immediately upon arrival is computed by

$$P_{imm} = \frac{\sum\limits_{i=0}^{\infty}\sum\limits_{n=0}^{N-1} \boldsymbol{\pi}(i,n)(\hat{D}\otimes I_{T_n})\mathbf{e}}{\hat{\lambda}}$$

where the average arrival rate of primary customers $\hat{\lambda}$ is computed by

$$\hat{\lambda} = \sum_{i=0}^{\infty}\sum_{n=0}^{N} \boldsymbol{\pi}(i,n)(\hat{D}\otimes I_{T_n})\mathbf{e}$$

and $\hat{D} = \sum\limits_{r=1}^{L} D_r$.

Remark 1. *If all the matrices $D^{(i)}$ are the diagonal, that is, the underlying process ψ_t of arrivals cannot make the jumps into other states at the moments of retrials, then the average arrival rate of primary customers $\hat{\lambda}$ is equal to the arrival rate λ of the MMAP that arrives at time intervals when the orbit is empty. The value λ is given by $\lambda = \boldsymbol{\theta}\hat{D}\mathbf{e}$ where the row vector $\boldsymbol{\theta}$ is the unique solution of the system $\boldsymbol{\theta}(D_0+\hat{D}) = \mathbf{0},\ \boldsymbol{\theta}\mathbf{e} = 1$. If some of the matrices $D^{(i)}$ are non-diagonal, then generally speaking $\hat{\lambda} \neq \lambda$.*

The average intensity $\lambda^{(l)}_{out-serve}$ of flow of customers who leave the network after successful service from the l-th node is computed by

$$\lambda^{(l)}_{out-serve} = \sum_{i=0}^{\infty}\sum_{n=1}^{N} \boldsymbol{\pi}(i,n)(I_{\bar{W}}\otimes S_{N-n}(\mathbf{q}^{(l)}))\mathbf{e},\ l=\overline{1,L},$$

where $\mathbf{q}^{(l)}$ is a column vector of size L with all zero components except the component $(\mathbf{q}^{(l)})_l = q_l$.

The average intensity $\lambda_{out-serve}$ of flow of customers who leave the network after successful service is computed by

$$\lambda_{out-serve} = \sum_{i=0}^{\infty}\sum_{n=1}^{N} \boldsymbol{\pi}(i,n)(I_{\bar{W}}\otimes S_{N-n}(\mathbf{q}))\mathbf{e}.$$

The average intensity $\lambda^{(l)}_{out-imp}$ of flow of customers who leave the network due to impatience from the l-th node is computed by

$$\lambda^{(l)}_{out-imp} = \sum_{i=0}^{\infty}\sum_{n=2}^{N} \boldsymbol{\pi}(i,n)(I_{\bar{W}}\otimes Z_n(\boldsymbol{\beta}^{(l)}))\mathbf{e},\ l=\overline{1,L},$$

where $\boldsymbol{\beta}^{(l)}$ is a column vector of size L with all zero components except the component $(\boldsymbol{\beta}^{(l)})_l = \beta_l$.

The average intensity $\lambda_{out-imp}$ of flow of customers who leave the network due to impatience is computed by

$$\lambda_{out-imp} = \sum_{i=0}^{\infty}\sum_{n=2}^{N} \boldsymbol{\pi}(i,n)(I_{\bar{W}}\otimes Z_n(\boldsymbol{\beta}))\mathbf{e}.$$

The probability of an arbitrary customer loss due to impatience from the orbit is computed by

$$P^{imp-orbit}_{loss} = \hat{\lambda}^{-1}\gamma N_{orbit}.$$

The probability of an arbitrary customer loss due to non-persistence from the orbit is computed by

$$P_{loss}^{nonpersist-orbit} = \hat{\lambda}^{-1} h \sum_{i=1}^{\infty} \boldsymbol{\pi}(i,N)(D^{(i)} \otimes I_{T_N})\mathbf{e}.$$

The probability of an arbitrary customer loss due to impatience from the network is computed by

$$P_{loss}^{imp-net} = \hat{\lambda}^{-1} \lambda_{out-imp}.$$

The probability of an arbitrary customer loss due to impatience from the lth node of the network is computed by

$$P_{loss}^{imp-net,l} = \hat{\lambda}^{-1} \lambda_{out-imp}^{(l)}, \; l = \overline{1,L}.$$

The probability of an arbitrary customer loss is computed by

$$P_{loss} = 1 - \frac{\lambda_{out-serve}}{\hat{\lambda}} = P_{loss}^{imp-orbit} + P_{loss}^{imp-net} + P_{loss}^{nonpersist-orbit}.$$

The average intensity $\mu_{out}^{(l)}$ of output flow of successfully served customers from the l-th node is computed by

$$\mu_{out}^{(l)} = \sum_{i=0}^{\infty} \sum_{n=1}^{N} \boldsymbol{\pi}(i,n)(I_{\bar{W}} \otimes S_{N-n}(\mathbf{m}^{(l)}))\mathbf{e}, \; l = \overline{1,L},$$

where $\mathbf{m}^{(l)}$ is the column vector of size L with all zero components except the component $(\mathbf{m}^{(l)})_l = \mu_l$.

The load of the l-th node ρ_l can be found as follows:

$$\rho_l = \frac{\mu_{out}^{(l)}}{\mu_l(1 - P_{loss}^{imp-net,l})}, \; l = \overline{1,L}.$$

This characteristic of the node operation is very important because knowledge of its value is helpful to recognize the so-called bottlenecks in the network and to make certain managerial updates.

5. Numerical Examples

We present the results of three numerical experiments. In the first experiment, we illustrate the importance on account of possible dependence of arrival processes of primary and retrial customers. The aim of the second experiment is the numerical investigation of the dependence of the main performance measures of the system on the threshold N and illustration of the importance of account the impatience of customers staying in the network. In the third experiment, we show how our results can be used for localization of the bottlenecks of the network and improvement of the performance of the network via upgrading the bottleneck node.

Remark 2. *The importance of correlation in the arrival process for a similar queuing network without retrials was shown in Reference [10]. In this paper, we do not present the results illustrating the importance of the correlation. We mention only that the correlation in the arrival process of primary customers has an essential impact on the networks with retrials as well.*

Example 1. *Let us consider a queuing network consisting of $L = 3$ nodes. The mean service rates at these nodes are $\mu_1 = 2$, $\mu_2 = 1.5$, $\mu_3 = 2$, respectively.*

The transition probabilities of the customers in the network $q_{l,k}$, $l = \overline{1,L}$, $k = \overline{0,L}$, $k \neq l$, are defined as the corresponding entries of Table 1.

The probabilities defining the choice of the node at the moment of admission of a customer from the orbit are chosen as: $p_1 = 0.2$, $p_2 = 0.3$, and $p_3 = 0.5$. The intensity of impatience of a customer from the orbit is assumed to be $\gamma = 0.02$. The probability of a customer departure from the orbit after an unsuccessful retrial is $h = 0.3$. The intensities of impatience of customers from the nodes are given as follows: $\beta_1 = 0.05$, $\beta_2 = 0.01$, $\beta_2 = 0.03$.

Table 1. Transition probabilities $q_{l,k}$.

$q_{l,k}$	$k = 0$	$k = 1$	$k = 2$	$k = 3$
$l = 1$	$\frac{1}{2}$	-	$\frac{1}{4}$	$\frac{1}{4}$
$l = 2$	$\frac{1}{3}$	$\frac{2}{15}$	-	$\frac{8}{15}$
$l = 3$	$\frac{1}{2}$	$\frac{1}{4}$	$\frac{1}{4}$	-

We assume that the $MMAP$ arrival flow of primary customers when the orbit is empty is defined by the following matrices:

$$D_0 = \begin{pmatrix} -1.764 & 0.014 \\ 0.07 & -0.42 \end{pmatrix}, \; D_1 = \begin{pmatrix} 0.07 & 0.007 \\ 0 & 0.14 \end{pmatrix},$$

$$D_2 = \begin{pmatrix} 0.028 & 0.035 \\ 0.0042 & 0.203 \end{pmatrix}, \; D_3 = \begin{pmatrix} 1.603 & 0.007 \\ 0.0021 & 0.0007 \end{pmatrix}.$$

The retrial rates of customers from the orbit when i customers are staying in the orbit are defined by the entries of the matrix

$$D^{(i)} = i\tilde{D},$$

where

$$\tilde{D} = \begin{pmatrix} 0.2 & 0 \\ 0 & 0.02 \end{pmatrix}.$$

Remark 3. *We intentionally choose the matrix $D^{(i)}$ in the simple diagonal form. This allows us to calculate the average individual retrial rate α of a customer from the orbit as*

$$\alpha = \sigma \tilde{D} \mathbf{e},$$

where σ is the unique solution of the following system

$$\sigma(D_0 + \sum_{l=1}^{L} D_l - \text{diag}\{\tilde{D}\mathbf{e}\}) = \mathbf{0}, \quad \sigma \mathbf{e} = 1.$$

In this example, α is equal to 0.1185929648.

In the considered model, the arrival flows of primary and retrials customers are dependent. In the existing literature, such dependence was not analyzed. In this example, we clarify whether or not this dependence is essential. To this end, we also consider the case when customers in the orbit retry to obtain access to the network independently on primary customers as assumed in the existing literature. We suppose that each customer from the orbit makes repeated attempts with the intensity $\alpha = 0.1185929648$.

It is easy to see that the results for the model with independent flows of primary and retrial customers are obtained as the particular case of our results if we assume that the matrix $\tilde{D}$ has the following form:

$$\tilde{D} = \begin{pmatrix} \alpha & 0 \\ 0 & \alpha \end{pmatrix}.$$

Let us vary the number N of customers, which can be serviced in the network simultaneously, over the interval $[1, 15]$. The computations were performed on a PC with an Intel Core i7-8700 CPU and 16 GB RAM. The computation time for all N from 1 to 15 is about 4 min and 50 s.

Figures 2–4 illustrate the dependence on N of the probability of an arbitrary customer loss due to non-persistence from the orbit $P_{loss}^{nonpersist-orbit}$, due to impatience from the orbit $P_{loss}^{imp-orbit}$ and the loss probability of an arbitrary customer P_{loss} for the systems with dependent and independent arrivals of primary and retrial customers.

As it is seen from these figures, an account of the dependence of arrivals of primary and retrial customers is important for the precious prediction of the system performance measures. Under the chosen values of the network parameters, this dependence deteriorates the performance of the network. All loss probabilities are greater when the flows are dependent. In this example, the error in the estimation the loss probabilities can be up to 20% on their real values.

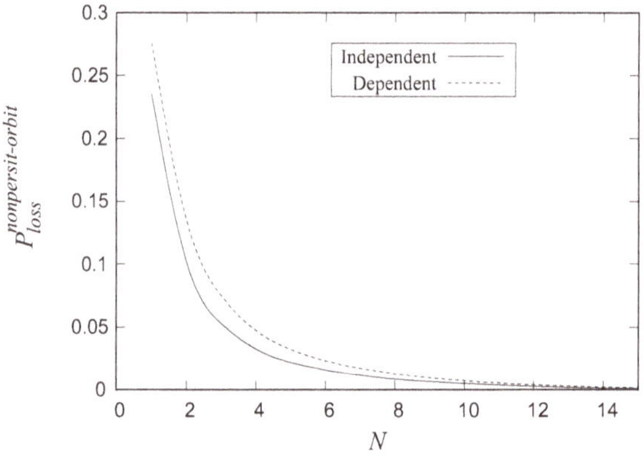

Figure 2. Dependence of the probability $P_{loss}^{nonpersist-orbit}$ on the number N.

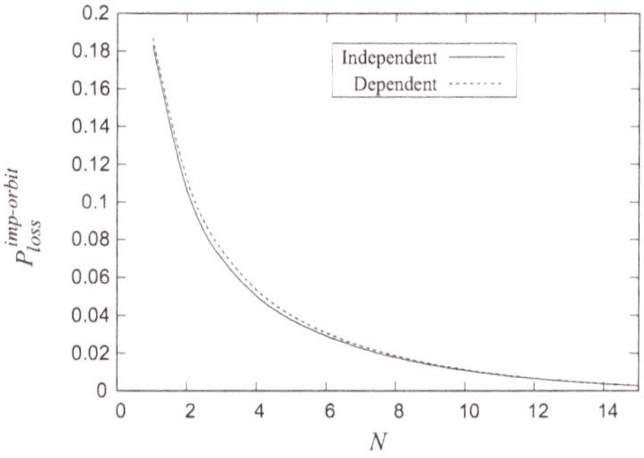

Figure 3. Dependence of the probability $P_{loss}^{imp-orbit}$ on the number N.

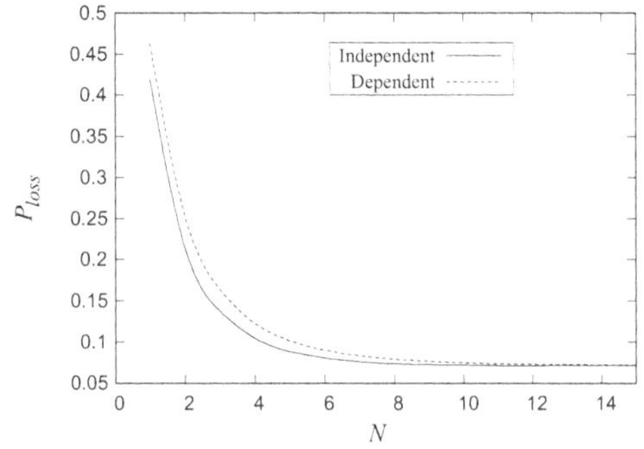

Figure 4. Dependence of the probability P_{loss} on the number N.

Example 2. *In this example, we numerically investigate the dependence of the main performance measures of the system on the threshold N. We also investigate the importance of accounting the impatience of customers in the nodes of the network. Let us assume that all system parameters are the same as in the previous example except for the matrix*

$$\tilde{D} = \begin{pmatrix} 0.2 & 0.002 \\ 0.001 & 0.02 \end{pmatrix}.$$

In the previous example, we chose relatively small intensities of impatience of customers from the nodes: $\beta_1 = 0.05$, $\beta_2 = 0.01$, $\beta_2 = 0.03$. *The question arises, whether or not it is possible to ignore such small intensities of impatience and assume that the customers are patient, that is,* $\beta_l = 0$, $l = \overline{1, L}$.

Figures 5–7 illustrate the dependence on N of the average number N_{orbit} of customers in the orbit, the probability P_{imm} that an arbitrary customer is admitted to the network immediately upon arrival and the probability of an arbitrary customer loss due to impatience from the network $P_{loss}^{imp-net}$ for the cases of the patient and impatient customers in the nodes of the network.

The probability of an arbitrary customer loss due to impatience from the network $P_{loss}^{imp-net}$ is equal to 0 when the customers in the nodes are patient and essentially increases with the growth of N when the customers are impatient. It is worth to note that the loss of the customers due to impatience in the nodes causes the reduction of the average number of customers in the orbit and the increase of the probability P_{imm}.

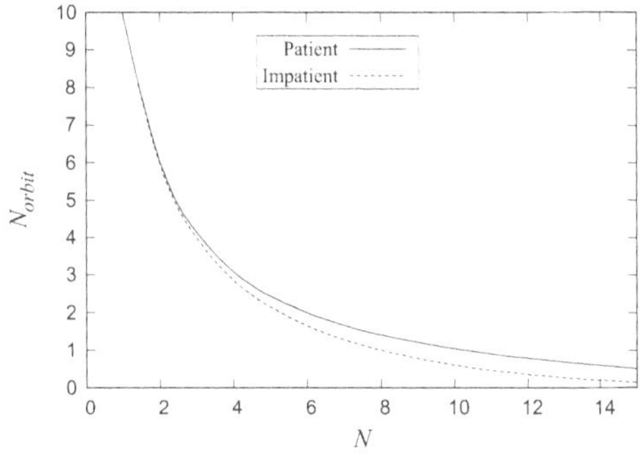

Figure 5. Dependence of the average number N_{orbit} on the number N.

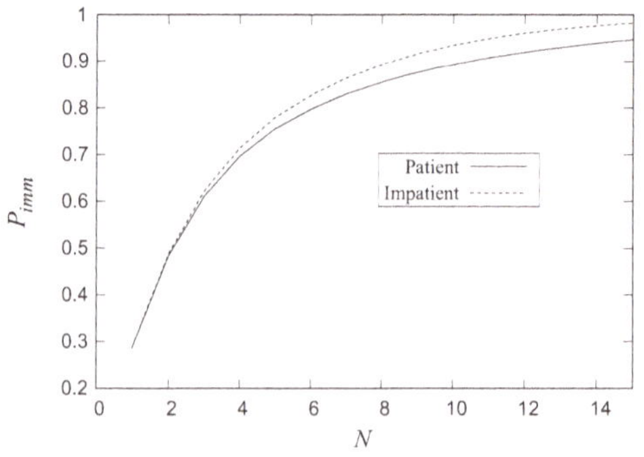

Figure 6. Dependence of the probability P_{imm} on the number N.

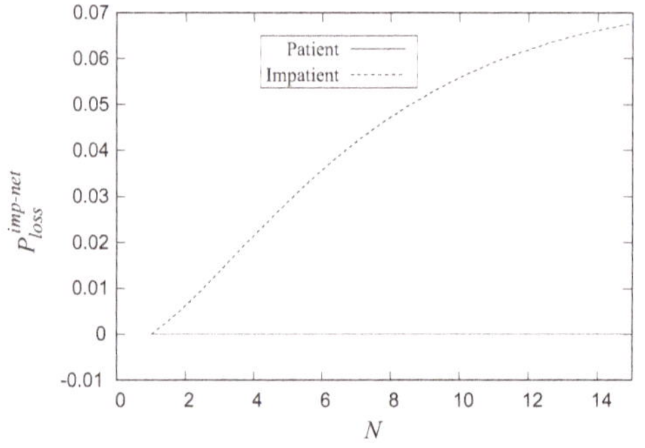

Figure 7. Dependence of the probability $P_{loss}^{imp-net}$ on the number N.

Analogously, Figures 8–10 illustrate the dependence on N of the probability of an arbitrary customer loss due to impatience from the orbit $P_{loss}^{imp-orbit}$, the probability of an arbitrary customer loss due to non-persistence from the orbit $P_{loss}^{nonpersist-orbit}$, and the loss probability of an arbitrary customer P_{loss} for the case of the patient and impatient customers in the network.

It can be seen that the impatience in the nodes reduced the probabilities of customers loss from the orbit (due to impatience or non-persistence). However, the total loss probability P_{loss} is essentially higher when the customers in the nodes are impatient. With the growth of N, the difference of values of this probability for the cases of the impatient and patient customers becomes more significant.

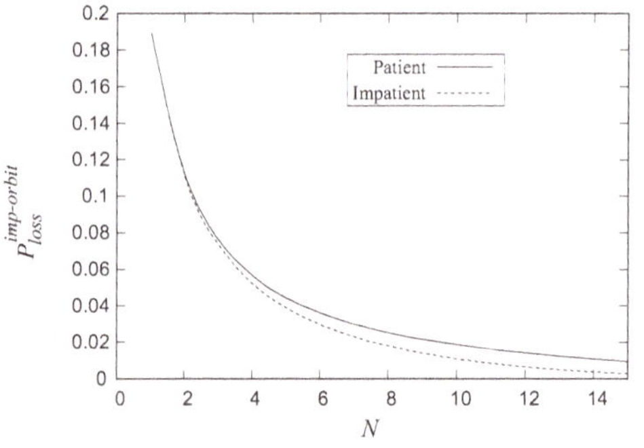

Figure 8. Dependence of the probability $P_{loss}^{imp-orbit}$ on the number N.

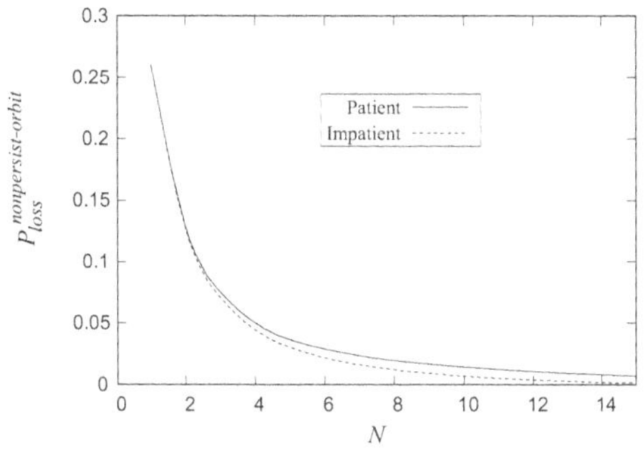

Figure 9. Dependence of the probability $P_{loss}^{nonpersist-orbit}$ on the number N.

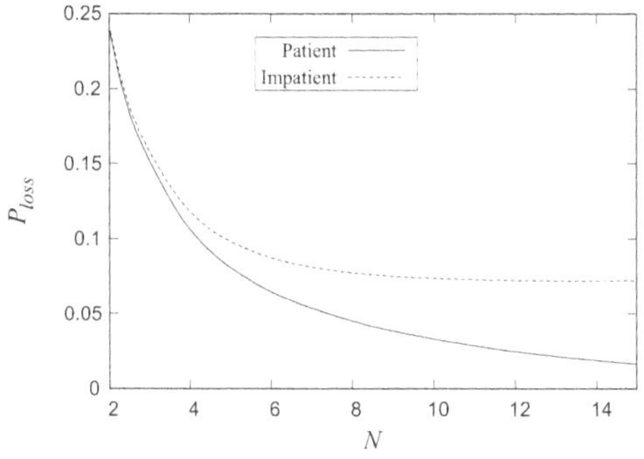

Figure 10. Dependence of the probability P_{loss} on the number N.

Let us introduce the following economical criterion of the quality of the network operation:

$$E(N) = \hat{\lambda}(a(P_{loss}^{imp-orbit} + P_{loss}^{nonpersist-orbit}) + bP_{loss}^{imp-net}),$$

where a is a charge paid for the loss of one customer from the orbit and b is a charge paid for the loss of one customer from the network per unit time. It is evident that the loss of a customer from the network is more painful than the loss of a customer from the orbit because the lost from the network customer possibly have already been serviced in some nodes, that is, the system has spent some resources for providing service to such a customer. Thus, in this example, we assume that $a = 1$ and $b = 3$.

Figure 11 illustrates the dependence of the economic criterion $E(N)$ on N for the cases of the patient and impatient customers in the network.

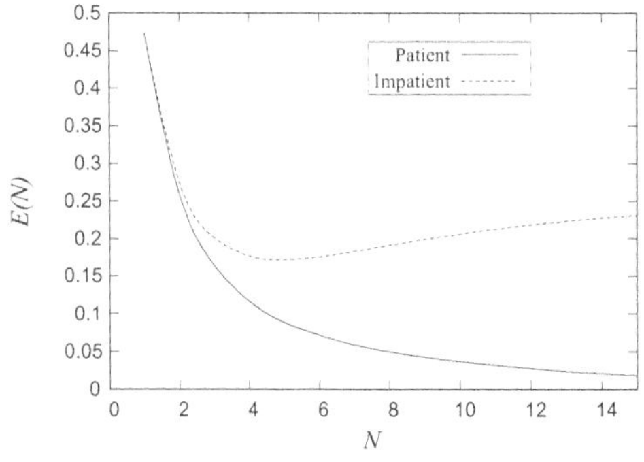

Figure 11. Dependence of the economical criterion $E(N)$ on the number N.

The optimal value for the case of patient customers is $E(N) = 0.0184$ when $N = 15$. This means that in the case of patient customers it is reasonable to accept as many customers as possible. The optimal value for the case of impatient in the network customers is $E(N) = 0.171865$ when $N = 5$. Presented in this paper results can be helpful for the optimal choice of the limit N imposed on the number of customers that can be admitted to the network simultaneously.

Example 3. *In this experiment, we show how our results can be used for identification of the bottleneck of the network and further improving the performance of the network via the proper upgrade of the bottleneck node. Let us choose the parameters of the network the same as in the previous example in the case of impatient customers. One can see that the loss probability of customers is quite high even for a large value of N ($N = 15$). About 7.2% of customers are lost due to different reasons. To understand the reason for such not very good operation of the network, let us compute the average load of each network's node.*

Figure 12 illustrates the dependence of loads of the nodes on the parameter N.

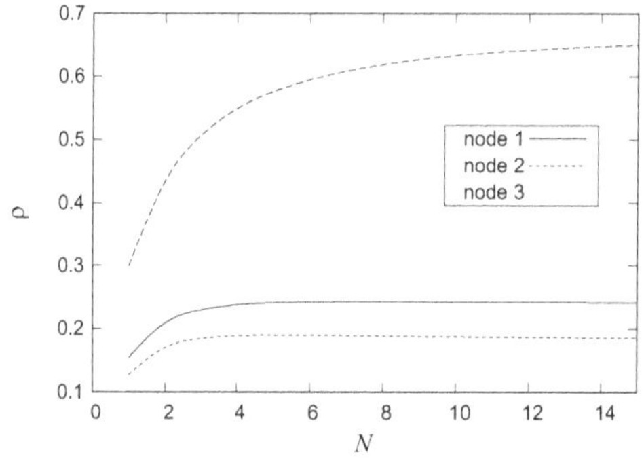

Figure 12. Dependence of the load of the nodes on the parameter N.

As it is seen from this Figure, the load of the third node is much higher than loads of other nodes. Let us make an upgrade of the third node in such a way that after upgrade the service rate in this node increases from 2 to 4 and compute the main performance measures of the network.

Figures 13 and 14 illustrate the dependence of the probability P_{imm} that an arbitrary customer is admitted to the network immediately upon arrival and the loss probability of an arbitrary customer P_{loss} on the parameter N before and after upgrade.

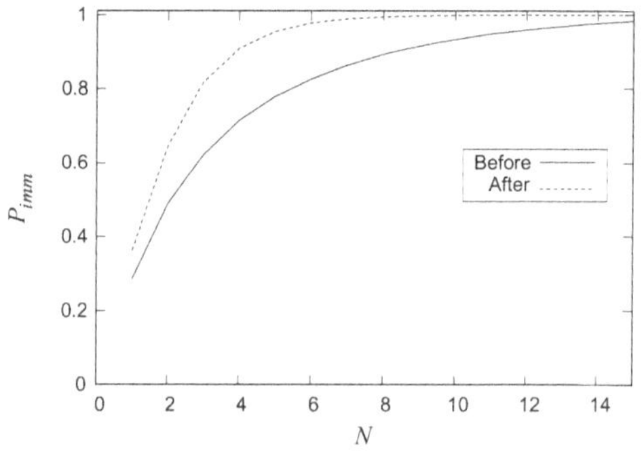

Figure 13. Dependence of the probability P_{imm} on the number N.

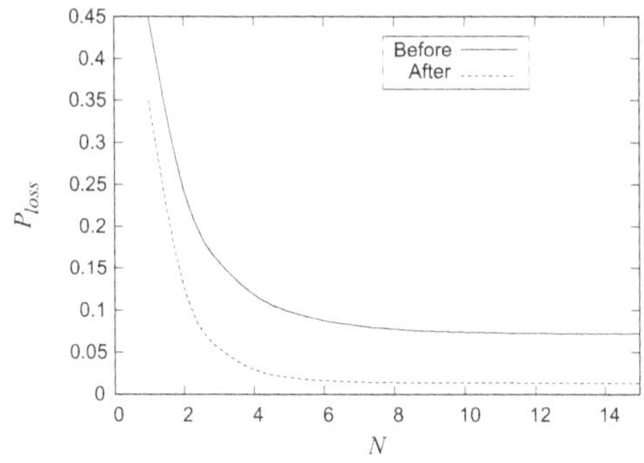

Figure 14. Dependence of the probability P_{loss} on the number N.

One can see from these Figures that after upgrade the loss probability P_{loss} essentially decreases and the probability P_{imm} of immediate access essentially increases. Figure 15 illustrates the dependence of the economic criterion $E(N)$ on N before and after the upgrade.

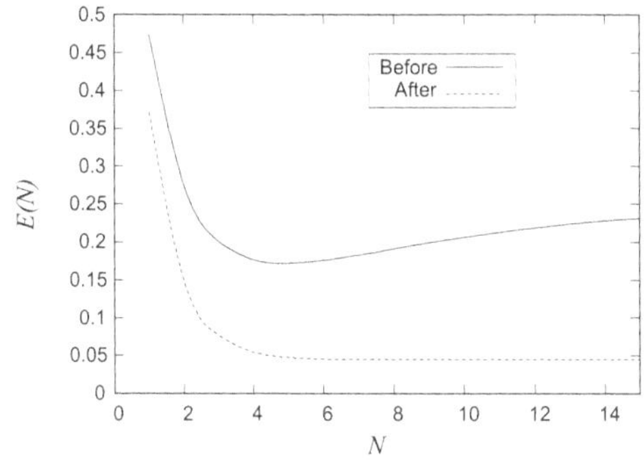

Figure 15. Dependence of the economic criterion $E(N)$ on the number N before and after upgrade.

The optimal value $E(N)$ after upgrade is $E(N) = 0.0448422$ when $N = 7$ what is almost four times smaller than the optimal value $E(N) = 0.171865$ before the upgrade (that was achieved when $N = 5$). Definitely, the increase of the service rate in the third node can cost some money. However, it allows increasing the optimal number of the limit N from 5 to 7 with the essential improvement of the quality of operation of the network.

It is worth to note that the improvement of the quality of operation of the network via the account of loads of the nodes can be achieved also via the modification of the routing of the customers in the node and the choices of the target node by the primary and retrial customers.

6. Conclusions

In this paper, we analyzed a semi-open queuing network with customers retrial. The number of customers in the network must not exceed the fixed threshold (capacity of the network). An arriving primary customer is admitted to the network and starts processing only if the current number of customers in the network is less than the network capacity. Otherwise, the customer moves to the orbit having an infinite capacity and tries to enter the network after random time intervals. The arrivals of primary customers and retrials of customers from the orbit depend on the same underlying process. This allows more adequately model real-world arrival processes than it can be done using known in the literature models of the arrival and retrial processes. Customers are impatient both in the orbit and infinite buffers of the nodes of the network. The behavior of the network is described by a multidimensional Markov chain. The generator of this chain is derived. The problem of computing the stationary state distribution is discussed. The expressions for computing the main performance measures of the network are derived. Numerical results illustrate the importance of the account of the dependency of arrivals of the primary and retrial customers as well as the importance of account of impatience of customers. Possibilities of optimization of the quality of the network operation by means of the optimal choice of the network capacity and identification and elimination of bottlenecks in the network are demonstrated.

Author Contributions: Conceptualization C.K. and A.D.; methodology C.K. and S.D.; software S.D.; validation C.K., A.D. and K.S.; formal analysis C.K., S.D. and A.D.; investigation C.K., K.S. and A.D.; Writing—Original draft preparation, C.K. and S.D.; Writing—Review and editing, K.S. and A.D.; supervision, C.K.; project administration C.K. and K.S.

Funding: The work by Chesoong Kim has been supported by Basic Science Research Program through the National Research Foundation of Korea (NRF) funded by the Ministry of Education (Grant No. NRF-2017R1D1A3A03000523). The work by S. Dudin, A. Dudin and K. Samouylov has been supported by RUDN University Program 5-100.

Conflicts of Interest: The authors declare no conflict of interest.

References

1. Bolch, G.; Greiner, S.; De Meer, H.; Trivedi, K.S. *Queuing Networks and Markov Chains: Modeling and Performance Evaluation with Computer Science Applications*; John Wiley & Sons: Hoboken, NJ, USA, 2006.
2. Boucherie, R.J.; van Dijk, N.M. *Queuing Networks*; A Fundamental Approach; Springer: Cham, Switzerland, 2011.
3. Shortle, J.F.; Thompson, J.M.; Gross, D.; Harris, C.M. *Fundamentals of Queuing Theory*; John Wiley & Sons: Hoboken, NJ, USA, 2018.
4. Smith, J.M. *Introduction to Queuing Networks: Theory & Practice*; Springer: Cham, Switzerland, 2018.
5. Roy, D. Semi-open queuing networks: A review of stochastic models, solution methods and new research areas. *Int. J. Prod. Res.* **2016**, *54*, 1735–1752. [CrossRef]
6. Dallery, Y. Approximate analysis of general open queuing networks with restricted capacity. *Perform. Eval.* **1990**, *11*, 209–222. [CrossRef]
7. Dhingra, V.; Kumawat, G.L.; Roy, D.; de Koster, R. Solving semi-open queuing networks with time-varying arrivals: An application in container terminal landside operations. *Eur. J. Oper. Res.* **2018**, *267*, 855–876. [CrossRef]
8. Ekren, B.Y.; Heragu, S.S.; Krishnamurthy, A.; Malmborg, C.J. Matrix-geometric solution for semi-open queuing network model of autonomous vehicle storage and retrieval system. *Comput. Ind. Eng.* **2014**, *68*, 78–86. [CrossRef]
9. Jia, J.; Heragu, S.S. Solving semi-open queuing networks. *Oper. Res.* **2009**, *57*, 391–401. [CrossRef]
10. Kim, J.; Dudin, A.; Dudin, S.; Kim, C. Analysis of a Semi-Open queuing Network with Markovian Arrival Process. *Perform. Eval.* **2018**, *120*, 1–19. [CrossRef]
11. Palmer, G.I.; Harper, P.R.; Knight, V.A. Modelling deadlock in open restricted queuing networks. *Eur. J. Oper. Res.* **2018**, *266*, 609–621. [CrossRef]
12. Artalejo, J.R.; Gomez-Corral, A. *Retrial Queuing Systems: A Computational Approach*; Springer: Berlin/Heidelberg, Germany, 2008.
13. Falin, G.I.; Templeton, J.G.C. *Retrial Queues*; Chapman&Hall: London, UK, 1997.
14. Kim, C.S.; Dudin, S. Analysis of Semi-Open queuing Network with Customer Retrials. *J. Korean Inst. Ind.* **2019**, *45*, 193–202.
15. Mandelbaum, A.; Massey, W.A.; Reiman, M.I. Strong approximations for Markovian service networks. *Queuing Syst.* **1998**, *30*, 149–201. [CrossRef]
16. Chakravarthy, S.R. The batch Markovian arrival process: A review and future work. In *Advances in Probability Theory and Stochastic Processes*; Krishnamoorthy, A., Raju, N., Ramaswami, V., Eds.; Notable Publications Inc.: Branchburg, NJ, USA, 2001; pp. 21–29.
17. Lucantoni, D. New results on the single server queue with a batch Markovian arrival process. *Commun. Stat. Stoch. Models* **1991**, *7*, 1–46. [CrossRef]
18. Vishnevski, V.M.; Dudin, A.N. queuing systems with correlated arrival flows and their applications to modeling telecommunication networks. *Autom. Remote Control* **2017**, *78*, 1361–1403. [CrossRef]

19. Strelen, J.C. Approximate analysis of queuing networks with Markovian arrival processes and phase type service times. In *Modellierung und Bewertung von Rechen- und Kommunikationssystemen*; Irmscher, K., Mittasch, C., Richter, K., Eds.; VDE- Verlag GmbH.: Berlin, Germany; Offenbach, Germany, 1997; pp. 55–70.
20. Gomez-Corral, A. A tandem queue with blocking and Markovian arrival process. *Queuing Syst.* **2002**, *41*, 343–370. [CrossRef]
21. Kim, C.S.; Dudin, A.; Dudin, S.; Dudina, O. Tandem queuing system with impatient customers as a model of call center with Interactive Voice Response. *Perform. Eval.* **2013**, *70*, 440–453. [CrossRef]
22. Kim, C.S.; Klimenok, V.; Taramin, O. A tandem retrial queuing system with two Markovian flows and reservation of channels. *Comput. Oper. Res.* **2010**, *37*, 1238–1246. [CrossRef]
23. Kim, C.S.; Park, S.H.; Dudin, A.; Klimenok, V.; Tsarenkov, G. Investigaton of the $BMAP/G/1 \to \bullet/PH/1/M$ tandem queue with retrials and losses. *Appl. Math. Model.* **2010**, *34*, 2926–2940. [CrossRef]
24. He, Q.M. Queues with marked calls. *Adv. Appl. Probab.* **1996**, *28*, 567–587. [CrossRef]
25. Dudin, A.N.; Klimenok, V.I. $BMAP/SM/1$ model with Markov modulated retrials. *Top* **1999**, *7*, 267–278. [CrossRef]
26. Graham, A. *Kronecker Products and Matrix Calculus with Applications*; Ellis Horwood: Cichester, UK, 1981.
27. Klimenok, V.I.; Dudin, A.N. Multi-dimensional asymptotically quasi-Toeplitz Markov chains and their application in queuing theory. *Queuing Syst.* **2006**, *54*, 245–259. [CrossRef]
28. Dudina, O.; Kim, C.; Dudin, S. Retrial queuing system with Markovian arrival flow and phase-type service time distribution. *Comput. Ind. Eng.* **2013**, *66*, 360–373. [CrossRef]
29. Dudin, S.; Dudina, O. Retrial multi-server queuing system with PHF service time distribution as a model of a channel with unreliable transmission of information. *Appl. Math. Model.* **2019**, *65*, 676–695. [CrossRef]

mathematics

MDPI

Article

On the Rate of Convergence and Limiting Characteristics for a Nonstationary Queueing Model

Yacov Satin [1], Alexander Zeifman [1,2,3,*] and Anastasia Kryukova [1]

1 Department of Applied Mathematics, Vologda State University, 160000 Vologda, Russia
2 Institute of Informatics Problems of the Federal Research Center "Computer Science and Control" of the Russian Academy of Sciences, 119333 Moscow, Russia
3 Vologda Research Center of the Russian Academy of Sciences, 160014 Vologda, Russia
* Correspondence: a_zeifman@mail.ru

Received: 29 June 2019; Accepted: 28 July 2019; Published: 30 July 2019

Abstract: Consideration is given to the nonstationary analogue of $M/M/1$ queueing model in which the service happens only in batches of size 2, with the arrival rate $\lambda(t)$ and the service rate $\mu(t)$. One proposes a new and simple method for the study of the queue-length process. The main probability characteristics of the queue-length process are computed. A numerical example is provided.

Keywords: queueing systems; rate of convergence; non-stationary; Markovian queueing models; limiting characteristics

1. Introduction

Non-stationary Markovian queueing models have been the subject of extensive research for the past few decades. It is well-known that the direct computation of time-dependent characteristics for arbitrary (in)homogeneous continuous-time Markov chains, which show up in the analysis of various queueing models, is a difficult problem. Thus, usually the alternative way is taken: one resorts to different types of approximations. One can find an overview of the approaches for the performance evaluation of time-dependent queueing systems up to 2016 in [1]. The papers [2–4] are devoted to the construction of the main performance characteristics and papers [5–8] deal with the estimation of the convergence rate and approximations. The general framework for the study of time-dependent queueing system systems is described in detail in the recent paper [8]. It consists of several steps, among which the most important one is the estimation of the upper bounds for the rate of convergence to the limiting regime. Having such bound allows one to find (compute) the time instant, say t^*, starting from which probabilistic properties of $X(t)$ do not depend on the value of $X(0)$ (assuming that the process starts at time $t = 0$). Thus, for example, if the transition intensities are periodic (say, 1-time-periodic), one can truncate the process on the interval $[t^*, t^* + 1]$ and solve the forward Kolmogorov system of differential equations on this interval with $X(0) = 0$. In such a way, one may build approximations for any limiting probability characteristics of $X(t)$ and estimate stability (perturbation) bounds. For the details regarding the stability bounds, one can refer to [9–15] and references therein. If the reduced intensity matrix of a Markov chain (see the next section for the definition) is essentially positive, then the approach for the computation of the upper bounds on the rate of convergence l_1 metric is available: one may use the method of logarithmic norm of a linear operator function and use the bounds for the Cauchy operator of the (reduced) forward Kolmogorov system (see [6,7]). Note that such bounds may be sharp if the difference between the two initial conditions is nonnegative.

However, the method of the logarithmic norm is not always applicable, i.e., there are Markov chains with such transition intensities, for which it does not yield upper bounds for the rate of convergence. Such a Markov chain is the topic of this paper. Specifically, one considers an inhomogeneous analogue of the classical $M/M/1$ queue, in which the service happens only in batches of size 2. For this queue, we propose the simple new method, based on the direct application of differential inequalities, for the estimation of the queue-size probability characteristics.

2. Model Description and Basic Transformations

Consideration is given to the Markov chain $X(t)$ being the queue-length (including a customer in server) at time t in the $M_t/M_t/1$ queuing system with batch service (only). It is assumed that customers enter the system only by one and the arrival intensity does not depend on the number of customers in the system, but depends on time, and is equal to $\lambda(t)$. Customers can be served only in batches of size 2 and the service intensity does not depend on the number of customers in the system, but depends on time, and is equal to $\mu(t)$. The transition diagram for $X(t)$ is given in Figure 1.

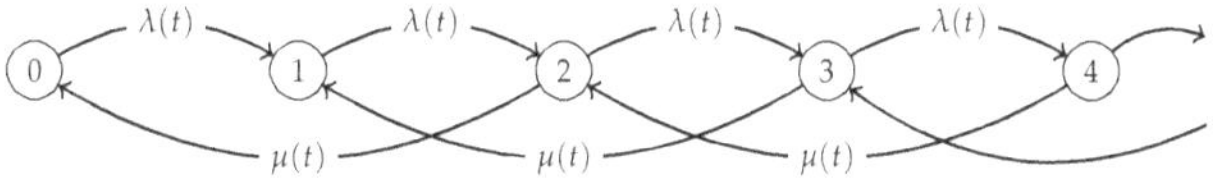

Figure 1. Transition diagram for the Markov chain $X(t)$.

Denote by $p_{ij}(s,t) = Pr\left\{X(t) = j\,/X(s) = i\right\}$, $i,j \geq 0$, $0 \leq s \leq t$ the transition probabilities and by $p_i(t) = P\{X(t) = i\}$—the probability that $X(t)$ is in state i at time t. Let $\mathbf{p}(t) = (p_0(t), p_1(t), \ldots, p_S(t))^T$ be probability distribution vector at instant t. Throughout the paper, it is assumed that

$$\Pr\left(X\left(t+h\right) = j | X\left(t\right) = i\right) = \begin{cases} \lambda\left(t\right)h + \alpha_{ij}\left(t,h\right) & \text{if } j = i+1,\ i \geq 0, \\ \mu\left(t\right)h + \alpha_{ij}\left(t,h\right) & \text{if } j = i-2,\ i \geq 2, \\ 1 - \lambda(t)h + \alpha_i\left(t,h\right) & \text{if } j = i,\ 0 \leq i \leq 1, \\ 1 - (\lambda(t) + \mu(t))h + \alpha_i\left(t,h\right) & \text{if } j = i,\ i \geq 2, \\ \alpha_{ij}\left(t,h\right) & \text{otherwise,} \end{cases} \tag{1}$$

where all $\alpha_{ij}(t,h)$ are $o(h)$ and $\alpha_i(t,h)$ are $o(h)$ uniformly in i for any $t \geq 0$. In addition, it is assumed that the intensity functions $\lambda(t)$ and $\mu(t)$ are nonnegative, continuous and bounded on the interval $[0,\infty)$, $\lambda(t) + \mu(t) \leq L < \infty$ for any $t \geq 0$. Then, the probabilistic dynamics of the process is represented by the forward Kolmogorov system of differential equations:

$$\frac{d}{dt}\mathbf{p}(t) = A(t)\mathbf{p}(t), \tag{2}$$

where $A(t)$ is the transposed intensity matrix of the process, having the following form:

$$A(t) = \begin{pmatrix} -\lambda(t) & 0 & \mu(t) & 0 & 0 & \cdots \\ \lambda(t) & -\lambda(t) & 0 & \mu(t) & 0 & \cdots \\ 0 & \lambda(t) & -\left(\lambda(t)+\mu(t)\right) & 0 & \mu(t) & \cdots \\ 0 & 0 & \lambda(t) & -\left(\lambda(t)+\mu(t)\right) & 0 & \cdots \\ \ddots & \ddots & \ddots & \ddots & \ddots & \ddots \\ 0 & 0 & 0 & 0 & 0 & \cdots \\ \cdots & \cdots & \cdots & \cdots & \cdots & \cdots \end{pmatrix}. \tag{3}$$

Throughout the paper, by $\|\cdot\|$, we denote the l_1-norm, i.e., $\|\mathbf{p}(t)\| = \sum_{k\geq 0}|p_k(t)|$, and $\|A(t)\| = \sup_{j\geq 1}\sum_{i\geq 1}|a_{ij}|$. Let Ω be a set all stochastic vectors, i.e., l_1 vectors with non-negative coordinates and unit norm. Hence, we have $\|A(t)\| = 2\sup_{k\geq 1}|q_{kk}(t)| \leq 2L$ for almost all $t \geq 0$. Hence, the operator function $A(t)$ from l_1 into itself is bounded and continuous for all $t \geq 0$ and thus Label (2) is a differential equation in the space l_1 with bounded operator, which has a unique solution for any arbitrary initial condition (see [16]).

The method that is being proposed in this paper relies on the following transformation (referring to [6]) of the intensity matrix $A(t)$. Since $p_0(t) = 1 - \sum_{i \geq 1} p_i(t)$ due to the normalization condition, one can rewrite the system (2) as follows:

$$\frac{d}{dt}\mathbf{z}(t) = B(t)\mathbf{z}(t) + \mathbf{f}(t), \tag{4}$$

where

$$\mathbf{f}(t) = (\lambda(t), 0, 0, \dots)^T, \; \mathbf{z}(t) = (p_1(t), p_2(t), \dots)^T,$$

$$B(t) = \begin{pmatrix} -2\cdot\lambda(t) & -\lambda(t) & \mu(t)-\lambda(t) & -\lambda(t) & \cdots \\ \lambda(t) & -(\lambda(t)+\mu(t)) & 0 & \mu(t) & \cdots \\ 0 & \lambda(t) & -(\lambda(t)+\mu(t)) & 0 & \cdots \\ \ddots & \ddots & \ddots & \ddots & \ddots \\ 0 & 0 & 0 & 0 & \cdots \\ \cdots & \cdots & \cdots & \cdots & \cdots \end{pmatrix}. \tag{5}$$

Note that the bounds on the rate of convergence of the solutions of the system of differential equations

$$\frac{d}{dt}\mathbf{y}(t) = B(t)\mathbf{y}(t) \tag{6}$$

correspond to the same bounds of $X(t)$.

Denote by T the upper triangular matrix of ones i.e., $t_{ij} = 1$ for $j \geq i$ and 0, otherwise. Then,

$$T^{-1} = \begin{pmatrix} 1 & -1 & 0 & 0 & \cdots \\ 0 & 1 & -1 & 0 & \cdots \\ 0 & 0 & 1 & -1 & \cdots \\ 0 & 0 & 0 & 1 & \cdots \\ \vdots & \vdots & \vdots & \vdots & \ddots \end{pmatrix}.$$

Put $\mathbf{u}(t) = T\mathbf{y}(t)$. Then, we have

$$\frac{d}{dt}\mathbf{u}(t) = B^*(t)\mathbf{u}(t), \tag{7}$$

where

$$B^*(t) = \begin{pmatrix} -\lambda(t) & -\mu(t) & \mu(t) & 0 & 0 & 0 & \cdots \\ \lambda(t) & -(\lambda(t)+\mu(t)) & 0 & \mu(t) & 0 & 0 & \cdots \\ 0 & \lambda(t) & -(\lambda(t)+\mu(t)) & 0 & \mu(t) & 0 & \cdots \\ 0 & 0 & \lambda(t) & -(\lambda(t)+\mu(t)) & 0 & \mu(t) & \cdots \\ 0 & 0 & 0 & \lambda(t) & -(\lambda(t)+\mu(t)) & 0 & \cdots \\ \ddots & \ddots & \ddots & \ddots & \ddots & \ddots & \ddots \\ \cdots & \cdots & \cdots & \cdots & \cdots & \cdots & \cdots \end{pmatrix}.$$

Such transformation has been applied in a series of papers for general Markovian queueing models (see, for example, [7]). As it was mentioned above, the analysis of the rate of convergence to the limiting regime (a detailed description of this approach and its generalization can be found, for example, in [8,17,18]) was based on the logarithmic norm of an operator function from l_1 to itself, which can be computed by the simple formula:

$$\gamma(B(t)) = \sup_{j \geq 1} \left(b_{jj}(t) + \sum_{i \geq 1, i \neq j} |b_{ij}(t)| \right). \tag{8}$$

For the considered Markov chain $X(t)$, the method based on the logarithmic norm no longer applied. This is due to the fact that all column sums in $B^*(t)$ are equal to zero. In the next section, one outlines another approach, which is based on the direct applications of the differential inequalities. It was firstly considered for a finite Markovian queueing model in [19].

3. Bounds on the Rate of Convergence

Let $\{d_i, i \geq 0\}$ be a sequence such that $\inf_{i \geq 0} |d_i| = d > 0$. Denote by $D = diag\,(d_0, d_1, d_2, \dots)$ the diagonal matrix. By putting $\mathbf{w}(t) = D\mathbf{u}(t)$ from (7), one obtains

$$\frac{d}{dt}\mathbf{w}(t) = B^{**}(t)\mathbf{w}(t), \tag{9}$$

where the matrix $B^{**}(t) = (b^{**}(t))_{i,j=1}^{\infty} = DB^*(t)D^{-1}$ has the following form:

$$B^{**}(t) = \begin{pmatrix} -\lambda(t) & -\mu(t)\cdot\frac{d_1}{d_2} & \mu(t)\cdot\frac{d_1}{d_3} & 0 & 0 & \cdots \\ \lambda(t)\cdot\frac{d_2}{d_1} & -(\lambda(t)+\mu(t)) & 0 & \mu(t)\cdot\frac{d_2}{d_4} & 0 & \cdots \\ 0 & \lambda(t)\cdot\frac{d_3}{d_2} & -(\lambda(t)+\mu(t)) & 0 & \mu(t)\cdot\frac{d_3}{d_5} & \cdots \\ 0 & 0 & \lambda(t)\cdot\frac{d_4}{d_3} & -(\lambda(t)+\mu(t)) & 0 & \cdots \\ \ddots & \ddots & \ddots & \ddots & \ddots & \ddots \\ \cdots & \cdots & \cdots & \cdots & \cdots & \cdots \end{pmatrix}.$$

Let $\mathbf{u}(t)$ be an arbitrary solution of (7). Consider an interval (t_1, t_2) with fixed signs of coordinates of $\mathbf{u}(t)$. Let now signs of the entries d_i coincide with signs of the corresponding coordinates $u_i(t)$ of $\mathbf{u}(t)$. Then, $d_i u_i(t) > 0$ for all $i \geq 1$ on the time interval (t_1, t_2) and hence $\sum_{k=1}^{\infty} d_k u_k(t) = \|\mathbf{w}(t)\|$ can be considered as the corresponding norm. Put $\alpha_j(t) = -\sum_i b_{ij}^{**}(t)$ and assume that

$$\alpha_j(t) \geq \alpha_D(t),\ j \geq 1. \tag{10}$$

Consider now the system (9) on the interval (t_1, t_2). Then, the following bound holds:

$$\frac{d}{dt}\|\mathbf{w}(t)\| = \frac{d}{dt}\left(\sum_{k \geq 1} w_k(t)\right) = \sum_{j \geq 1}\sum_{i \geq 1} b_{ij}^{**}(t) w_j(t) \leq -\alpha_D(t)\|\mathbf{w}(t)\|. \tag{11}$$

If one puts $\alpha^*(t) = \inf \alpha_D(t)$, where the infimum is taken over all intervals with different combinations of coordinate signs of the solution, then for any such interval one has *in the own corresponding norm*, the inequality $\|\mathbf{w}(t)\| \leq e^{-\int_s^t \alpha^*(\tau)\,d\tau}\|\mathbf{w}(s)\|$.

Let firstly all coordinates of $\mathbf{u}(t)$ be positive. Put $d_1 = 1$, $d_2 = 1/\delta$, $d_3 = \delta$ and $d_{k+1} = \delta d_k$, for $k \geq 3$, where $\delta > 1$. Then, one has:

$$\alpha_1^1(t) = \lambda(t)\left(1 - \delta^{-1}\right),$$

$$\alpha_2^1(t) = \mu(t)\,(1+\delta) - \lambda(t)\left(\delta^2 - 1\right),$$

$$\alpha_3^1(t) = \mu(t)\left(1-\delta^{-1}\right) - \lambda(t)(\delta-1),$$

$$\alpha_4^1(t) = \mu(t)\left(1-\delta^{-3}\right) - \lambda(t)(\delta-1),$$

$$\alpha_k^1(t) = \mu(t)\left(1-\delta^{-2}\right) - \lambda(t)(\delta-1), \quad k \geq 5.$$

Therefore, one can take on the corresponding interval $\alpha_D^1(t) = \min_{1\leq i\leq 4} \alpha_i^1(t)$ and $d^1 = \inf_i |d_i| = \delta^{-1}$.

Let now $u_1(t) < 0$ and $u_k(t) > 0$ for $k \geq 1$. In this case, we put $d_1 = -1$, $d_2 = \delta$ and $d_{k+1} = \delta d_k$, for $k \geq 2$, for the same $\delta > 1$. Then, one has:

$$\alpha_1^2(t) = \lambda(t)(1+\delta),$$

$$\alpha_2^2(t) = \mu(t)\left(1-\delta^{-1}\right) - \lambda(t)(\delta-1),$$

$$\alpha_3^2(t) = \mu(t)\left(1+\delta^{-2}\right) - \lambda(t)(\delta-1),$$

$$\alpha_k^2(t) = \mu(t)\left(1-\delta^{-2}\right) - \lambda(t)(\delta-1), \quad k \geq 4.$$

Therefore, one can take on the corresponding interval $\alpha_D^2(t) = \min_{1\leq i\leq 3} \alpha_i^2(t)$ and $d^2 = \inf_i |d_i| = 1$. Moreover, it can be noted that, in any other case, a number of negative elements will be added to the column sums in (10). Hence, all values of $\alpha_k(t)$ in the other situations can only increase, and therefore the corresponding values of $\alpha_D(t)$ for the same d_k will be even greater.

Finally, if one takes $\alpha^*(t) = \inf \alpha_D(t)$, where the infimum is taken over all intervals with different combinations of coordinate signs of the solution, then the following bounds hold:

$$\alpha^*(t) \geq \min\left[\lambda(t)\left(1-\delta^{-1}\right), \mu(t)(1+\delta) - \lambda(t)\left(\delta^2-1\right), \mu(t)\left(1-\delta^{-1}\right) - \lambda(t)(\delta-1)\right], \tag{12}$$

and the corresponding 'absolute infimum'

$$d^* = \min\left(d^1, d^2\right) = \delta^{-1}. \tag{13}$$

By applying the comparison of norms, as it was done in [7], one obtains the following theorem.

Theorem 1. *Let*

$$\int_0^\infty \alpha^*(t)\,dt = +\infty \tag{14}$$

for some $\delta > 1$. Then, $X(t)$ is weakly ergodic and the following bounds on the rate of convergence hold:

$$\|\mathbf{u}(t)\| \leq \delta e^{-\int_0^t \alpha^*(\tau)\,d\tau}\|\mathbf{w}(0)\|, \tag{15}$$

$$\|\mathbf{p}^*(t) - \mathbf{p}^{**}(t)\| \leq 4\delta e^{-\int_0^t \alpha^*(\tau)\,d\tau}\|\mathbf{w}(0)\|, \tag{16}$$

for any initial conditions.

Note that the inequality $W = \inf_{k\geq 1} \frac{d_k}{k} > 0$ holds for both sequences. It implies an existence of the limiting mean for the process and the corresponding bounds on the rate of convergence (see, for example, [6,7]).

Let the process $X(t)$ be homogeneous i.e., let $\lambda(t) = \lambda$ and $\mu(t) = \mu$ be positive numbers. Then, (14) is equivalent to $\alpha^* > 0$ and this is equivalent to $0 < \lambda < \mu$. Put $\delta = \sqrt{\frac{\mu}{\lambda}}$. Hence,

$$\alpha_0^* = \min\left[\left(\sqrt{\mu} - \sqrt{\lambda}\right)^2, \lambda\left(1 - \sqrt{\frac{\lambda}{\mu}}\right)\right], \tag{17}$$

and the following is the corollary to Theorem 1.

Corollary 1. *Let $X(t)$ be the queue-length process in $M/M/1$ queue with service in batches of size 2. Let $0 < \lambda < \mu$. Then, $X(t)$ is ergodic and the following bounds hold:*

$$\|\mathbf{u}(t)\| \leq \delta e^{-\alpha_0^* t}\|\mathbf{w}(0)\|, \tag{18}$$

and

$$\|\mathbf{p}^*(t) - \pi\| \leq 4\delta e^{-\alpha^* t}\|\mathbf{w}(0)\|, \tag{19}$$

for any initial condition $X(0)$.

Note that the inequality $W = \inf_{k\geq 1} \frac{d_k}{k} > 0$ implies an existence of the constant limiting mean for the process and the corresponding bounds on the rate of convergence.

4. Numerical Example

There exists a number of investigations of queueing models with service in batches (or group services) (see, for example, [20,21]). Consider one example of such a queueing model with periodic arrival and service rates. We will be interested in the following quantities: $p_i(t)$ the probability that the total number of customers in the system at time t is i and the mean number $E(t,k) = E(X(t)|X(0) = k)$ of customers in the system at time t, provided that initially (at instant $t = 0$), there were k customers in the system.

Let $\lambda(t) = 2 + \sin 2\pi t$ and $\mu(t) = 4 - \cos 2\pi t$. Put $\delta = \frac{11}{10}$. Then, $\int_0^1 \alpha^*(t)\,dt \geq \frac{1}{22} > 0$ and the assumptions of Theorem 1 are fulfilled. Hence, $X(t)$ is exponentially weakly ergodic (i.e., $\lim_{t\to\infty} \|\mathbf{p}^*(t) - \mathbf{p}^{**}(t)\| \to 0$ for any initial conditions $\mathbf{p}^*(0)$ and $\mathbf{p}^{**}(0)$, where $\mathbf{p}^*(t)$ and $\mathbf{p}^{**}(t)$ are the solutions of (2).) and has the 1-periodic limiting mean (A Markov chain has the limiting mean $m(t)$, if $\lim_{t\to\infty} (m(t) - E(t,k)) = 0$ for any k.) $m(t)$. Now, applying the known truncation technique (See the detailed discussion and bounds in [22]), one can compute all probability characteristics of the queue-length process $X(t)$. The corresponding graphs are shown in Figures 2–9. To ensure that the truncation error is less than 10^{-3}, one can truncate the process $X(t)$ at the level $N = 100$. Then, one can compute any probability characteristic using the "extreme" initial conditions $X(0) = 0$ and $X(0) = N = 100$. Inequality (16) gives the corresponding (very rough) upper bounds on the rate of convergence for the state probabilities and for the mean number of customers in the system. Therefore, one can compute all characteristics on the intervals $[0, t^*]$ and $[t^*, t^* + 1]$, and obtain the limiting state probabilities and the limiting mean with error less 2×10^{-3}. In Figures 2–9 below, it can be seen that in fact it suffices to set $t^* = 28$. Figures 2, 4, 6 and 8 show the mean number of customers in the system and the probabilities $p_0(t)$, $p_1(t)$ and $p_2(t)$ converge to their limiting values. One can explicitly see how they approach the time t^*, starting from which the characteristics do not longer depend on the initial conditions. Other figures show their approximate limiting values.

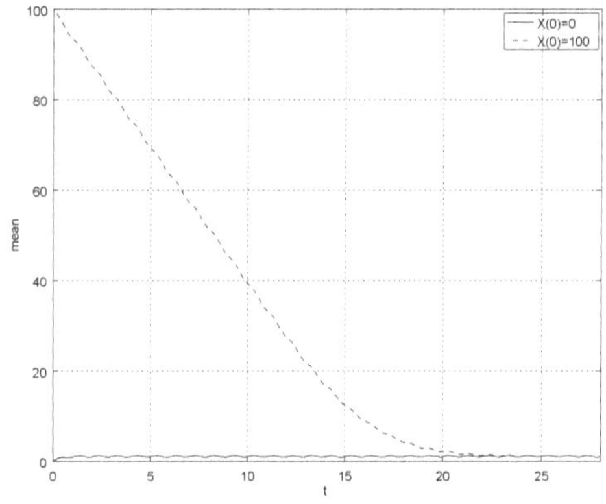

Figure 2. Example. The mean $E(t,0)$ and $E(t,100)$ for $t \in [0,28]$, this figure shows the rate of convergence.

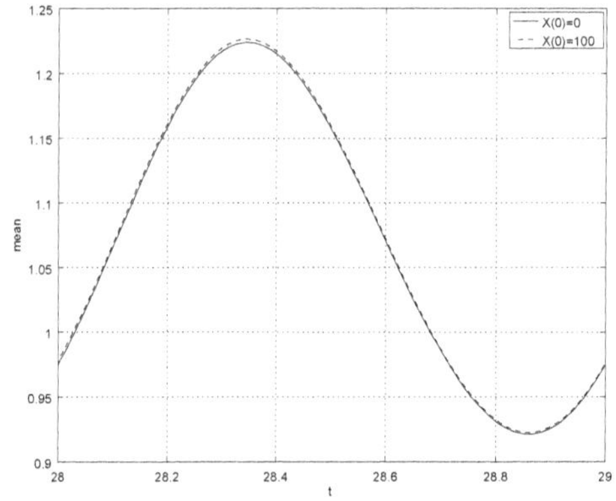

Figure 3. Example. The mean $E(t,0)$ and $E(t,100)$ for $t \in [28,29]$, this figure shows approximation of the limiting mean.

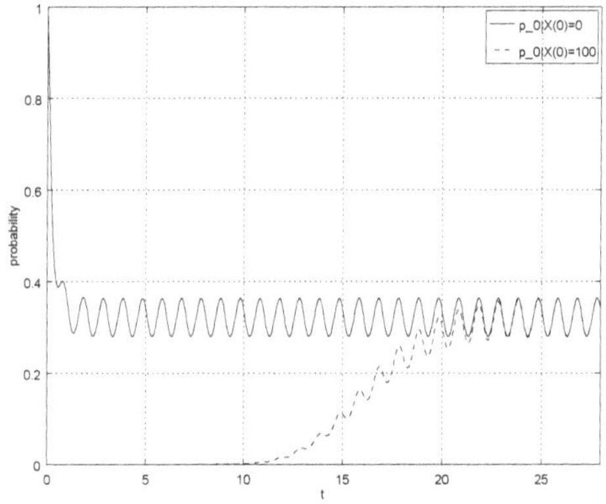

Figure 4. Example. Probability of the empty queue $p_0(t)$ for $t \in [0,28]$ and initial conditions $X(0) = 0$ and $X(0) = 100$, this figure shows the rate of convergence.

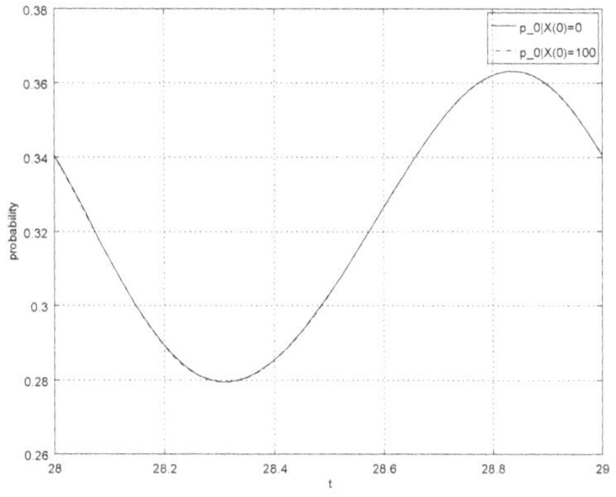

Figure 5. Example. Probability of the empty queue $p_0(t)$ for $t \in [28, 29]$ and initial conditions $X(0) = 0$ and $X(0) = 100$, this figure shows approximation of the limiting probability $p_0(t)$.

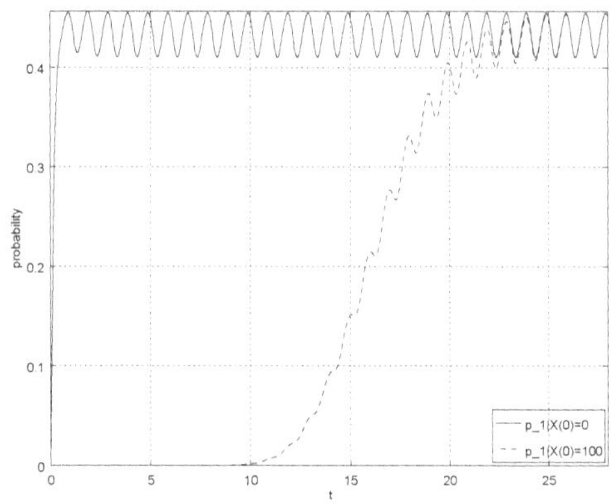

Figure 6. Example. Probability $p_1(t)$ for $t \in [0, 28]$ and initial conditions $X(0) = 0$ and $X(0) = 100$, this figure shows the rate of convergence.

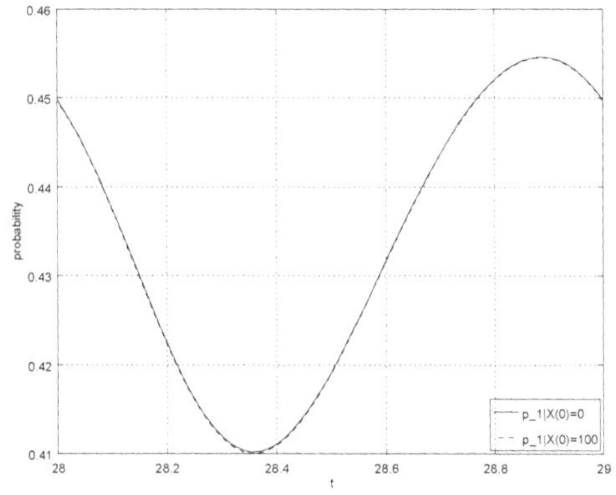

Figure 7. Example. Probability $p_1(t)$ for $t \in [28, 29]$ and initial conditions $X(0) = 0$ and $X(0) = 100$, this figure shows approximation of the limiting probability $p_1(t)$.

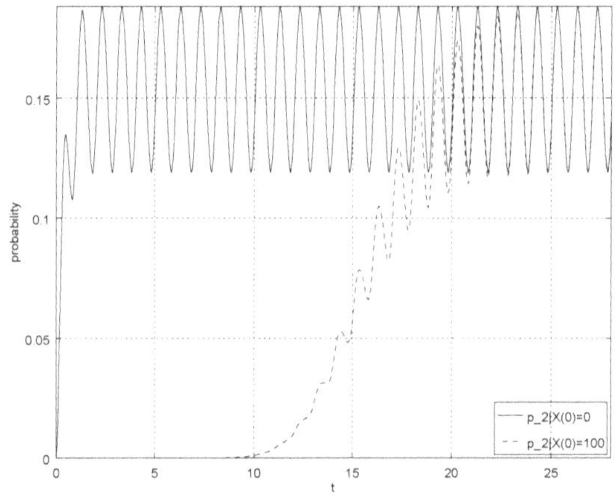

Figure 8. Example. Probability $p_2(t)$ for $t \in [0, 28]$ and initial conditions $X(0) = 0$ and $X(0) = 100$, this figure shows the rate of convergence.

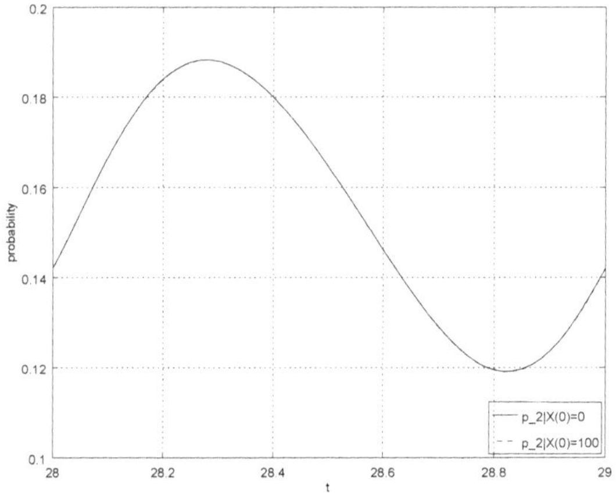

Figure 9. Example. Probability $p_2(t)$ for $t \in [28, 29]$ and initial conditions $X(0) = 0$ and $X(0) = 100$, this figure shows approximation of the limiting probability $p_2(t)$.

5. Conclusions

In the paper, some estimates of the rate of convergence and the corresponding approach were discussed for an inhomogeneous countable state continuous-time Markov chain. This chain is considered as the queue-length process of a simple nonstationary model of a queue with single arrivals and batch service (only in batches of size 2). The applied approach allows for studying new classes of the continuous-time Markov chain such that the corresponding reduced intensity matrix is not essentially nonnegative.

Author Contributions: Conceptualization, A.Z. and Y.S.; Methodology, A.Z., Y.S. and A.K.; Software, Y.S.; Validation, Y.S., A.Z. and A.K.; Investigation, A.Z., Y.S. and A.K.; Writing—Original Draft Preparation,Y.S., A.Z. and A.K.; Writing—Review and Editing, Y.S., A.Z. and A.K.; Supervision, A.Z.; Project Administration, A.Z.

Funding: This research was supported by Russian Science Foundation under grant 19-11-00020.

Acknowledgments: The authors thank the referees for useful comments that improved the paper and Rostislav Razumchik for some helpful discussions.

Conflicts of Interest: The authors declare no conflict of interest.

References

1. Schwarz, J.A.; Selinka, G.; Stolletz, R. Performance analysis of time-dependent queueing systems: Survey and classification. *Omega* **2016**, *63*, 170–189. [CrossRef]
2. Di Crescenzo, A.; Giorno, V.; Krishna Kumar, B.; Nobile, A.G. A Time-Non-Homogeneous Double-Ended Queue with Failures and Repairs and Its Continuous Approximation. *Mathematics* **2018**, *6*, 81. [CrossRef]
3. Giorno, V.; Nobile, A.G.; Spina, S. On some time non-homogeneous queueing systems with catastrophes. *Appl. Math. Comp.* **2014**, *245*, 220–234. [CrossRef]
4. Ammar, S.I.; Alharbi, Y.F. Time-dependent analysis for a two-processor heterogeneous system with time-varying arrival and service rates. *Appl. Math. Model.* **2018**, *54*, 743–751. [CrossRef]
5. Meyn, S.P.; Tweedie, R.L. Stability of Markovian processes III: Foster- Lyapunov criteria for continuous time processes. *Adv. Appl. Probab.* **1993**, *25*, 518–548. [CrossRef]
6. Zeifman, A.; Leorato, S.; Orsingher, E.; Satin Ya Shilova, G. Some universal limits for nonhomogeneous birth and death processes. *Queueing Syst.* **2006**, *52*, 139–151. [CrossRef]
7. Zeifman, A.; Razumchik, R.; Satin, Y.; Kiseleva, K.; Korotysheva, A.; Korolev, V. Bounds on the Rate of Convergence for One Class of Inhomogeneous Markovian Queueing Models with Possible Batch Arrivals and Services. *Int. J. Appl. Math. Comp. Sci.* **2018**, *28*, 141–154. [CrossRef]
8. Zeifman, A.; Satin, Y.; Kiseleva, K.; Korolev, V.; Panfilova, T. On limiting characteristics for a non-stationary two-processor heterogeneous system. *Appl. Math. Comput.* **2019**, *351*, 48–65. [CrossRef]
9. Kartashov, N.V. Criteria for uniform ergodicity and strong stability of Markov chains with a common phase space. *Theory Probab. Appl.* **1985**, *30*, 71–89.
10. Liu, Y. Perturbation bounds for the stationary distributions of Markov chains. *SIAM J. Matrix Anal. Appl.* **2012**, *33*, 1057–1074. [CrossRef]
11. Mitrophanov, A.Y. Stability and exponential convergence of continuous-time Markov chains. *J. Appl. Probab.* **2003**, *40*, 970–979. [CrossRef]
12. Mitrophanov, A.Y. The spectral gap and perturbation bounds for reversible continuous-time Markov chains. *J. Appl. Probab.* **2004**, *41*, 1219–1222. [CrossRef]
13. Zeifman, A.I.; Korolev, V.Y. On perturbation bounds for continuous-time Markov chains. *Stat. Probab. Lett.* **2014**, *88*, 66–72. [CrossRef]
14. Mitrophanov, A.Y. Connection between the Rate of Convergence to Stationarity and Stability to Perturbations for Stochastic and Deterministic Systems. In Proceedings of the 38th International Conference Dynamics Days Europe (DDE 2018), Loughborough, UK, 3–7 September 2018. Available online: http://alexmitr.com/talk_DDE2018_Mitrophanov_FIN_post_sm.pdf (accessed on 29 June 2019).
15. Rudolf, D.; Schweizer, N. Perturbation theory for Markov chains via Wasserstein distance. *Bernoulli* **2018**, *24*, 2610–2639. [CrossRef]
16. Daleckij, J.L.; Krein, M.G. *Stability of Solutions of Differential Equations in Banach Space*; American Mathematical Society: Providence, RI, USA, 2002; Volume 43.
17. Sinitcina, A.; Satin, Y.; Zeifman, A.; Shilova, G.; Sipin, A.; Kiseleva, K.; Panfilova, T.; Kryukova, A.; Gudkova, I.; Fokicheva, E. On the Bounds for a Two-Dimensional Birth-Death Process with Catastrophes. *Mathematics* **2018**, *6*, 80. [CrossRef]
18. Zeifman, A.; Satin, Y.; Kiseleva, K.; Korolev, V. On the Rate of Convergence for a Characteristic of Multidimensional Birth-Death Process. *Mathematics* **2019**, *7*, 477. [CrossRef]
19. Zeifman, A.; Satin, Y.; Kiseleva, K.; Kryukova, A. Applications of Differential Inequalities to Bounding the Rate of Convergence for Continuous-time Markov Chains. *AIP Conf. Proc.* **2019**, *2116*, 090009.
20. Brugno, A.; D'Apice, C.; Dudin, A.; Manzo, R. Analysis of an MAP/PH/1 queue with flexible group service. *Int. J. Appl. Math. Comp. Sci.* **2017**, *27*, 119–131. [CrossRef]

21. Lee, H.W.; Baek, J.W.; Jeon, J. Analysis of the MX/G/1 queue under D-policy. *Stoch. Anal. Appl.* **2005**, *23*, 785–808. [CrossRef]
22. Zeifman, A.; Satin Ya Korolev, V.; Shorgin, S. On truncations for weakly ergodic inhomogeneous birth and death processes. *Int. J. Appl. Math. Comp. Sci.* **2014**, *24*, 503–518. [CrossRef]

mathematics

MDPI

Article

Statistical Tests for Extreme Precipitation Volumes

Victor Korolev [1,2,3], Andrey Gorshenin [1,2,*] and Konstatin Belyaev [1,4]

1 Faculty of Computational Mathematics and Cybernetics, Lomonosov Moscow State University, Moscow 119991, Russia
2 Federal Research Center "Computer Science and Control" of the Russian Academy of Sciences, Moscow 119333, Russia
3 Hangzhou Dianzi University, Hangzhou 310018, China
4 P. P. Shirshov Institute of Oceanology of the Russian Academy of Sciences, Moscow 117997, Russia
* Correspondence: agorshenin@frccsc.ru

Received: 24 June 2019; Accepted: 17 July 2019; Published: 19 July 2019

Abstract: The analysis of the real observations of precipitation based on the novel statistical approach using the negative binomial distribution as a model for describing the random duration of a wet period is considered and discussed. The study shows that this distribution fits very well to the real observations and generalized standard methods used in meteorology to detect an extreme volume of precipitation. It also provides a theoretical base for the determination of asymptotic approximations to the distributions of the maximum daily precipitation volume within a wet period, as well as the total precipitation volume over a wet period. The paper demonstrates that the relation of the unique precipitation volume, having the gamma distribution, divided by the total precipitation volume taken over the wet period is given by the Snedecor–Fisher or beta distributions. It allows us to construct statistical tests to determine the extreme precipitations. Within this approach, it is possible to introduce the notions of relatively and absolutely extreme precipitation volumes. An alternative method to determine an extreme daily precipitation volume based on a certain quantile of the tempered Snedecor–Fisher distribution is also suggested. The results of the application of these methods to real data are presented.

Keywords: wet periods; total precipitation volume; asymptotic approximation; extreme order statistics; random sample size; testing statistical hypotheses

MSC: 62G30; 62E20; 62P12; 65C20

1. Introduction

Estimates of regularities and trends in heavy and extreme daily precipitation are important for understanding climate variability and change at relatively small or medium time horizons [1–3]. However, such estimates are much more uncertain compared to those derived for mean precipitation or total precipitation during a wet period [4]. This uncertainty is due to the fact that, first, estimates of heavy precipitation depend closely on the accuracy of the daily records; they are more sensitive to missing values. Second, uncertainties in the estimates of heavy and extreme precipitation are caused by the inadequacy of the mathematical models used for the corresponding calculations. Third, these uncertainties are boosted by the lack of reasonable means for the unambiguous (algorithmic) determination of extreme or anomalously heavy precipitation amplified by some statistical significance problems owing to the low occurrence of such events. As a consequence, continental-scale estimates of the variability and trends in heavy precipitation based on daily precipitation might generally agree qualitatively but may exhibit significant quantitative differences. In Reference [5] a detailed review of this phenomenon is presented where it is noted that for the European continent, most results hint at a growing intensity of heavy precipitation over the last five decades. The changes in extreme precipitation volumes are also among the relevant problems of such areas as climate [6,7], shallow landslides [8], debris flows [9], and so forth. Therefore, tools for analyzing such observations are being improved, including precipitation reanalysis [10].

There are several schemes on how to detect the extreme precipitation volume and what can be understood as the extreme precipitation. It can be selected from the total precipitation as unique heavy rain or show, for instance, which contains the precipitation volume more then 95% or more volume among total precipitation for given period independently of whether rain was observed or not. Alternatively, the precipitation and its volume can be considered only during the wet period ignoring all dry spells. In this case the concept of extreme precipitation and its statistics will differ from its counterpart in the first case. The details are discussed for instance in Reference [11]. The classical extreme value approach based on Pickands-Balkema-de Haan theorem [12,13] (the POT method) is used for solving such problems for precipitation data in References [14–16]. But sometimes it can lead to less accurate estimates of extreme values [17] than the methods proposed in the present paper (one possible alternative for estimating precipitation extremes based on the log-histospline was considered in Reference [18]). In the present paper, a rather reasonable approach to the unambiguous (algorithmic) determination of extreme or abnormally heavy daily and total precipitation within a wet period is proposed.

It is traditionally assumed that the duration of a wet period (the number of subsequent wet days) follows the geometric distribution (for example, see Reference [5]). But the sequence of dry and wet days is not only independent, it is also devoid of the Markov property [19]. Our approach introduces the negative binomial model for the duration of wet periods measured in days. This model demonstrates excellent fiting the numbers of successive wet days with the negative binomial distribution with shape parameter less than one, see References [20,21]. Numerous examples for dry and wet periods for different stations in Europe and Russia have been presented in Reference [22]. It provides a theoretical base for the determination of asymptotic approximations to the distributions of the maximum daily precipitation volume within a wet period and of the total precipitation volume for a wet period. The asymptotic distribution of the maximum daily precipitation volume within a wet period turns out to be a tempered Snedecor-Fisher distribution (i.e., the distribution of a positive power of a random variable with the Snedecor-Fisher distribution) whereas the total precipitation volume for a wet period turns out to be the gamma distribution. Both approximations appear to be very accurate. These asymptotic approximations are deduced using limit theorems for statistics constructed from samples with random sizes.

In this paper, two approaches are proposed to the definition of anomalously extremal precipitation. The first approach to the definition (and determination) of abnormally heavy daily precipitation is based on the tempered Snedecor-Fisher distribution (some methods of statistical estimation of its parameters will also be discussed). The second approach is based on the assumption that the total precipitation volume over a wet period has the gamma distribution. This assumption is theoretically justified by a version of the law of large numbers for sums of a random number of random variables in which the number of summands has the negative binomial distribution and is empirically substantiated by the statistical analysis of real data. Hence, the hypothesis that the total precipitation volume during a certain wet period is anomalously large can be formulated as the homogeneity hypothesis of a sample from the gamma distribution. Two equivalent tests are proposed for testing this hypothesis. One of them is based on the beta distribution whereas the second is based on the Snedecor–Fisher distribution. Both of these tests deal with the relative contribution of the total precipitation volume for a wet period to the considered set (sample) of successive wet periods. Within the second approach it is possible to introduce the notions of relatively abnormal and absolutely anomalous precipitation volumes. The results of the application of these tests to real data are presented yielding the conclusion that the intensity of wet periods with anomalously large precipitation volume increases.

The proposed approaches have several important benefits First, estimates of total precipitation are weakly affected by the accuracy of the daily records and are less sensitive to missing values. Second, they are based on limit theorems of probability theorems that yield unambiguous asymptotic approximations which are used as adequate mathematical models. Third, these approaches provide unambiguous algorithms for the determination of extreme or anomalously heavy daily or total precipitation that do not involve statistical significance problems owing to the low occurrence of such (relatively rare) events.

Suggested methods improve the approach described in Reference [11], where an estimate of the fractional contribution from the wettest days to the total was developed which is less hampered by the limited number of wet days. This paper demonstrates that the proposed methods have a theoretical background and at the same time they are perfectly consistent with the real data. Our methods are also compared with previously suggested in Reference [11] on the same database. The comparison clearly showed that in many case the new approach gives more precise assessment for extremes and allows distinguishing the ordinary and extreme volume for precipitation more detailed. It is worth noting that all algorithms have been implemented as MATLAB software tools.

The paper is organized as follows. In Section 2 mathematical models to derive statistical tests for the extreme precipitation events are introduced. The tempered Snedecor-Fisher distribution as an asymptotic approximation to the maximum daily precipitation volume within a wet period and some corresponding analytic properties are presented. Two equivalent statistical tests for a total precipitation volume over a wet period to be abnormally large based on testing the homogeneity hypothesis of a sample from the gamma distribution are introduced. Section 3 describes the application of the proposed models and algorithms to real data using precipitation observations in Potsdam and Elista in about 60 years. Section 4 is devoted to the main conclusions of the work.

2. Mathematical Models to Derive Statistical Tests for Precipitation Volume to Be Anomalous Large

At the beginning of this section some notations that will be used below are introduced. All the random variables under consideration are defined on the same probability space $(\Omega, \mathfrak{F}, \mathbb{P})$. The results are expounded in terms of random variables with the corresponding distributions. The symbol $\stackrel{d}{=}$ denotes the coincidence of distributions.

Let $G_{r,\lambda}$ be a random variable having the gamma distribution with shape parameter $r > 0$ and scale parameter $\lambda > 0$:

$$\mathbb{P}(G_{r,\lambda} < x) = \int_0^x \frac{\lambda^r}{\Gamma(r)} z^{r-1} e^{-\lambda z} dz, \quad x \geqslant 0,$$

Let W_γ be a random variable with the Weibull distribution with the distribution function $[1 - e^{-x^\gamma}]\mathbf{1}(x \geqslant 0)$ ($\mathbf{1}(A)$ is the indicator function of a set A). The distribution of the random variable $|X|$, where X is a random variable with the standard normal distribution function, is a folded normal ($x \geqslant 0$):

$$\mathbf{P}(|X| < x) = 2\Phi(x) - 1. \tag{1}$$

Let $S_{\alpha,1}$ and $S'_{\alpha,1}$ $(0 < \alpha < 1)$ be independent and identically-distributed random variables with the same strictly stable distribution [23]. So, the density $v_\alpha(x)$ of the random variable $R_\alpha = S_{\alpha,1}/S'_{\alpha,1}$ can be represented [24] as follows ($x > 0$):

$$v_\alpha(x) = \frac{\sin(\pi\alpha)x^{\alpha-1}}{\pi[1 + x^{2\alpha} + 2x^\alpha \cos(\pi\alpha)]}. \tag{2}$$

2.1. The Tempered Snedecor–Fisher Distribution as an Asymptotic Approximation to the Maximum Daily Precipitation Volume Within a Wet Period

As it has been demonstrated in Reference [25], the asymptotic probability distribution of extremal daily precipitation within a wet period can be represented as follows (here $r > 0$, $\lambda > 0$, and $\gamma > 0$):

$$F(x;r,\lambda,\gamma) = \left(\frac{\lambda x^{\gamma}}{1+\lambda x^{\gamma}}\right)^{r}, \quad x \geqslant 0. \tag{3}$$

Moreover, the theoretical conditions of limit theorems correspond with the real data (in sense of fitting Pareto distribution, see Reference [26]). The function (3) is a scale mixture of the Fréchet (inverse Weibull) distribution. It can be demonstrated for a random variable $M_{r,\gamma,\lambda}$ with a distribution function $F(x;r,\lambda,\gamma)$ that

$$M_{r,\gamma,\lambda} \stackrel{d}{=} \left(\frac{Q_{r,1}}{\lambda r}\right)^{1/\gamma}.$$

that is, the distribution of the random variable $M_{r,\gamma,\lambda}$ up to a non-random scale factor coincides with that of the positive power of a random variable with the Snedecor–Fisher distribution. In other words, the distribution function $F(x;r,\lambda,\gamma)$ (3) up to a power transformation of the argument x coincides with the Snedecor–Fisher distribution function. In statistics, distributions with arguments subjected to the power transformation are conventionally called tempered. Therefore, the distribution $F(x;r,\lambda,\gamma)$ can be called tempered Snedecor–Fisher distribution. Some properties of the distribution of the random value $M_{r,\gamma,\lambda}$ were discussed in [26]. In particular, it was shown that the limit distribution (3) can be represented as a scale mixture of exponential or stable or Weibull or Pareto or folded normal laws ($r \in (0,1]$, $\gamma \in (0,1]$, $\lambda > 0$):

$$M_{r,\gamma,\lambda} \stackrel{d}{=} \frac{G_{r,\lambda}^{1/\gamma} S_{\gamma,1}}{W_1} \stackrel{d}{=} \frac{W_{\gamma}}{W'_{\gamma}} \cdot \frac{1}{Z_{r,\lambda}^{1/\gamma}} \stackrel{d}{=} W_1 \cdot \frac{R_{\gamma}}{W'_1 Z_{r,\lambda}^{1/\gamma}} \stackrel{d}{=} \frac{\Pi R_{\gamma}}{Z_{r,\lambda}^{1/\gamma}} \stackrel{d}{=} \frac{|X|\sqrt{2W_1}R_{\gamma}}{W'_1 Z_{r,\lambda}^{1/\gamma}},$$

where $W_{\gamma} \stackrel{d}{=} W'_{\gamma}$, $W_1 \stackrel{d}{=} W'_1$, the random variable R_{γ} has the density (2), the random variable Π has the Pareto distribution ($\mathbb{P}(\Pi > x) = (x+1)^{-1}$, $x \geqslant 0$), and in each term the involved random variables are independent.

It should be mentioned that the same mathematical reasoning can be used for the determination of the asymptotic distribution of the maximum daily precipitation within m wet periods with arbitrary finite $m \in \mathbb{N}$. Indeed, fix arbitrary positive $r_1, \ldots, r_m$ and $p \in (0,1)$. Let $N_{r_1,p}^{(1)}, \ldots, N_{r_m,p}^{(m)}$ be independent random variables having the negative binomial distributions with parameters r_j, p, $j = 1, \ldots, m$, respectively. By the consideration of characteristic functions it can be easily verified that

$$N_{r_1,p}^{(1)} + \ldots + N_{r_m,p}^{(m)} \stackrel{d}{=} N_{r,p}, \tag{4}$$

where $r = r_1 + \ldots + r_m$. If all r_j coincide, then $r = mr_1$ and in accordance with relation (4), the asymptotic distribution of the maximum daily precipitation within m wet periods has the form ($x \geqslant 0$)

$$F^{(m)}(x;r,\lambda,\gamma) = F(x;mr_1,\lambda,\gamma) = \left(\frac{\lambda x^{\gamma}}{1+\lambda x^{\gamma}}\right)^{mr_1}.$$

And if now m infinitely increases and simultaneously λ changes as $\lambda = cm$, $c \in (0,\infty)$, then, obviously,

$$\lim_{m\to\infty} F^{(m)}(x;r,\lambda,\gamma) = \lim_{m\to\infty} F(x;mr_1,cm,\gamma) = e^{-\mu x^{-\gamma}}$$

with $\mu = (cr_1)^{-1}$, that is, the distribution function
$F^{(m)}(x;r,\lambda,\gamma)$ of the maximum daily precipitation within m wet periods turns into the classical Fréchet distribution.

This model makes it possible to propose the following approach to the definition (and determination) of an anomalously heavy daily precipitation volume. The grounds for this approach is an obvious observation that if $X_1, X_2, \ldots, X_N$ is a sample of N positive observations, then with finite (possibly, random) N, among X_i's there is always an extreme observation, say, X_1, such that $X_1 \geqslant X_i$, $i = 1, 2, \ldots, N$. Two cases are possible: X_1 is a 'typical' observation and its extreme character is conditioned by purely stochastic circumstances (there must be an extreme observation within a finite homogeneous sample) and X_1 is abnormally large so that it is an outlier and its extreme character is due to some exogenous factors. It follows from (3) that the distribution of X_1 in the first case is the tempered Snedecor–Fisher distribution. Therefore, if X_1 exceeds a certain (pre-defined) quantile of this distribution, then it is regarded as suspicious to be an outlier. The examples will be demonstrated in Section 3.2.

2.2. The Algorithms of Statistical Fitting of the Tempered Snedecor-Fisher Distribution Model

In this section the algorithms and corresponding formulas for a statistical estimation of the parameters r, λ and γ of the tempered Snedecor–Fisher distribution (3) are briefly given (for details, see Reference [26]).

Let $\{X_{i,j}\}$, $i = 1, \ldots, m$, $j = 1, \ldots, m_i$, be the precipitation volumes on the jth day of the ith wet sequence. Let $X^*_{(1)}, \ldots, X^*_{(m)}$ be order statistics constructed from the sample $X^*_1, \ldots, X^*_m$, where $X^*_k = \max\{X_{k,1}, \ldots, X_{k,m_k}\}$. The unknown parameters r, λ and γ can be found as a solution of a following system of equations (for fixed values p_1, p_2 and p_3, $0 < p_1 < p_2 < p_3 < 1$):

$$X^*_{([mp_k])} = \left(\frac{p_k^{1/r}}{\lambda - \lambda p_k^{1/r}} \right)^{1/\gamma}, \quad k = 1, 2, 3 \tag{5}$$

(here the symbol $[a]$ denotes the integer part of a number a).

Proposition 1. *The values of parameters γ and λ can be estimated as follows:*

$$\widetilde{\gamma}_q = \frac{\frac{1}{r}(\log p_1 - \log p_3) + \log(1 - p_3^{\frac{1}{r}}) - \log(1 - p_1^{\frac{1}{r}})}{\log X^*_{([mp_1])} - \log X^*_{([mp_3])}}, \tag{6}$$

$$\widetilde{\lambda}_q = \frac{p_2^{\frac{1}{r}}}{(1 - p_2^{\frac{1}{r}})(X^*_{([mp_2])})^{\gamma}}. \tag{7}$$

The expressions (6) and (7) are obtained as solutions to the system of Equation (5) while the parameter r is determined numerically.

Proposition 2. *If the value of parameter r is estimated as a corresponding parameter of the negative binomial distribution, least squares estimates of parameters γ and λ are as follows:*

$$\widehat{\gamma}_{LS} = \sum_{j=1}^{m-1} \log X^*_{(j)} \left(\left(\log \frac{j^{1/r}}{m^{1/r} - j^{1/r}} \right)^{m-1} - \sum_{k=1}^{m-1} \log \frac{k^{1/r}}{m^{1/r} - k^{1/r}} \right) \times$$
$$\times \left((m-1) \sum_{j=1}^{m-1} \left(\log X^*_{(j)} \right)^2 - \left(\sum_{j=1}^{m-1} \log X^*_{(j)} \right)^2 \right)^{-1}, \tag{8}$$

$$\widehat{\lambda}_{LS} = \exp\left\{ \frac{1}{m-1} \left(\sum_{j=1}^{m-1} \log \frac{j^{1/r}}{m^{1/r} - j^{1/r}} - \widehat{\gamma}_{LS} \sum_{j=1}^{m-1} \log X^*_{(j)} \right) \right\}. \tag{9}$$

The derivation of the Formulas (8) and (9) is based on a minimization of the discrepancy between the empirical and model distribution functions using the least squares techniques.

The examples of fitting tempered Snedecor–Fisher distribution to the real data will be discussed in Section 3.1. The examples of determining an extreme daily precipitation volume based on quantiles of this distribution will be demonstrated in Section 3.2 using precipitation observations in Potsdam and Elista from 1950 to 2007.

2.3. The Tests for a Total Precipitation Volume to Be Anomalously Extremal Based on the Homogeneity Test of a Sample From the Gamma Distribution

Here some algorithms of testing the hypotheses that a total precipitation volume during a wet period is anomalously extremal within a certain time horizon are proposed. Moreover, this approach makes it possible to consider relatively anomalously extremal volumes and absolutely anomalously extremal volumes for a given time horizon.

Let $m \in \mathbb{N}$ and $G_{r,\mu}^{(1)}, G_{r,\mu}^{(2)}, \ldots, G_{r,\mu}^{(m)}$—be independent random variables having the same gamma distribution with shape parameter $r > 0$ and scale parameter $\mu > 0$. In Reference [11] it was suggested to use the distribution of the ratio

$$R^* = \frac{G_{r,\mu}^{(1)}}{G_{r,\mu}^{(1)} + G_{r,\mu}^{(2)} + \ldots + G_{r,\mu}^{(m)}} \stackrel{d}{=} \frac{G_{r,1}^{(1)}}{G_{r,1}^{(1)} + G_{r,1}^{(2)} + \ldots + G_{r,1}^{(m)}}$$

as a heuristic model of the distribution of the extremely large precipitation volume based on the assumption that fluctuations of daily precipitation follow the gamma distribution. The gamma model for the distribution of daily precipitation volume is less adequate than the Pareto one [26]. Here we will modify the technique proposed in Reference [11] and make it more adequate and justified.

Let $X_1, X_2, \ldots$ be daily precipitation volumes on wet days. For $k \in \mathbb{N}$ denote $S_k = X_1 + \ldots + X_k$. The statistical analysis of the observed data shows that the average daily precipitation volume on wet days is finite:

$$\frac{1}{n}\sum_{j=1}^{n} X_j \Longrightarrow a \in (0, \infty). \tag{10}$$

Here the symbol $\Longrightarrow$ denotes the convergence in distribution.

Figure 1 illustrates the stabilization of the cumulative averages of daily precipitation volumes as n grows in Potsdam (continuous line) and Elista (dash line), and thus, the practical validity of assumption (10). It should be emphasized that $X_1, X_2, \ldots$ can be dependent.

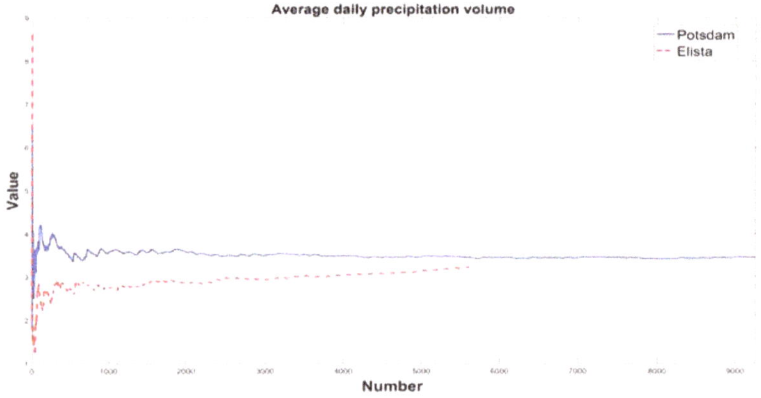

Figure 1. Stabilization of the cumulative averages of daily precipitation volumes as n grows in Potsdam (continuous line) and Elista (dash line).

Let $r > 0$, $\mu > 0$, $q \in (0,1)$, $n \in \mathbb{N}$. Let the random variable N_{r,p_n} have the negative binomial distribution with parameters r and $p_n = \min\{q, \mu/n\}$. Using the properties of characteristic functions it is easy to make sure that

$$n^{-1}N_{r,p_n} \Longrightarrow G_{r,\mu} \stackrel{d}{=} \tfrac{1}{\mu}G_{r,1} \tag{11}$$

as $n \to \infty$. So, the following analog of the law of large numbers for negative binomial random sums can be obtained. It can be actually regarded as a generalization of the Rényi theorem concerning the rarefaction of renewal processes.

Theorem 1. *Assume that the daily precipitation volumes on wet days $X_1, X_2, ...$ satisfy condition (10). Let the numbers $r > 0$, $q \in (0,1)$ and $\mu > 0$ be arbitrary. For each $n \in \mathbb{N}$, let the random variable N_{r,p_n} have the negative binomial distribution with parameters r and $p_n = \min\{q, \mu/n\}$. Assume that the random variables N_{r,p_n} are independent of the sequence $X_1, X_2, ...$ Then*

$$n^{-1}\sum_{j=1}^{N_{r,p_n}} X_j \Longrightarrow aG_{r,\mu} \stackrel{d}{=} \tfrac{a}{\mu}G_{r,1} \quad \textit{as } n \to \infty. \tag{12}$$

Proof. Consider a sequence of random variables $\left\{W_n = \sum_{j=1}^{n} X_j\right\}$, $n \in \mathbb{N}$. Let us introduce the following notations:

$$b_n = d_n = n, \quad N_n = N_{r,p_n}, \quad W = a, \quad N = \frac{1}{\mu}G_{r,1},$$

where for every $n \in \mathbb{N}$ the random variable N_n is independent of the sequence $\{W_n\}$. Then, the expressions

$$b_n^{-1}W_n \Longrightarrow W, \quad d_n^{-1}b_{N_n} \Longrightarrow N$$

take place as $n \to \infty$ (see (10) and (11), respectively). As all the conditions of the theorem for random sequences with independent random indices [27,28] are satisfied, we obtain the following expression:

$$d_n^{-1}W_{N_n} \Longrightarrow W \cdot N \quad \text{as } n \to \infty.$$

Taking into account above-mentioned notations, one can conclude that relation (12) holds. □

Therefore, with the account of the excellent fit of the negative binomial model for the duration of a wet period [26] with rather small p_n, the gamma distribution can be regarded as an adequate and theoretically well-based model for the total precipitation volume during a (long enough) wet period. This theoretical conclusion based on the negative binomial model for the distribution of duration of a wet period is vividly illustrated by the empirical data as shown on Figure 2 where the histograms of total precipitation volumes in Potsdam and Elista and the fitted gamma distributions are shown. For comparison, the densities of the best generalized Pareto distributions are also presented. It can be seen that even the best fitted Pareto distributions demonstrate worse fit than the gamma distribution.

Let $m \in \mathbb{N}$ and $G_{r,\mu}^{(1)}, G_{r,\mu}^{(2)}, \ldots, G_{r,\mu}^{(m)}$ be independent random variables having the same gamma distribution with parameters $r > 0$ and $\mu > 0$. Consider the relative contribution of the random variable $G_{r,\mu}^{(1)}$ to the sum $G_{r,\mu}^{(1)} + G_{r,\mu}^{(2)} + \ldots + G_{r,\mu}^{(m)}$:

$$\begin{aligned} R = \frac{G_{r,\mu}^{(1)}}{G_{r,\mu}^{(1)} + G_{r,\mu}^{(2)} + \ldots + G_{r,\mu}^{(m)}} &\stackrel{d}{=} \frac{G_{r,1}^{(1)}}{G_{r,1}^{(1)} + G_{r,1}^{(2)} + \ldots + G_{r,1}^{(m)}} \stackrel{d}{=} \\ \stackrel{d}{=} \left(1 + \frac{1}{G_{r,1}^{(1)}}(G_{r,1}^{(2)} + \ldots + G_{r,1}^{(m)})\right)^{-1} &\stackrel{d}{=} \left(1 + \frac{G_{(m-1)r,1}}{G_{r,1}}\right)^{-1}, \end{aligned} \tag{13}$$

where the gamma-distributed random variables on the right hand side are independent. So, the random variable R characterizes the relative precipitation volume for one (long enough) wet period with respect to the total precipitation volume registered for m wet periods.

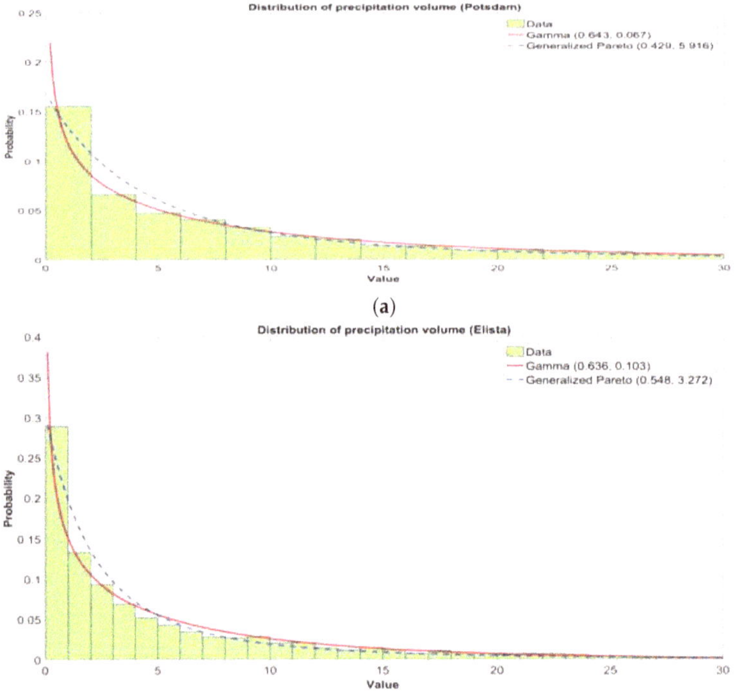

(a) (b)

Figure 2. The histograms of total precipitation volumes in Potsdam (**a**) and Elista (**b**) and the fitted gamma and generalized Pareto distributions.

The distribution of the random variable R is completely determined by the distribution of the ratio of two independent gamma-distributed random variables. To find the latter, denote $k = (m-1)r$ and obtain

$$\frac{G_{k,1}}{G_{r,1}} = \frac{k}{r} \cdot \left(\frac{r}{k} \cdot \frac{G_{k,1}}{G_{r,1}} \right) \stackrel{d}{=} \frac{k}{r} \cdot Q_{k,r},$$

where $Q_{k,r}$ is the random variable having the Snedecor–Fisher distribution determined for $k > 0$, $r > 0$ by the Lebesgue density

$$f_{k,r}(x) = \frac{\Gamma(k+r)}{\Gamma(k)\Gamma(r)} \left(\frac{k}{r}\right)^k \frac{x^{k-1}}{(1+\frac{k}{r}x)^{k+r}}, \quad x \geqslant 0, \tag{14}$$

(as is known, $Q_{k,r} \stackrel{d}{=} rG_{k,1}(kG_{r,1})^{-1}$, where the random variables $G_{k,1}$ and $G_{r,1}$ are independent (see, e.g., Reference [29], p. 32)). It is worth noting that the particular value of the scale parameter is insignificant. For convenience, it is assumed equal to one.

So, $R \stackrel{d}{=} \left(1 + \frac{k}{r}Q_{k,r}\right)^{-1}$, and, as is easily made sure by standard calculation using (14), the distribution of the random variable R is determined by the density

$$p(x;k,r) = \frac{\Gamma(k+r)}{\Gamma(r)\Gamma(k)}(1-x)^{k-1}x^{r-1}, \quad 0 \leqslant x \leqslant 1,$$

that is, it is the beta distribution with parameters $k = (m-1)r$ and r.

Then the test for the homogeneity of an independent sample of size m consisting of the gamma-distributed observations of total precipitation volumes during m wet periods with known γ based on the random variable R looks as follows. Let $V_1, \dots, V_m$ be the total precipitation volumes during m wet periods and, moreover, $V_1 \geqslant V_j$ for all $j \geqslant 2$. Calculate the quantity

$$SR = \frac{V_1}{V_1 + \ldots + V_m}$$

(SR means "Sample R"). From what was said above it follows that under the hypothesis H_0: "the precipitation volume V_1 under consideration is not anomalously large" the random variable SR has the beta distribution with parameters $k = (m-1)r$ and r. Let $\alpha \in (0,1)$ be a small number, $\beta_{k,r}(1-\alpha)$ be the $(1-\alpha)$-quantile of the beta distribution with parameters $k = (m-1)r$ and r. If $SR > \beta_{k,r}(1-\alpha)$, then the hypothesis H_0 must be rejected, that is, the volume V_1 of precipitation during one wet period must be regarded as anomalously large. Moreover, the probability of erroneous rejection of H_0 is equal to α.

Instead of R (13), the quantity

$$R_0 = \frac{(m-1)G_{r,\mu}^{(1)}}{G_{r,\mu}^{(2)} + \ldots + G_{r,\mu}^{(m)}} \stackrel{d}{=} \frac{k}{r}\frac{G_{r,\mu}}{G_{k,\mu}} \stackrel{d}{=} \frac{k}{r}\frac{G_{r,1}}{G_{k,1}} \stackrel{d}{=} Q_{r,k}$$

can be considered. Then, as is easily seen, the random variables R and R_0 are related by the one-to-one correspondence

$$R = \frac{R_0}{m-1+R_0} \quad \text{or} \quad R_0 = \frac{(m-1)R}{1-R},$$

so that the homogeneity test for a sample from the gamma distribution equivalent to the one described above and, correspondingly, the test for a precipitation volume during a wet period to be anomalously large, can be based on the random variable R_0 which has the Snedecor-Fisher distribution with parameters r and $k = (m-1)r$.

Namely, again let $V_1, \dots, V_m$ be the total precipitation volumes during m wet periods and, moreover, $V_1 \geqslant V_j$ for all $j \geqslant 2$. Calculate the quantity

$$SR_0 = \frac{(m-1)V_1}{V_2 + \ldots + V_m} \tag{15}$$

(SR_0 means "Sample R_0"). From what was said above it follows that under the hypothesis H_0: "the precipitation volume V_1 under consideration is not anomalously large" the random variable SR has the Snedecor–Fisher distribution with parameters r and $k = (m-1)r$. Let $\alpha \in (0,1)$ be a small number, $q_{r,k}(1-\alpha)$ be the $(1-\alpha)$-quantile of the Snedecor–Fisher distribution with parameters r and $k = (m-1)r$. If $SR_0 > q_{r,k}(1-\alpha)$, then the hypothesis H_0 must be rejected, that is, the volume V_1 of precipitation during one wet period must be regarded as anomalously large. Moreover, the probability of erroneous rejection of H_0 is equal to α.

Let l be a natural number, $1 \leqslant l < m$. It is worth noting that, unlike the test based on the statistic R, the test based on R_0 can be modified for testing the hypothesis H_0': "the precipitation volumes $V_{i_1}, V_{i_2}, \dots, V_{i_l}$ do not make an anomalously large cumulative contribution to the total precipitation volume $V_1 + \ldots + V_m$". For this purpose denote

$$T_l = V_{i_1} + V_{i_2} + \ldots + V_{i_l}, \quad T = V_1 + V_2 + \ldots + V_m$$

and consider the quantity

$$SR_0' = \frac{(m-l)T_l}{l(T-T_l)}.$$

In the same way as it was done above, it is easy to make sure that

$$SR_0' \stackrel{d}{=} \frac{(m-l)G_{lr,l}}{lG_{(m-l)r,1}} \stackrel{d}{=} Q_{lr,(m-l)r}.$$

Let $\alpha \in (0,1)$ be a small number, $q_{lr,(m-1)r}(1-\alpha)$ be the $(1-\alpha)$-quantile of the Snedecor-Fisher distribution with parameters lr and $k=(m-l)r$. If $SR_0' > q_{lr,(m-l)r}(1-\alpha)$, then the hypothesis H_0' must be rejected, that is, the cumulative contribution of the precipitation volumes $V_{i_1}, V_{i_2}, \ldots, V_{i_l}$ into the total precipitation volume $V_1 + \ldots + V_m$ must be regarded as anomalously large. Moreover, the probability of erroneous rejection of H_0' is equal to α.

The examples of application of the test for a total precipitation volume within a wet period to be anomalously large will be discussed in Section 3.4 using precipitation observations in Potsdam and Elista from 1950 to 2007.

3. The Results of the Analysis of Real Data

3.1. Statistical Fitting of the Tempered Snedecor-Fisher Distribution Model to Real Data

The numerical results of estimation of the parameters of daily precipitation in Potsdam and Elista from 1950 to 2009 using both algorithms described in Propositions 1 and 2 (see Section 2.2) are presented in Tables 1 and 2. The first column indicates the censoring threshold: since the tempered Snedecor–Fisher distribution is an asymptotic model which is assumed to be more adequate with small "success probability," the estimates were constructed from differently censored samples which contain only those wet periods whose duration is no less than the specified threshold.

Table 1. Estimation of the parameters of daily precipitation in Potsdam ($r = 0.847$).

Minimum Duration	Sample Size	D_q	D_{LS}	$\widetilde{\lambda}_q$	$\widehat{\lambda}_q$	$\widetilde{\gamma}_{LS}$	$\widehat{\gamma}_{LS}$
1	3323	0.09	0.092	0.169	0.212	1.18	1.29
2	2066	0.045	0.065	0.0383	0.054	1.755	1.71
3	1282	0.031	0.041	0.01	0.013	2.261	2.183
4	862	0.026	0.027	0.0049	0.0045	2.449	2.524
6	384	0.025	0.026	0.0015	0.0012	2.822	2.949
8	163	0.04	0.045	0.0007	0.0005	3.174	3.255
10	73	0.041	0.042	0.0003	0.0003	3.385	3.352
15	12	0.13	0.09	0.0014	0.0009	2.667	2.973

Table 2. Estimation of the parameters of daily precipitation in Elista ($r = 0.876$).

Minimum Duration	Sample Size	D_q	D_{LS}	$\widetilde{\lambda}_q$	$\widehat{\lambda}_q$	$\widetilde{\gamma}_{LS}$	$\widehat{\gamma}_{LS}$
1	2937	0.06	0.06	0.361	0.349	1.053	1.263
2	1374	0.049	0.055	0.108	0.101	1.424	1.574
3	656	0.041	0.045	0.0454	0.0376	1.707	1.9
4	319	0.051	0.06	0.0234	0.0273	1.891	1.94
6	77	0.07	0.075	0.0181	0.0144	2.011	2.186
7	42	0.15	0.01	0.0197	0.0207	1.983	2.179
8	22	0.12	0.14	0.014	0.0358	2.01	1.764
10	10	0.17	0.16	0.0136	0.0375	2.163	1.802

The second column contains the correspondingly censored sample size. The third and fourth columns contain the sup-norm discrepancies D_q and D_{LS} between the empirical $\widehat{F}(x)$ and fitted tempered Snedecor–Fisher distributions $F_{SF}^{(q)}(x)$ and $F_{SF}^{(LS)}(x)$ for two types of estimators (quantile (6), (7) and least squares (8), (9)) described above. The quantities D_q and D_{LS} are as follows:

$$D_q = \max_{x \in \mathbf{X}} \left| \widehat{F}(x) - F_{SF}^{(q)}(x) \right|, \quad D_{LS} = \max_{x \in \mathbf{X}} \left| \widehat{F}(x) - F_{SF}^{(LS)}(x) \right|,$$

where $\mathbf{X} = (X_1, \ldots, X_n)$ is a sample. The rest columns contain the corresponding values of the parameters estimated by these two methods.

According to Tables 1 and 2, the best accuracy is attained when the censoring threshold equals 3 days for Elista and 5–6 days for Potsdam. The least squares method leads to the more accurate estimates. Figures 3 and 4 demonstrate examples of fitted tempered Snedecor–Fisher distribution to precipitation in Potsdam and Elista, respectively.

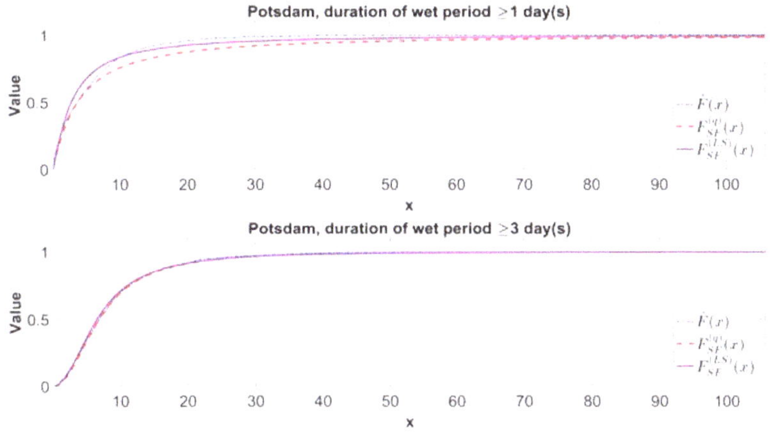

Figure 3. Fitted tempered Snedecor–Fisher distribution to precipitation in Potsdam.

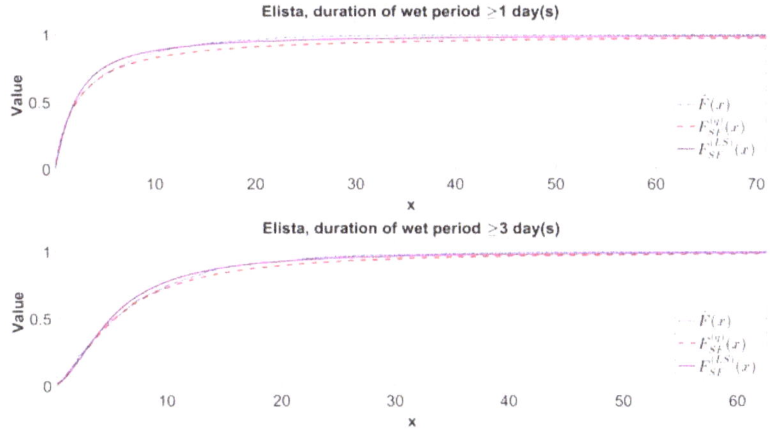

Figure 4. Fitted tempered Snedecor–Fisher distribution to precipitation in Elista.

3.2. Determining an Extreme Daily Precipitation Volume Based on Quantiles of the Tempered Snedecor-Fisher Distribution

In applied problems dealing with extreme values there is a common tradition which, possibly, has already become a prejudice, that statistical regularities in the behavior of extreme values necessarily obey one of well-known three types of extreme value distributions. In general, this is certainly so, if the sample size is very large, that is, the time horizon under consideration is very wide. In other words, the models based on the extreme value distributions have asymptotic character. However, in real practice, when the sample size is finite and the extreme values of the process under consideration are studied on the time horizon of a moderate length, the classical extreme value distributions may turn out to be inadequate models. In these situations a more thorough analysis may generate other models which appear to be considerably more adequate. This is exactly the case discussed in the present paper. Methodically, this approach is similar to the classical techniques of dealing with extreme observations [30]. The novelty of the proposed method is in a more accurate specification of the distribution of extreme daily precipitation which turned out to be the tempered Snedecor-Fisher distribution. Figure 5 demonstrates the algorithm of determination of an anomalously heavy daily precipitation.

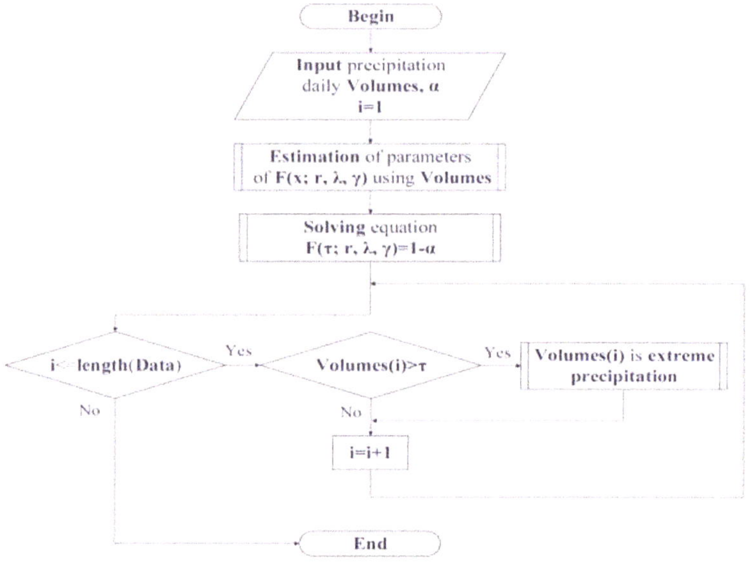

Figure 5. The algorithm of determination of an anomalously heavy daily precipitation.

It is easy to see that the the probability of the error of the first kind (occurring in the case where a regularly large maximum value is erroneously recognized as an anomalously large outlier) for this test is approximately equal to α, it is a small fixed positive number.

The application of this test to real data is illustrated by Figures 6 and 7. On these figures the lower horizontal line corresponds to the threshold equal to the quantile of the fitted tempered Snedecor-Fisher distribution of order 0.9. The middle and upper lines correspond to the quantiles of orders 0.95 and 0.99, respectively. Figure 6 contains all data. For the sake of vividness, on Figure 7 only one, maximum, daily precipitation is exposed for each wet period.

From Figure 7 it is seen that during 58 years (from 1950 to 2007) in Potsdam there were 13 wet periods containing anomalously heavy maximum daily precipitation volumes (at 99% threshold) and 69 wet periods containing anomalously heavy maximum daily precipitation volumes (at 95% threshold). Other maxima were 'regular.'

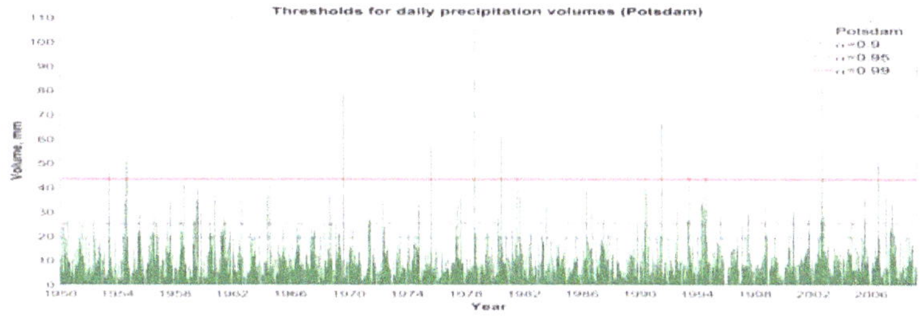

(a)

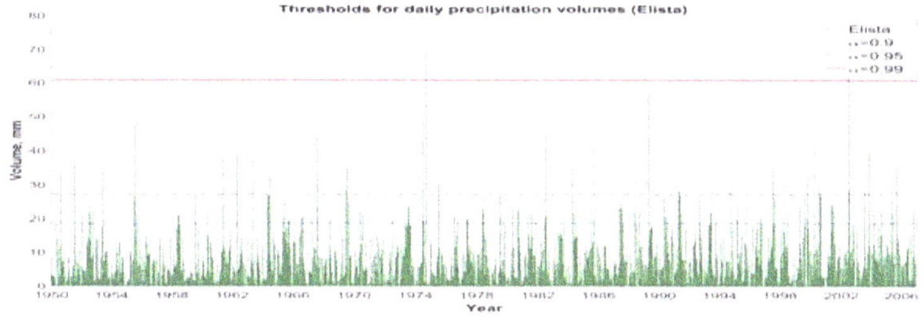

(b)

Figure 6. Testing daily precipitation within a wet period for abnormal heaviness: (**a**) Potsdam; (**b**) Elista, all data.

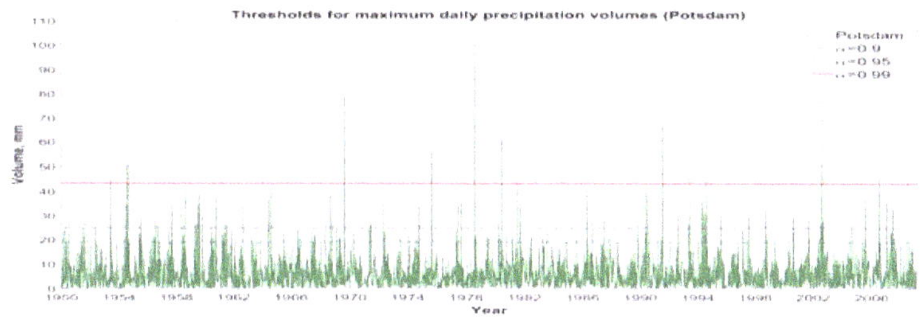

(a)

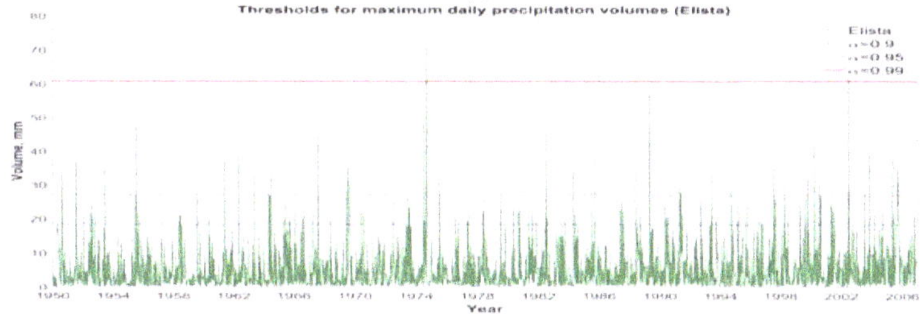

(b)

Figure 7. Testing maximum daily precipitation within a wet period for abnormal heaviness: (**a**) Potsdam; (**b**) Elista, data containing only maximum daily precipitation for every wet period.

During the same period in Elista there were only 2 wet periods containing anomalously heavy maximum daily precipitation volumes (at 99% threshold) and 40 wet periods containing anomalously heavy maximum daily precipitation volumes (at 95% threshold). Other maxima were 'regular.' The proportion of abnormal maxima exceeding 99% and 95% thresholds in Potsdam is quite adequate (the latter is approximately five times greater than the former) whereas in Elista this proportion is noticeably different. Perhaps, this can be explained by the fact that, for Elista, heavy rains are rare events.

3.3. Comparison With the Extreme Precipitation Detected by the Beta-Distributed Tests

At this section we present the results of comparison of the precipitation events from two samples both having the length 20 rains over the selected time-period (Elista and Potsdam). The extreme rain is detected as it passes the 95% quantile in the corresponding statistics. The first approach exploits the Beta distribution and it is based on the ratio between the selected precipitation and total precipitation sum over the sample. This test has been previously suggested in the paper [11]. It leads to the threshold TH_{beta} where all necessary parameters of the Beta distribution have been estimated directly from the samples. The second threshold TH_{SF} is determined using the tempered Snedecor–Fisher distribution (3) (see Section 2.1). The results of the comparison are presented in Table 3 where only two maximum rains are accounted. The column *Decision* indicates whether the corresponding volume is considered to be extreme according to the threshold values.

Table 3. An example of comparing the results with the previously proposed test.

City	Volume	Ratio	TH_{beta}	TH_{SF}	Decision
Potsdam	28.7	0.45	0.25	28.66	Yes/Yes
	16	0.25	0.25	28.66	Yes/No
Elista	21.9	0.59	0.23	21.19	Yes/Yes
	5.2	0.14	0.23	21.19	No/No

It is clearly seen that the Snedecor-Fisher test detects the extremal rains more accurately. Unlike the beta test, it selects only the rains that substantially extend the average or other heavy rains but which, however, are not extremely heavy. This means that the supposed scheme really can be applied to detect and forecast the potentially dangerous phenomena and to separate them from the substantial but not extremally big precipitations.

3.4. Determination Of Abnormalities Types Based on the Results of the Statistical Analysis

This section demonstrates the application of the statistical methodology based on the test (15) to determine extremeness of the precipitation volumes within wet periods. This method allows us to improve the quality of detecting events of such type in comparison with approaches based on thresholds.

It should be emphasized that the parameter m of the Snedecor-Fisher distribution of the test statistic SR_0 (15) is tightly connected with the time horizon, the abnormality of precipitation within which is studied. Indeed, the average duration of a wet/dry period (or the average distance between the first days of successive wet periods) in Potsdam turns out to be $5.804 \approx 6$ days. So, one observation of a total precipitation during a wet period, on the average, corresponds to approximately 6 days. This means, that, for example, the value $m = 5$ corresponds to approximately one month on the time axis, the value $m = 15$ corresponds to approximately 3 months, the value $m = 60$ corresponds to approximately one year. Figure 8 presents a flowchart illustrating the proposed algorithm.

It is important that the test for whether a total precipitation volume during one wet period is anomalously large can be applied to the observed time series in a moving mode. For this purpose, a window (a set of successive observations) should be determined. The number of observations in this set, say, m, is called the window width. The observations within a window constitute the sample to be analyzed. After the test has been performed for a given position of the window, the window moves rightward by one observation so that the leftmost observation at the previous position of the window is excluded from the sample and the observation next to the rightmost observation is added to the sample. The test is performed once more and so on. It is clear that each fixed observation falls in exactly m successive windows (from mth to $N - m + 1$, where N denotes the number of wet periods). Two cases are possible: (i) the fixed observation is recognized as anomalously large within each of m windows containing this observation and (ii) the fixed observation is recognized as anomalously large within at least one of m windows containing this observation. In the case (i) the observation will be called absolutely anomalously large with respect to a given time horizon (approximately equal to $m \cdot 5.804 \approx 6m$ days). In the case (ii) the observation will be called relatively anomalously large with respect to a given time horizon. Of course, these definitions admit intermediate cases where the observation is recognized as anomalously large for $q \cdot m$ windows with $q \in [\frac{1}{m}, 1]$.

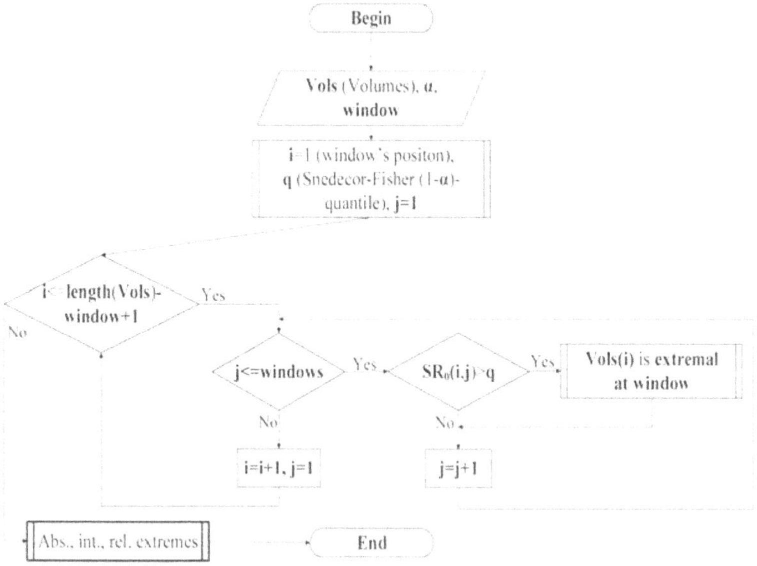

Figure 8. The algorithm of determination of extremeness of the precipitation volumes.

The results of the application of the test for a total precipitation volume during one wet period to be anomalously large based on SR_0 in the moving mode are shown on Figures 9–11 (Potsdam) and Figures 12–14 (Elista) for different time horizons (30, 90 and 360 days). The notation $Extr_{int}$ corresponds to the intermediate extremes (the fixed observation is recognized as anomalously large within at least $\lceil m/2 \rceil$ windows containing this observation, here the symbol $\lceil \cdot \rceil$ denotes the next larger integer). They are marked by circles on the figures. The absolutely extreme volumes are shown as triangles, and the relatively ones are marked by squares.

It is seen that at relatively small time horizons, the test yields non-trivial and unobvious conclusions. However, as the time horizon increases, the results of the test become more expected. At small time horizons there are some big precipitation volumes that are not recognized as abnormal. At large time horizons there are almost no regular big precipitation volumes at significance level $\alpha = 0.05$ whereas at the smaller significance level $\alpha = 0.01$ there are some regular big precipitation volumes which are thus not recognized as abnormal.

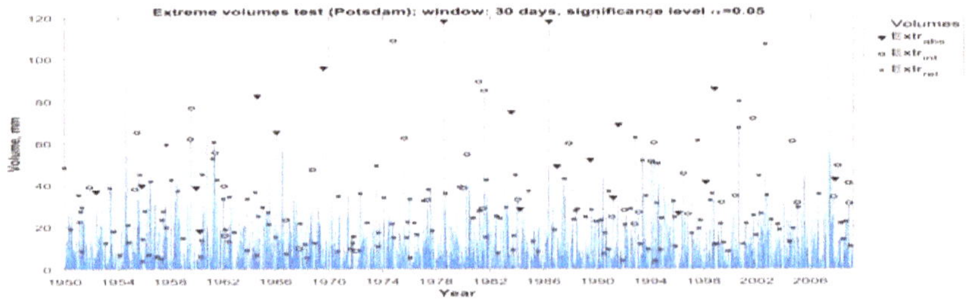

(a)

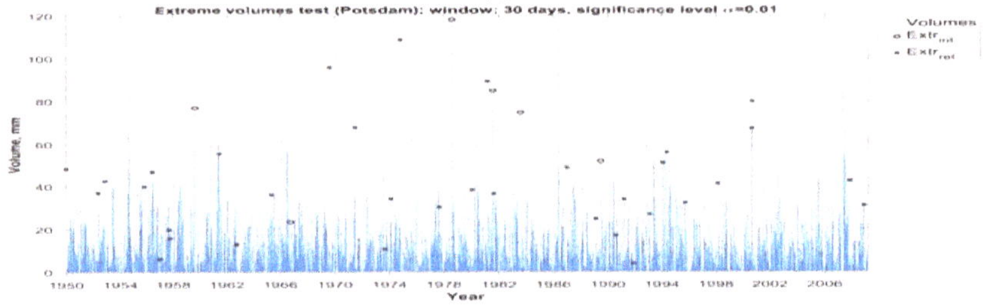

(b)

Figure 9. Abnormal precipitation volumes, Potsdam, time horizon = 30 days, significance levels $\alpha = 0.05$ (**a**) and $\alpha = 0.01$ (**b**).

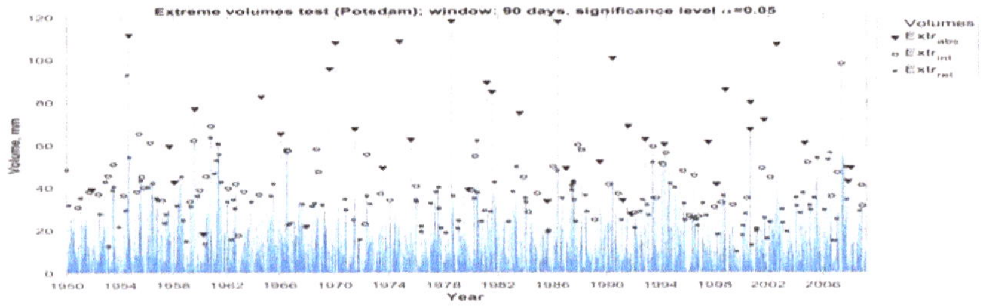

(a)

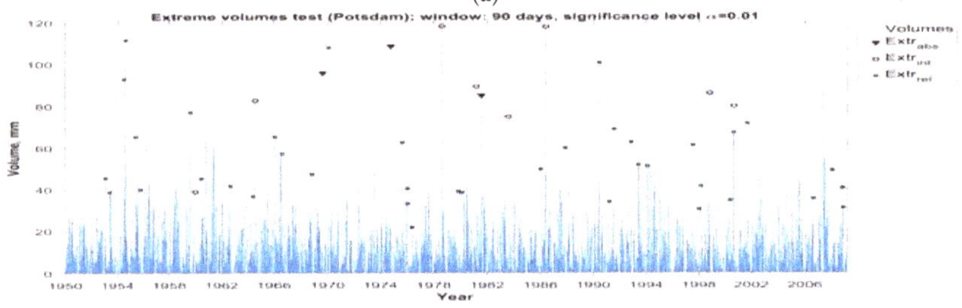

(b)

Figure 10. Abnormal precipitation volumes, Potsdam, time horizon = 90 days, significance levels $\alpha = 0.05$ (**a**) and $\alpha = 0.01$ (**b**).

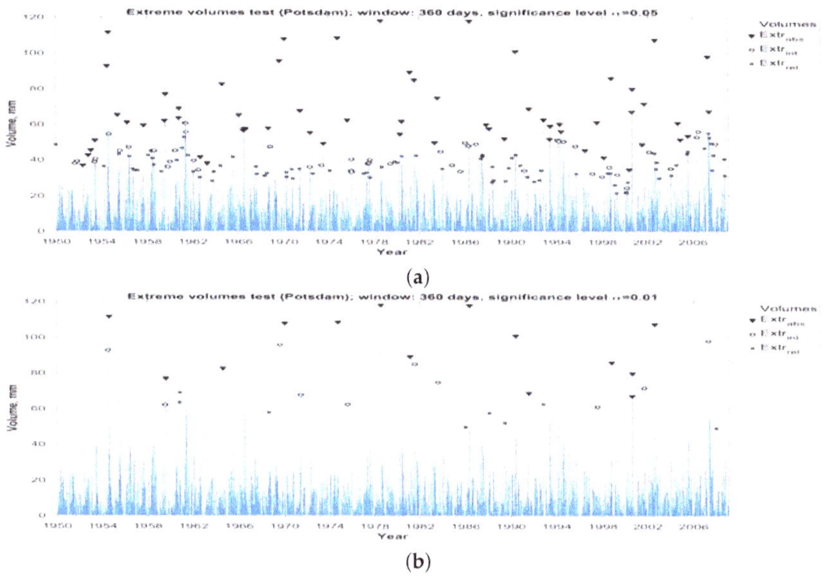

Figure 11. Abnormal precipitation volumes, Potsdam, time horizon = 360 days, significance levels $\alpha = 0.05$ (**a**) and $\alpha = 0.01$ (**b**).

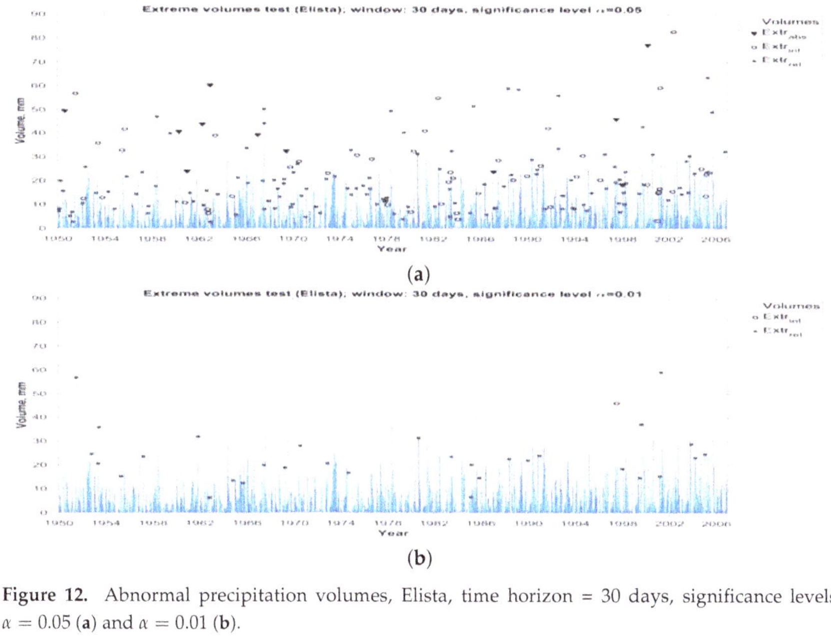

Figure 12. Abnormal precipitation volumes, Elista, time horizon = 30 days, significance levels $\alpha = 0.05$ (**a**) and $\alpha = 0.01$ (**b**).

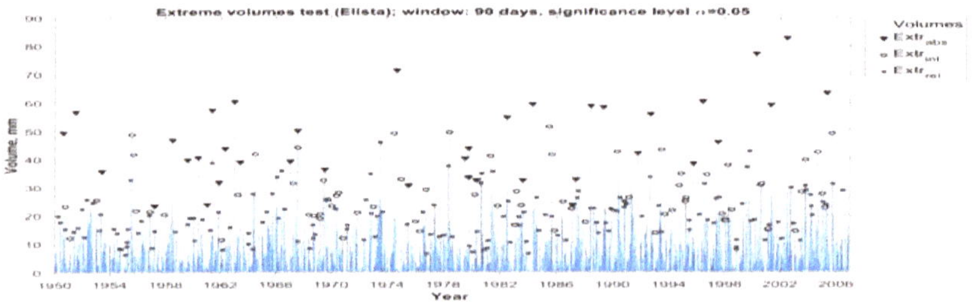

(a)

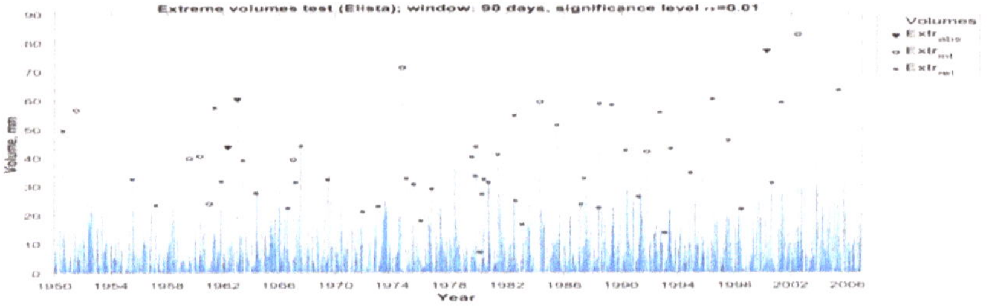

(b)

Figure 13. Abnormal precipitation volumes, Elista, time horizon = 90 days, significance levels $\alpha = 0.05$ (**a**) and $\alpha = 0.01$ (**b**).

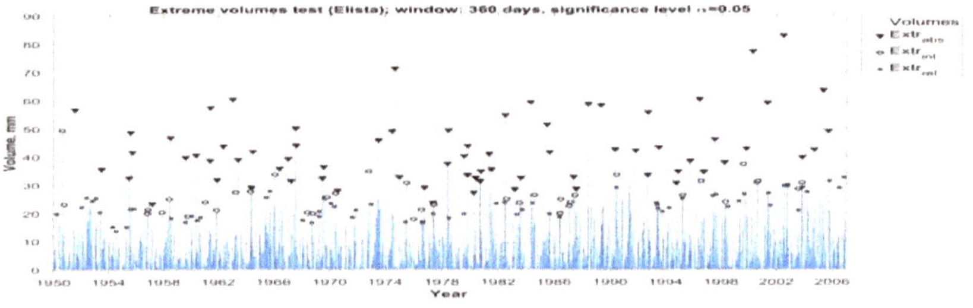

(a)

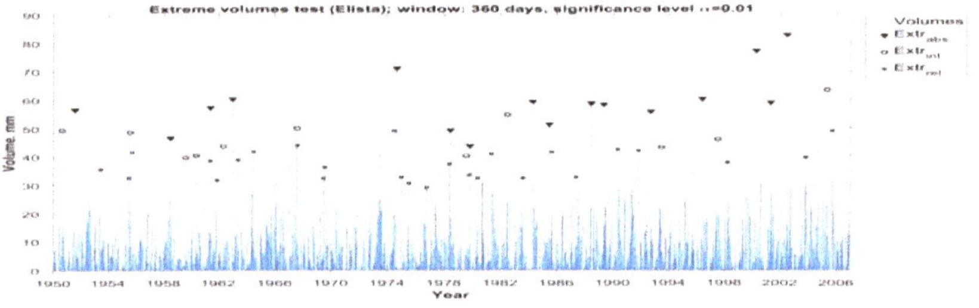

(b)

Figure 14. Abnormal precipitation volumes, Elista, time horizon = 360 days, significance levels $\alpha = 0.05$ (**a**) and $\alpha = 0.01$ (**b**).

4. Conclusions and Discussion

In this paper, it is shown that the negative binomial distribution can be fruitful for the description of the statistical regularities in the duration of wet periods observed in practice and can be used as a base for obtaining the model for the extreme precipitation per wet period and, consequently, for testing the hypothesis that the specific precipitation volume considered over a given wet period is anomalously extreme. Several approaches to the definition of an anomalously extreme precipitation are proposed. This is an important issue, since up until now there has not been a single-valued criterion for which precipitation volume should be regarded as anomalous and which is not. Obviously, one and the same volume can be regarded normal in a region where precipitations are quite frequent, for instance in tropical zones and at the time it can be considered absolutely anomalous in a dry region, say, in a desert. The proposed tests consider the relative part of precipitation and the time scale and, therefore, are free from the aforementioned disadvantage. Moreover, the proposed techniques of testing the corresponding hypotheses can be easily numerically implemented.

The models and methods described in the paper can be applied to various time series in order to detect anomalously extreme values. For example, the considered algorithms can be used for other geophysical time series such as wind speed, heat fluxes and so forth, both univariate or multivariate. This can be important for global climate prediction models, for forecasting and evaluation of dangerous phenomena and processes. One more possible useful application of these models is filling the gaps in precipitation time series due to missing observations. Indeed, the statistically estimated parameters of the proposed models can be used as additional attributes in training samples when machine learning techniques or Artificial Intelligence systems are used for the simulation of missing data without actual extension of available information.

Author Contributions: Conceptualization, V.K., A.G., K.B.; Formal analysis, V.K., A.G.; Funding acquisition, A.G.; Investigation, V.K., A.G., K.B.; Methodology, V.K., A.G.; Project administration, V.K., A.G.; Resources, A.G., K.B.; Software, A.G.; Supervision, V.K.; Validation, A.G.; Visualization, A.G.; Writing—original draft, V.K., A.G.; Writing—review & editing, V.K., A.G., K.B.

Funding: This research was funded by the Russian Foundation for Basic Research (grant number 17-07-00851) and the RF Presidential scholarship program (grant number 538.2018.5).

Acknowledgments: Authors thank the reviewers for their valuable comments that helped to improve the presentation of the material.

Conflicts of Interest: The authors declare no conflict of interest.

References

1. Groisman, P.; Legates, D. Documenting and detecting long-term precipitation trends: Where we are and what should be done. *Clim. Chang.* **1995**, *31*, 601–622. [CrossRef]
2. Groisman, P.; Karl, T.; Easterling, D.; Knight, R.; Jamason, P.; Hennessy, K.; Suppiah, R.; Page, C.; Wibig, J.; Fortuniak, K.; et al. Changes in the probability of heavy precipitation: Important indicators of climatic change. *Clim. Chang.* **1999**, *42*, 243–283.:1005432803188. [CrossRef]
3. Lockhoff, M.; Zolina, O.; Simmer, C.; Schulz, J. Evaluation of Satellite-Retrieved Extreme Precipitation over Europe using Gauge Observations. *J. Clim.* **2014**, *27*, 607–623. [CrossRef]
4. Zolina, O.; Simmer, C.; Belyaev, K.; Kapala, A.; Gulev, S.; Koltermann, P. Multidecadal trends in the duration of wet spells and associated intensity of precipitation as revealed by a very dense observational German network. *Environ. Res. Lett.* **2014**, *9*. [CrossRef]
5. Zolina, O.; Simmer, C.; Belyaev, K.; Kapala, A.; Gulev, S.; Koltermann, P. Changes in the duration of European wet and dry spells during the last 60 years. *J. Clim.* **2013**, *26*, 2022–2047. [CrossRef]
6. Mo, C.; Ruan, Y.; He, J.; Jin, J.; Liu, P.; Sun, G. Frequency analysis of precipitation extremes under climate change. *Int. J. Climatol.* **2019**, *39*, 1373–1387. [CrossRef]
7. Donat, M.; Angelil, O.; Ukkola, A. Intensification of precipitation extremes in the world's humid and water-limited regions. *Environ. Res. Lett.* **2019**, *14*. [CrossRef]
8. Bezak, N.; Auflic, M.J.; Mikos, M. Application of hydrological modelling for temporal prediction of rainfall-induced shallow landslides. *Landslides* **2019**, *16*, 1273–1283. [CrossRef]
9. Huang, J.; van Asch, T.; Wang, C.; Li, Q. Study on the combined threshold for gully-type debris flow early warning. *Nat. Hazards Earth Syst. Sci.* **2019**, *19*, 41–51. [CrossRef]
10. Bliznak, V.; Kaspar, M.; Muller, M.; Zacharov, P. Sub-daily temporal reconstruction of extreme precipitation events using NWP model simulations. *Atmos. Res.* **2019**, *224*, 65–80. [CrossRef]

11. Zolina, O.; Simmer, C.; Belyaev, K.; Kapala, A.; Gulev, S. Improving estimates of heavy and extreme precipitation using daily records from European rain gauges. *J. Hydrometeorol.* **2009**, *10*, 701–716. [CrossRef]
12. Balkema, A.; de Haan, L. Residual life time at great age. *Ann. Probab.* **1974**, *2*, 792–804. [CrossRef]
13. Pickands, J. Statistical inference using extreme order statistics. *Ann. Stat.* **1975**, *3*, 119–131. [CrossRef]
14. Begueria, S.; Vicente-Serrano, S. Mapping the hazard of extreme rainfall by peaks over threshold extreme value analysis and spatial regression techniques. *J. Appl. Meteorol. Climatol.* **2006**, *45*, 108–124. [CrossRef]
15. Kyselý, J.; Picek, J.; Beranova, R. Estimating extremes in climate change simulations using the peaks-over-threshold method with a non-stationary threshold. *Glob. Planet. Chang.* **2010**, *72*, 55–68. [CrossRef]
16. Begueria, S.; Angulo-Martinez, M.; Vicente-Serrano, S.; Lopez-Moreno, I.; El-Kenawy, A. Assessing trends in extreme precipitation events intensity and magnitude using non-stationary peaks-over-threshold analysis: A case study in northeast Spain from 1930 to 2006. *Int. J. Climatol.* **2011**, *31*, 2102–2114. [CrossRef]
17. Gorshenin, A.; Korolev, V. Determining the extremes of precipitation volumes based on a modified "Peaks over Threshold". *Inform. Primenen.* **2018**, *12*, 16–24. [CrossRef]
18. Huang, W.; Nychka, D.; Zhang, H. Estimating precipitation extremes using the log-histospline. *Environmetrics* **2019**, *30*. [CrossRef]
19. Gorshenin, A. Pattern-based analysis of probabilistic and statistical characteristics of precipitations. *Inform. Primenen.* **2017**, *11*, 38–46. [CrossRef]
20. Gorshenin, A. On some mathematical and programming methods for construction of structural models of information flows. *Inform. Primenen.* **2017**, *11*, 58–68. [CrossRef]
21. Korolev, V.; Gorshenin, A.; Gulev, S.; Belyaev, K.; Grusho, A. Statistical Analysis of Precipitation Events. *AIP Conf. Proc.* **2017**, *1863*. [CrossRef]
22. Vasilieva, M.; Gorshenin, A.; Korolev, V. Statistical analysis of probability characteristics of precipitation in different geographical regions. *Adv. Intell. Syst. Comput.* **2020**, *902*, 629–639._5. [CrossRef]
23. Zolotarev, V. *One-Dimensional Stable Distributions*; American Mathematical Society: Providence, RI, USA, 1986.
24. Kotz, S.; Ostrovskii, I. A mixture representation of the Linnik distribution. *Stat. Probab. Lett.* **1996**, *26*, 61–64. [CrossRef]
25. Korolev, V.; Gorshenin, A. The probability distribution of extreme precipitation. *Dokl. Earth Sci.* **2017**, *477*, 1461–1466. [CrossRef]
26. Gorshenin, A.; Korolev, V. Scale mixtures of Frechet distributions as asymptotic approximations of extreme precipitation. *J. Math. Sci.* **2018**, *234*, 886–903. [CrossRef]
27. Korolev, V. Convergence of random sequences with independent random indexes. I. *Theory Probab. Appl.* **1994**, *39*, 313–333. [CrossRef]
28. Korolev, V. Convergence of random sequences with independent random indexes. II. *Theory Probab. Appl.* **1995**, *40*, 770–772. [CrossRef]
29. Johnson, N.; Kotz, S.; Balakrishnan, N. *Continuous Univariate Distributions*, 2nd ed.; Wiley: New York, NY, USA, 1995; Volume 2.
30. Embrechts, P.; Klüppelberg, K.; Mikosch, T. *Modeling Extremal Events*; Springer: Berlin, Germany, 1998.

mathematics

Article

Non-Parametric Threshold Estimation for the Wiener–Poisson Risk Model

Honglong You [1,*] **and Yuan Gao** [2]

[1] School of Statistics, Qufu Normal University, Qufu 273165, China
[2] School of Mathematics, Qufu Normal University, Qufu 273165, China; gyyql413@163.com
* Correspondence: yougaoyou815@qfnu.edu.cn

Received: 15 April 2019; Accepted: 27 May 2019; Published: 3 June 2019

Abstract: In this paper, we consider the Wiener–Poisson risk model, which consists of a Wiener process and a compound Poisson process. Given the discrete record of observations, we use a threshold method and a regularized Laplace inversion technique to estimate the survival probability. In addition, we also construct an estimator for the distribution function of jump size and study its consistency and asymptotic normality. Finally, we give some simulations to verify our results.

Keywords: Wiener–Poisson risk model; survival probability; Nonparametric threshold estimation

1. Introduction

Let $S = \{S_t\}_{t\geq 0}$ with $S_0 = 0$ be a compound Poisson process defined as

$$S_t = \sum_{i=1}^{N_t} \gamma_i, \quad t \geq 0,$$

where $\{N_t\}_{t\geq 0}$ is a Poisson process with unknown intensity $\lambda > 0$, and $\gamma_1, \gamma_2, \gamma_3, \ldots$ are independent and identically distributed positive sequence of random variables with unknown distribution function F supported on $(0, \infty)$.

The Wiener–Poisson risk process is defined by

$$X_t = x + ct + \sigma W_t - \sum_{i=1}^{N_t} \gamma_i, \quad t \geq 0, \tag{1}$$

where x is a given positive constant, $\sigma > 0$ is an unknown constant, the corresponding process $\{N_t, t \geq 0\}$ is called the claim number process, $\{\gamma_i\}_{i=1,2,\ldots}$ is a sequence of claims, and $\{W_t\}_{t\geq 0}$ is a standard Brownian motion. Suppose that $\{N_t\}_{t\geq 0}$, $\{W_t\}_{t\geq 0}$ and $\{\gamma_i\}_{i=1,2,\ldots}$ are independent of each other, and the mean and variance of claim are finite, i.e., $\mu_\gamma = \int_0^\infty xF(dx) < \infty$, $\sigma_\gamma^2 = \int_0^\infty x^2F(dx) - \mu_\gamma^2 < \infty$. Further, we assume that the risk model in Equation (1) has a relative safety loading $\omega = \frac{c}{\lambda\mu_\gamma} - 1 > 0$. Let $\tau(x) = \inf\{t > 0; X_t \leq 0, X_0 = x\}$. The survival probability of the risk model in Equation (1) is defined by:

$$\Phi(x) = \mathbf{P}\left(\tau(x) = \infty\right), \tag{2}$$

and $\Psi(x) = 1 - \Phi(x)$, the probability of ruin.

In the last few decades, many works have been contributed to the survival probability for the risk model in Equation (1) and its extended risk model. In [1], the author first introduced the risk model in Equation (1) and established an asymptotic estimate for $\Psi(x)$. In [2], the authors showed that $\Phi(x)$ satisfies a defective renewal equation. By renewal theory, they obtained the Pollaczeck–Khinchin formula of $\Phi(x)$. Accurate calculation and approximation for $\Psi(x)$ has always been an inspiration and an important source of technological development for actuarial mathematics (see, e.g., [3–9]). Although various approximations to the probability of ruin (e.g., importance sampling or saddle-point approximations) are now available, developing alternative approximations of different nature is still an interesting and practical problem.

In recent years, many authors studied the ruin probability by using statistical methods (see, e.g., [10–18]). In [17], the author assumed that $\{X_{t_i^n}|t_i^n = ih_n; i = 0, 1, 2, ..., n\}$ and $\{\gamma_1, \gamma_2, ..., \gamma_{N_{t_n^n}}\}$ are observed, where $h_n = t_i^n - t_{i-1}^n$ is the sampling interval and the time of claims are known. The author constructed an estimator for Gerber–Shiu function and obtained its asymptotic property. Please refer to Equation (1.2) in [17] for the details of Gerber–Shiu function.

In our work, we suppose that a sample $\{X_{t_1^n}, X_{t_2^n}, X_{t_3^n}, ..., X_{t_n^n}\}$ can be observed, where $h_n = t_i^n - t_{i-1}^n$ is the sampling interval. However, we cannot observe the exact time and size of claims. To estimate $\Phi(x)$, we have to estimate F, λ, $\lambda\mu_\gamma$ and σ^2. Given the discrete record of observations, we need to judge whether a claim occurs in the interval $(t_{i-1}^n, t_i^n]$. The threshold method from [19–22] is to determine that a single jump has occurred within $(t_{i-1}^n, t_i^n]$ if and only if the increment $|\Delta_i X| = |X_{t_i^n} - X_{t_{i-1}^n}|$ is larger than a suitable threshold function. By the threshold method and the work in [21,22], we can estimate F, λ, μ_γ and σ^2.

In [14,17], the authors estimated the ruin probability and Gerber–Shiu function by a regularized Laplace inversion technique. Using the threshold method and the work in [23], it is easy to obtain an estimator for the Laplace transform of $\Phi(x)$. To estimate $\Phi(x)$, the regularized Laplace inversion technique is used. Finally, we also obtain a rate of convergence for the estimator of $\Phi(x)$ in a sense of the integrated squared error (ISE).

This paper is organized as follows. In Section 2, we give some estimators for σ^2, $\lambda\mu_\gamma$, λ, F and its Laplace–Stieltjes transform. In Section 3, we study the asymptotic properties for the estimators. Finally, we give some conclusions in Section 5. All the technical proofs are presented in Appendix A.

2. Estimation of Survival Probability

To give the estimators for σ^2, $\lambda\mu_\gamma$, λ, F and the Laplace transform of F, we introduce the following filter:

$$C_i^n(\vartheta(h_n)) = \{\omega \in \Omega; |\Delta_i X(\omega)| > \vartheta(h_n)\}, \tag{3}$$

where $\vartheta(h_n)$ is a threshold function and $\mathcal{D}_i^n(\vartheta(h_n))$ is a complement of $C_i^n(\vartheta(h_n))$. In [19,20], the threshold function $\vartheta(h_n)$ satisfies $\lim_{h_n\to 0}\vartheta(h_n) = 0$ and $\lim_{h_n\to 0}\frac{\sqrt{h_n\log(\frac{1}{h_n})}}{\vartheta(h_n)} = 0$. In [21], the author gave an expression of threshold function $\vartheta(h_n) = Lh_n^b$, where $L > 0$ is a constant and $b \in (0, \frac{1}{2})$. Obviously, the expression of $\vartheta(h_n)$ from [21] satisfies the two conditions. In our work, the expression of $\vartheta(h_n)$ is similar to that in [21].

We first estimate F. Using $\{|\Delta_i X|; 0 \le i \le n, \mathbf{I}_{C_i^n(\vartheta(h_n))} = 1\}$ and empirical distribution function, we can try to construct an estimator of F as follows:

$$\hat{F}_n(u) = \frac{1}{\sum_{i=1}^n \mathbf{I}_{C_i^n(\vartheta(h_n))}} \sum_{i=1}^n \mathbf{I}_{\{|\Delta_i X| \le u\}} \mathbf{I}_{C_i^n(\vartheta(h_n))}, \quad u \ge 0. \tag{4}$$

By Equations (3.4) and (3.6) in [21], the estimators of σ^2 and λ are

$$\widetilde{\sigma^2}_n = \frac{\sum_{i=1}^n |\Delta_i X - ch_n|^2 \mathbf{I}_{\mathcal{D}_i^n(\vartheta(h_n))}}{T_n}, \quad \widetilde{\lambda}_n = \frac{\sum_{i=1}^n \mathbf{I}_{C_i^n(\vartheta(h_n))}}{T_n}.$$

By Equation (3.10) in [21], an estimator of $\lambda\mu_\gamma$ is given by

$$\widetilde{\lambda\mu_{\gamma}}_n = \frac{\sum_{i=1}^n |\Delta_i X| \mathbf{I}_{C_i^n(\vartheta(h_n))}}{T_n}.$$

Let $\rho = \frac{\lambda\mu_\gamma}{c}$. Obviously, the estimator of ρ is given by

$$\widetilde{\rho}_n = \frac{1}{c} \frac{\sum_{i=1}^n |\Delta_i X| \mathbf{I}_{C_i^n(\vartheta(h_n))}}{T_n}.$$

The Laplace transform of F is defined by $l_F = E[e^{-s\gamma_1}] = \int_0^\infty e^{-su} F(du)$. An estimator of l_F is given by

$$\widetilde{l_F}_n(s) = \frac{\sum_{i=1}^n e^{-s|\Delta_i X|} \mathbf{I}_{C_i^n(\vartheta(h_n))}}{\sum_{i=1}^n \mathbf{I}_{C_i^n(\vartheta(h_n))}},$$

where $s \in \mathbb{E}$ and $\mathbb{E}$ is a compact subset of $(0, \infty)$.

By the work in [23], the Laplace transform of $\Phi(x)$ can be obtained as follows:

$$\begin{aligned} L_\Phi(s) &= \int_0^\infty e^{-sx} \Phi(x) dx \\ &= \frac{1-\rho}{D(s)}, \quad s > 0 \end{aligned} \tag{5}$$

where $\rho = \frac{\lambda\mu}{c}$ and $D(s) = s + \frac{\sigma^2}{2c}s^2 - \frac{\lambda}{c}(1 - l_F(s))$.

Let us define an estimator of $L_\Phi(s)$ as follows:

$$\widetilde{L_\Phi}(s) = \frac{1-\widetilde{\rho}_n}{\widetilde{D}(s)}, \quad \widetilde{D}(s) = s + \frac{\widetilde{\sigma^2}_n}{2c}s^2 - \frac{\widetilde{\lambda}_n}{c}(1 - \widetilde{l_F}_n(s)), \quad s > 0. \tag{6}$$

To estimate $\Phi(x)$, we use the L^2-inversion method proposed from [24]. Now, we give the L^2-inversion method by Definition 1. We say that $f \in L^2(0, \infty)$ if $(\int_0^\infty |f(t)|^2 dt)^{\frac{1}{2}} < \infty$.

Definition 1. *Let $m > 0$ be a constant. The regularized Laplace inversion $L_m^{-1} : L^2(0,\infty) \to L^2(0,\infty)$ is given by*

$$L_m^{-1} g(t) = \frac{1}{\pi^2} \int_0^\infty \int_0^\infty \Psi_m(y) y^{-\frac{1}{2}} e^{-tvy} g(v) dv dy \tag{7}$$

for a function $g \in L^2(0,\infty)$ and $t \in (0,\infty)$, where

$$\Psi_m(y) = \int_0^{a_m} \cosh(\pi x) \cos(x \log y) dx$$

and $a_m = \pi^{-1} \cosh^{-1}(\pi m) > 0$.

For further information, and details of L_m^{-1}, please refer to [24].

To use Definition 1, it requires to verify $\widetilde{L_\Phi}(s) \in L^2(0,\infty)$. As n is sufficiently large, for $\mathbf{P}$-almost all $\omega \in \Omega$ and $s > 0$, we have

$$\mathbf{P}(\{\omega \in \Omega; (1-\widetilde{\rho}_n)s \le \widetilde{D}(s) \le s + \frac{\widetilde{\sigma^2}_n}{2c}s^2\}) = 1. \tag{8}$$

From Equations (6) and (8), it is obvious that $\widetilde{L_\Phi}(s) \notin L^2(0,\infty)$. The L^2-inversion method in Definition 1 cannot be applied at once.

Therefore, to use Definition 1, we have to amend $\widetilde{L_\Phi}(s)$.

Let

$$\Phi_\theta(x) = e^{-\theta x}\Phi(x), \quad x > 0$$

for arbitrary fixed $\theta > 0$. It is obvious that

$$L_{\Phi_\theta}(s) = L_\Phi(s+\theta), \quad s > 0.$$

An estimator of L_{Φ_θ} is given by

$$\widetilde{L_{\Phi_\theta}}(s) = \widetilde{L_\Phi}(s+\theta), \quad s > 0.$$

Obviously, $\widetilde{L_{\Phi_\theta}} \in L^2(0,\infty)$.

Finally, an estimator of $\Phi(x)$ is given by

$$\widetilde{\Phi}_{m(n)}(x) = e^{\theta x}\widetilde{\Phi_{\theta,m(n)}}(x), \quad x > 0, \tag{9}$$

where $\widetilde{\Phi_{\theta,m(n)}}(x) = L^{-1}_{m(n)}\widetilde{L_{\Phi_\theta}}(s)$ and $m(n) > 0$.

3. Asymptotic Properties

According to Theorem 3.1 in [19], the author assumed that $\sigma < Q$ and $\gamma_i \geq \Gamma$ with $Q > 0, \Gamma > 0$. In our work, Assumption 1 is used to prove the asymptotic properties of estimators.

Assumption 1. *There exist two positive constants Q and Γ such that $\sigma < Q$ and $\mathbf{P}(\{\omega \in \Omega; \gamma_i \geq \Gamma\}) = 1$ for $i = 1, 2, \ldots$.*

Let $\bar{F} = 1 - F$. With Equation (4), an estimator of $\bar{F}$ is given by

$$\hat{\bar{F}}_n(u) \quad = \quad \frac{1}{\sum_{i=1}^n \mathbf{I}_{C_i^n(\vartheta(h_n))}} \sum_{i=1}^n \mathbf{I}_{\{|\Delta_i X| > u\}} \mathbf{I}_{C_i^n(\vartheta(h_n))}. \tag{10}$$

Let $\mathcal{N}(m,n)$ be a normal distribution with expectation m and variance n. Theorem 1 gives the asymptotic properties of $\hat{F}_n(u)$.

Theorem 1. *Suppose that $T_n = nh_n \to \infty$, $nh_n^2 \to 0$, $h_n \to 0$ as $n \to \infty$ and Assumption 1 is satisfied, then*

$$\sqrt{T_n}\left(\hat{\bar{F}}_n(u) - \bar{F}(u)\right) \xrightarrow{\mathbf{D}} \mathcal{N}(0, \frac{\overline{F}(u)(1-\overline{F}(u))}{\lambda}). \tag{11}$$

Obviously,

$$\sqrt{T_n}\left(\hat{F}_n(u) - F(u)\right) \xrightarrow{\mathbf{D}} \mathcal{N}(0, \frac{F(u)(1-F(u))}{\lambda}).$$

Remark 1. *By Dvoretzky–Kiefer–Wolfowitz inequality, we have*

$$\mathbf{P}(\sup_{u\in[0,\infty)} |\hat{F}_n(u) - F(u)| > x) \leq Ce^{-2\lambda T_n x^2}, \quad x > 0,$$

where C is a positive constant, not depending on F. Note that this inequality may be expression in the form :

$$\mathbf{P}(\sqrt{T_n} \sup_{u\in[0,\infty)} |\hat{F}_n(u) - F(u)| > x) \leq Ce^{-2\lambda x^2}, \quad x > 0,$$

which clearly demonstrate that

$$\sqrt{T_n} \sup_{u \in [0,\infty)} |\hat{F}_n(u) - F(u)| = O_{\mathbf{P}}(1).$$

The asymptotic properties of $\widetilde{\sigma^2}_n$ are given by the following Lemma 1.

Lemma 1. *Suppose that $T_n = nh_n \to \infty$, $nh_n^2 \to 0$ and $h_n \to 0$ as $n \to \infty$, then*

$$\widetilde{\sigma^2}_n \xrightarrow{\mathbf{P}} \sigma^2, \quad n \to \infty. \tag{12}$$

$$\sqrt{n}(\widetilde{\sigma^2}_n - \sigma^2) \xrightarrow{\mathbf{D}} \mathcal{N}(0, 2\sigma^4), \quad n \to \infty. \tag{13}$$

Lemma 2. *Suppose that $T_n = nh_n \to \infty$, $nh_n^\beta \to 0$ for some $\beta \in (1,2]$, $h_n \to 0$ as $n \to \infty$ and Assumption 1 is satisfied. Then,*

$$\widetilde{\lambda}_n \xrightarrow{\mathbf{P}} \lambda, \quad \sup_{\{s|s \in \mathbb{E}\}} |\widetilde{\lambda}_n \widetilde{l_F}_n(s) - \lambda l_F(s)| \xrightarrow{\mathbf{P}} 0, \tag{14}$$

$$\sqrt{T_n}(\widetilde{\lambda}_n - \lambda) \xrightarrow{\mathbf{D}} \mathcal{N}(0, \lambda), \tag{15}$$

$$\widetilde{\rho}_n \xrightarrow{\mathbf{P}} \rho \tag{16}$$

and

$$\sqrt{n}(\widetilde{\rho}_n - \rho) \xrightarrow{\mathbf{D}} \mathcal{N}(0, \frac{\lambda\sigma^2}{c^2}), \tag{17}$$

as $n \to \infty$.

Let $\|f\|_B^2 = \int_0^B |f(t)|^2 dt$ for any function f and $B > 0$. Theorem 2 gives a rate of convergence for $\widetilde{\Phi}_{m(n)}(x)$ in a sense of ISE.

Theorem 2. *Suppose that there exists a constant $K > 0$ such that $0 \leq \Phi'(x) = g(x) \leq K < \infty$ and the conditions in Lemma 2 are satisfied. Then, for $m(n) = \sqrt{\frac{T_n}{\log T_n}}$ and for any constant $B > 0$, we have*

$$\|\widetilde{\Phi}_{m(n)} - \Phi\|_B^2 = O_P((\log T_n)^{-1}), \quad n \to \infty.$$

Remark 2. *The explicit expression for $\widetilde{\Phi}_{m(n)}(x)$ is*

$$\widetilde{\Phi}_{m(n)}(x) = \frac{e^{x\theta}}{\pi^2} \int_0^\infty \int_0^\infty e^{-xsy} \widetilde{L_{\Phi_\theta}}(s) \Psi_{m(n)}(y) y^{-\frac{1}{2}} ds dy$$

where $\Psi_{m(n)}(y) = \int_0^{a_{m(n)}} \cosh(\pi x) \cos(x \log(y)) dx$ and $a_{m(n)} = \pi^{-1} \cosh^{-1}(\pi m(n)) > 0$ and $m(n) = \sqrt{\frac{T_n}{\log T_n}}$.

When c, λ, σ, F, θ, $\vartheta(h_n)$ and sample size n are known, $\widetilde{\Phi}_{m(n)}(x)$ can be evaluated with the command integral2$(f; 0; \infty; 0; \infty)$ of Matlab.

4. Simulation

If $F(x) = 1 - e^{-\frac{1}{\mu_\gamma}x}$, the survival probability is given by

$$\Phi(x) = 1 - \frac{r_1 + \frac{1}{\mu_\gamma} + \frac{2\lambda\mu_\gamma}{\sigma^2}}{r_1 - r_2} e^{r_1 x} - \frac{r_2 + \frac{1}{\mu_\gamma} + \frac{2\lambda\mu_\gamma}{\sigma^2}}{r_2 - r_1} e^{r_2 x}, \quad x \geq 0, \tag{18}$$

where $r_2 < r_1 < 0$ are negative roots of the following equation

$$\frac{1}{2}\sigma^2 s + c - \frac{\lambda}{s + \frac{1}{\mu_\gamma}} = 0.$$

By the work in [25], Equation (18) is obtained easily.

Let $c = \lambda = 10$, $\mu_\gamma = \frac{1}{2}$, $\sigma = 5$, $\theta = 0.075$, $\vartheta(h_n) = h_n^b$, $b = \frac{1}{4}$ and $h_n = n^{-\frac{4}{5}}$.

Firstly, we computed $\widetilde{L_{\Phi_\theta}}(s)$. In Figure 1, we plot the mean points with sample sizes n =5000, 10,000, 30,000, 50,000, 80,000, which were computed based on 5000 simulation experiments.

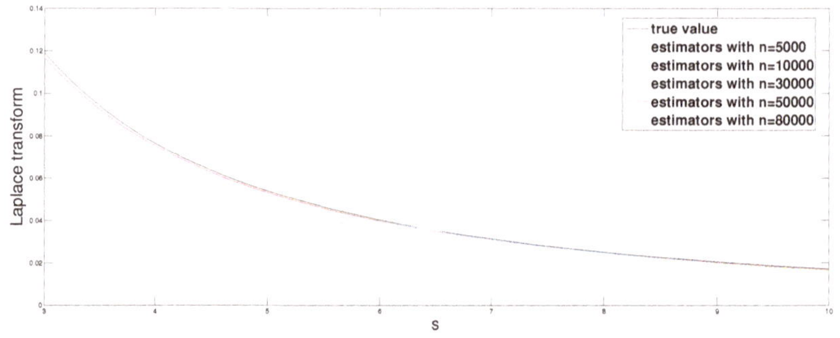

Figure 1. The estimator of $\mathcal{L}_{\Phi_\theta}$ with sample sizes n =5000, 10,000, 30,000, 50,000, 80,000.

In Remark 2, $\widetilde{\Phi}_{m(n)}(u)$ is a double complex integrals. Using Matlab to compute $\widetilde{\Phi}_{m(n)}(u)$ would take a long time. As shown in Figure 1, $\hat{\mathcal{L}}_{\Phi_\theta}$ is very close to $\mathcal{L}_{\Phi_\theta}$ as $n \geq$ 30,000. To improve computational efficiency, let

$$\Phi_p(x) = \frac{e^{x\theta}}{\pi^2} \int_0^\infty \int_0^\infty e^{-xsy} [\hat{\mathcal{L}}_{\Phi_\theta}(s)]_{n=30000} \Psi_p(y) y^{-\frac{1}{2}} ds dy,$$

where $[\hat{\mathcal{L}}_{\Phi_\theta}(s)]_{n=30000} = \frac{1 - \frac{1}{c30000(30000)^{-\frac{4}{5}}} \sum_{k=1}^{30000} (c(30000)^{-\frac{4}{5}} - Z_k) \mathbf{I}_{D_k^{30000}}}{\frac{1}{c(30000)^{-\frac{4}{5}}} (\frac{1}{30000} \sum_{k=1}^{30000} e^{sZ_k} - 1)}$, $\Psi_p(y) = \int_0^{a_p} \cosh(\pi x) \cos(x \log(y)) dx$ and $a_p = \pi^{-1} \cosh^{-1}(\pi p) > 0$.

In Figure 2, we plot the mean points with sample sizes n = 30,000 and $p = 100, 500, 800, 1000, 3000$, which were computed based on 5000 simulation experiments.

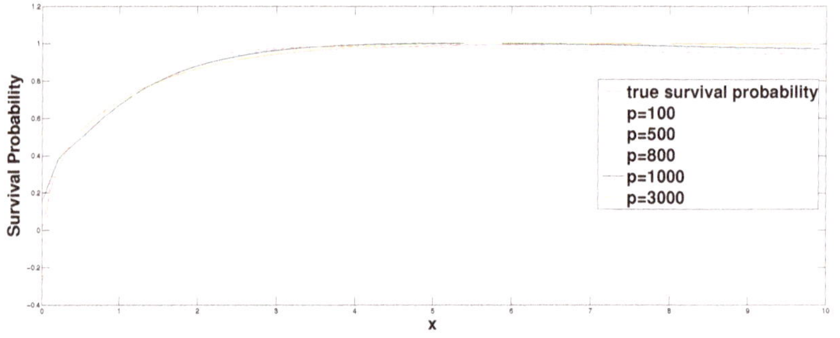

Figure 2. $\Phi_p(x)$ with sample size n = 30,000 and $p = 100, 500, 800, 1000, 3000$

5. Conclusions

In this paper, we use the threshold estimation technique and regularized Laplace inversion technique to constructed an estimator of survival probability for the Wiener–Poisson risk model. The rate of convergence for the estimator is a logarithmic rate. We adopt a method proposed by Cai et al. [26] to improve the speed in simulated calculation. The further work is to improve the speed of convergence for the estimator. We will combine the threshold estimation technique with Fourier transform (inversion) technique to construct an estimator of survival probability. We hope some further studies will be done when the risk model is the compound Poisson model with the barrier dividend strategy and investment. The Gerber–Shiu function and dividend function will be estimated by some statistical methods.

Author Contributions: Methodology, H.Y.; Formal analysis, H.Y. and Y.G.; Simulation, H.Y.; Writing–original draft, H.Y.

Funding: This research was partially supported by the National Natural Science Foundation of China (Grant Nos. 11571189, 11571198, 11501319 and 11701319), the Postdoctoral Science Foundation of China (Grant No. 2018M642634) and the Higher Educational Science and Technology Program of Shandong Province of China (Grant No. J15LI05).

Acknowledgments: The authors would like to express their thanks to three anonymous referees for their helpful comments and suggestions, which improved an earlier version of the paper.

Conflicts of Interest: The authors declare no conflicts of interest.

Appendix A. Proofs of Theorems

Proof of Lemma 1. The proof of Lemma 1 is easily obtained by Theorem 3.1 in [21]. □

Proof of Lemma 2. The proof of Equation (14) is given as Theorem 3.2 in [21]. By Theorem 3.1 in [19], we can get Equation (15). It is easy to get Equations (16) and (17) by Proposition 3.4 in [19] and Theorem 3 in [20]. □

To prove Theorem 1, we need the following Proposition, which can be easily obtained in Section 3.2 of [19].

Proposition A1. *Following from the condition of Theorem 1, for any $\epsilon > 0$,*

$$\lim_{n\to\infty} P\left(\left||\Delta_i X|\mathbf{I}_{C_i^n(\vartheta(h_n))} - \gamma_{\tau(i)}\mathbf{I}_{\{\Delta_i N\geq 1\}}\right| > \epsilon\right) = 0, \tag{A1}$$

where $\gamma_{\tau(i)}$ is the size of the eventual jump in time interval $(t_{i-1}^n, t_i^n]$.

Proof of Theorem 1. By Equation (10),

$$\sqrt{T_n}\left(\bar{\tilde{F}}_n(u) - \bar{F}(u)\right) = \frac{J}{\tilde{\lambda}_n}, \tag{A2}$$

where

$$J = \frac{\sum_{i=1}^n \left(\mathbf{I}_{\{|\Delta_i X|>u\}} - \bar{F}(u)\right)\mathbf{I}_{C_i^n(\vartheta(h_n))}}{\sqrt{T_n}}.$$

As $n \to \infty$, the expectation of J is

$$\begin{aligned}\lim_{n\to\infty}\mathbf{E}[J] &= \lim_{n\to\infty}\frac{n}{\sqrt{T_n}}\mathbf{E}\left[(\mathbf{I}_{\{|\Delta_i X|>u\}} - \bar{F}(u))\mathbf{I}_{C_i^n(\vartheta(h_n))}\right] \\ &= \lim_{n\to\infty}\frac{n}{\sqrt{T_n}}\left[\mathbf{P}(|\Delta_i X|\mathbf{I}_{C_i^n(\vartheta(h_n))} > u) - \bar{F}(u)\mathbf{P}(|\Delta_i X| > \vartheta(h_n))\right].\end{aligned}$$

By Proposition A1, we have

$$\lim_{n\to\infty} \mathbf{P}(|\Delta_i X|\mathbf{I}_{C_i^n(\vartheta(h_n))} > u) = \lim_{n\to\infty} \mathbf{P}(\gamma_{\tau^{(i)}}\mathbf{I}_{\{\Delta_i N\geq 1\}} > u) = \bar{F}(u)(\lambda h_n + o(h_n)). \tag{A3}$$

By the work in [19], it is obvious that

$$\lim_{n\to\infty} \mathbf{P}(|\Delta_i X| > \vartheta(h_n)) = \lambda h_n + o(h_n). \tag{A4}$$

Therefore,

$$\lim_{n\to\infty} \mathbf{E}[J] = 0.$$

As $n \to \infty$, the variance of J is

$$\begin{aligned}
\lim_{n\to\infty} \mathbf{Var}[J] &= \lim_{n\to\infty} \frac{n}{T_n}\mathbf{Var}\left[(\mathbf{I}_{\{|\Delta_i X|>u\}} - \bar{F}(u))\mathbf{I}_{C_i^n(\vartheta(h_n))}\right] \\
&= \lim_{n\to\infty} \frac{n}{T_n}\mathbf{E}[\mathbf{I}_{\{|\Delta_i X|\mathbf{I}_{C_i^n(\vartheta(h_n))}>u\}}] + \lim_{n\to\infty} \frac{n}{T_n}\mathbf{E}[(\bar{F}(u))^2\mathbf{I}_{C_i^n(\vartheta(h_n))}]. \\
&\quad - \lim_{n\to\infty} \frac{n}{T_n}\mathbf{E}[2\bar{F}(u)\mathbf{I}_{\{|\Delta_i X|\mathbf{I}_{C_i^n(\vartheta(h_n))}>u\}}].
\end{aligned}$$

By Equations (A3) and (A4),

$$\lim_{n\to\infty} \mathbf{Var}[J] = \lambda\bar{F}(u)(1-\bar{F}(u)).$$

With the central limit theorem, Slutsky's theorem and Lemma 2, we have

$$\sqrt{T_n}\left(\tilde{\bar{F}}_n(u) - \bar{F}(u)\right) \xrightarrow{\mathbf{D}} \mathcal{N}(0, \frac{\bar{F}(u)(1-\bar{F}(u))}{\lambda}), \quad n \to \infty.$$

□

To prove Theorem 2, we need the following Lemma A1.

Lemma A1. *Suppose that* $\int_0^\infty [t(t^{\frac{1}{2}}f(t))']^2 t^{-1} dt < \infty$ *for a function* $f \in L^2(0,\infty)$ *with the derivative* f'*. Then,*

$$\|L_n^{-1}L_f - f\| = O\left((\log n)^{-\frac{1}{2}}\right), \quad n \to \infty.$$

By Theorem 3.2 in [24], the proof of Lemma A1 can be found.

Proof of Theorem 2. By Equation (9),

$$\begin{aligned}
\|\widetilde{\Phi}_{m(n)} - \Phi\|_B^2 &\leq e^{2\theta B}\|\widetilde{\Phi}_{\theta,m(n)} - \Phi_\theta\|_B^2 \\
&\leq 2e^{2\theta B}\{\|L_{m(n)}^{-1}\widetilde{L_{\Phi_\theta}} - L_{m(n)}^{-1}L_{\Phi_\theta}\|^2 + \|\Phi_{\theta,m(n)} - \Phi_\theta\|^2\}.
\end{aligned} \tag{A5}$$

Let $\Phi'_\theta = g_\theta$. Now, we show that $\Phi_{\theta,m(n)}$ satisfies the condition of Lemma A1.

$$\begin{aligned}
\int_0^\infty [x(\sqrt{x}\Phi_\theta(x))']^2 \frac{1}{x}dx &= \int_0^\infty [x(\frac{1}{2\sqrt{x}}\Phi_\theta(x) + x\sqrt{x}g_\theta(x))]^2 \frac{1}{x}dx \\
&\leq \int_0^\infty 2\frac{1}{x}[x\frac{1}{2\sqrt{x}}\Phi_\theta(x)]^2 + \int_0^\infty 2\frac{1}{x}[x\sqrt{x}g_\theta(x)]^2 dx \\
&= \int_0^\infty \frac{1}{2}\Phi_\theta^2(x)dx + 2\int_0^\infty x^2 g_\theta^2(x)dx \\
&\leq \int_0^\infty \frac{1}{2}e^{-2\theta x}dx + 2\int_0^\infty x^2[g(x)e^{-\theta x} - \theta\Phi(x)e^{-\theta x}]^2 dx \\
&\leq \frac{1}{4\theta} + 4\int_0^\infty x^2 g^2(x)e^{-2\theta x}dx + 4\theta^2\int_0^\infty \Phi^2(x)x^2 e^{-2\theta x}dx \\
&\leq \frac{1}{4\theta} + 4(K^2+\theta^2)\int_0^\infty x^2 e^{-2\theta x}dx < \infty.
\end{aligned}$$

Therefore, by Lemma A1, we have

$$\|\Phi_{\theta,m(n)} - \Phi_\theta\|^2 = O(\frac{1}{\log m(n)}), \quad n \to \infty. \tag{A6}$$

By Equations (5) and (6),

$$\|\widetilde{L_{\Phi_\theta}} - L_\theta\|^2 = \int_0^\infty \left(\frac{(1-\rho)(\widetilde{D}(s+\theta) - D(s+\theta))}{\widetilde{D}(s+\theta)D(s+\theta)} + \frac{(\widetilde{\rho}_n - \rho)}{\widetilde{D}(s+\theta)}\right)^2 ds. \tag{A7}$$

Exploiting Equation (8) and $\mathbf{P}(\{\omega \in \Omega; \widetilde{\rho}_n = 1\}) = 0$, the right-hand side of Equation (A7) is bounded by

$$2\int_0^\infty \frac{(\widetilde{D}(s+\theta) - D(s+\theta))^2}{(s+\theta)^4(1-\widetilde{\rho}_n)^2}ds + 2\int_0^\infty (\frac{\widetilde{\rho}_n - \rho}{1-\widetilde{\rho}_n})^2 \frac{1}{(s+\theta)^2}ds. \tag{A8}$$

By Lemmas 1 and 2, the term

$$\begin{aligned}
\widetilde{D_n}(s+\theta) - D(s+\theta) &= \frac{(s+\theta)^2}{2c}(\widetilde{\sigma^2}_n - \sigma^2) + \frac{1}{c}\left((\lambda - \widetilde{\lambda}_n) + (\widetilde{\lambda}_n \widetilde{l_F}_n(s+\theta) - \lambda l_F(s+\theta))\right) \\
&= O_{\mathbf{P}}(T_n^{-\frac{1}{2}}) + \frac{1}{c}\frac{N_{T_n}}{T_n}\left(\int_0^\infty e^{-(s+\theta)x}(\hat{F}_n(dx) - F(dx))\right) \\
&= O_{\mathbf{P}}(T_n^{-\frac{1}{2}}) + \frac{1}{c}\frac{N_{T_n}}{T_n}\left(\int_0^\infty (s+\theta)e^{-(s+\theta)x}(\hat{F}_n(x) - F(x))dx\right) \\
&\leq O_{\mathbf{P}}(T_n^{-\frac{1}{2}}) + \frac{1}{c}\frac{N_{T_n}}{T_n}\sup_{x\in[0,\infty)}|\hat{F}_n(x) - F(x)| \\
&= O_{\mathbf{P}}(T_n^{-\frac{1}{2}})
\end{aligned} \tag{A9}$$

The last equality is obtained from Remark 1.

By Equation (A9) and Lemma 2, we have

$$2\int_0^\infty \frac{(\widetilde{D}(s+\theta) - D(s+\theta))^2}{(s+\theta)^4(1-\widetilde{\rho}_n)^2}ds = O_{\mathbf{P}}(T_n^{-1})$$

and

$$2\int_0^\infty (\frac{\widetilde{\rho}_n - \rho}{1-\widetilde{\rho}_n})^2 \frac{1}{(s+\theta)^2}ds = o_{\mathbf{P}}(T_n^{-1}).$$

Recall that $\|L^{-1}_{m(n)}\|^2 \leq \pi m^2(n)$ (see [24]), so

$$\|L^{-1}_{m(n)}\|^2 \|\widetilde{L_{\Phi_\theta}} - L_{\Phi_\theta}\|^2 \quad = \quad O_{\mathbf{P}}(\frac{m^2(n)}{T_n}). \tag{A10}$$

Combining Equations (A6) and (A10), we have

$$\|\widetilde{\Phi}_{m(n)} - \Phi\|_B^2 = O_{\mathbf{P}}(\frac{m^2(n)}{T_n}) + O_{\mathbf{P}}(\frac{1}{\log m(n)}). \tag{A11}$$

with an optimal $m(n) = \sqrt{\frac{T_n}{\log T_n}}$ balancing the the two right-hand terms in Equation (A11), the order becomes $O_{\mathbf{P}}((\log T_n)^{-1})$. $\square$

References

1. Gerber, H.U. An extension of the renewal equation and its application in the collective theory of risk. *Scand. Actuar. J.* **1970**, *1970*, 205–210. [CrossRef]
2. Dufresne, F.; Gerber, H.U. Risk theory for the compound Poisson process that is perturbed by diffusion. *Insur. Math. Econ.* **1991**, *10*, 51–59. [CrossRef]
3. Furrer, H.J.; Schimidli, H. Exponential inequalities for ruin probabilities of risk process perturbed by diffusion. *Insur. Math. Econ.* **1994**, *15*, 23–36. [CrossRef]
4. Gatto, R.; Mosimann, M. Four approaches to compute the probability of ruin in the compound Poisson risk process with diffusion. *Math. Comput. Modell.* **2012**, *55*, 1169–1185. [CrossRef]
5. Gatto, R.; Baumgartner, B. Saddlepoint Approximations to the Probability of Ruin in Finite Time for the Compound Poisson Risk Process Perturbed by Diffusion. *Methodol. Comput. Appl. Probab.* **2014**, *18*, 1–19. [CrossRef]
6. Gatto, R. Importance sampling approximations to various probabilities of ruin of spectrally negative Lévy risk processes. *Appl. Math. Comput.* **2014**, *243*, 91–104. [CrossRef]
7. Schimidli, H. Cramer-Lundberg approximations for ruin probabilites of risk processes perturbed by diffusion. *Insur. Math. Econ.* **1995**, *16*, 135–149. [CrossRef]
8. Veraverbeke, N. Asymptotic estimates for the probability of ruin in a Poisson model with diffusion. *Insur. Math. Econ.* **1993**, *13*, 57–62. [CrossRef]
9. Wang, Y.; Yin, C. Approximation for the ruin probabilities in a discrete time risk model with dependent risks. *Stat. Probab. Lett.* **2010**, *80*, 1335–1342. [CrossRef]
10. Bening, V.E.; Korolev, V.Y. Nonparametric estimation of the ruin probability for generalized risk processes. *Theory Probab. Its Appl.* **2002**, *47*, 1–16. [CrossRef]
11. Croux, K.; Veraverbeke, N. Non-parametric estimators for the probability of ruin. *Insur. Math. Econ.* **1990**, *9*, 127–130. [CrossRef]
12. Frees, EW. Nonparametric estimation of the probability of ruin. *Astin Bull.* **1986**, *16*, 81–90. [CrossRef]
13. Hipp, C. Estimators and bootstrap confidence intervals for ruin probabilities. *Astin Bull.* **1989**, *19*, 57–70. [CrossRef]
14. Mnatsakanov, R.; Ruymgaart, L.L.; Ruymgaart, F.H. Nonparametric estimation of ruin probabilities given a random sample of claims. *Math. Methods Stat.* **2008**, *17*, 35–43. [CrossRef]
15. Pitts, S.M. Nonparametric estimation of compound distributions with applications in insurance. *Ann. Inst. Stat. Math.* **1994**, *46*, 537–555.
16. Politis, K. Semiparametric estimation for non-ruin probabilities. *Scand. Actuar. J.* **2003**, *2003*, 75–96. [CrossRef]
17. Shimizu, Y. Non-parametric estimation of the Gerber-Shiu function for the Wiener-Poisson risk model. *Scand. Actuar. J.* **2012**, *2012*, 56–69. [CrossRef]
18. Zhang, Z.; Yang, H. Nonparametric estimate of the ruin probability in a pure-jump *lévy* risk model. *Insur. Math. Econ.* **2013**, *53*, 24–35. [CrossRef]
19. Mancini, C. Estimation of the characteristics of the jump of a general Poisson-diffusion model. *Scand. Actuar. J.* **2004**, *1*, 42–52. [CrossRef]

20. Mancini, C. Non-parametric threshold estimation for models with stochastic diffusion coefficient and jumps. *Scand. J. Stat.* **2009**, *36*, 270–296. [CrossRef]
21. Shimizu ,Y. A new aspect of a risk process and its statistical inference. *Insur. Math. Econ.* **2009**, *44*, 70–77. [CrossRef]
22. Shimizu, Y. Functional estimation for *lévy* measures of semimartingales with Poissonian jumps. *J. Multivar. Anal.* **2009**, *100*, 1073–1092. [CrossRef]
23. Morales, M. On the expected discounted penalty function for a perturbed risk process driven by a subordinator. *Insur. Math. Econ.* **2007**, *40*, 293–301. [CrossRef]
24. Chauveau, D.E.; Vanrooij, A.C.M.; Ruymgaart, F.H. Regularized inversion of noisy Laplace transforms. *Adv. Appl. Math.* **1994**, *15*, 186–201. [CrossRef]
25. Asmussen, S.; Albrecher, H. *Ruin Probabilities*, 2nd ed.; World Scientific: Singapore, 2010.
26. Cai, C.; Chen, N.; You, H. Nonparametric estimation for a spectrally negative Lévy process based on low-frequency observation. *J. Comput. Appl. Math.* **2018**, *328*, 432–442. [CrossRef]

mathematics

MDPI

Article

On the Rate of Convergence for a Characteristic of Multidimensional Birth-Death Process

Alexander Zeifman [1,*], Yacov Satin [2], Ksenia Kiseleva [3] and Victor Korolev [4,5]

1 Department of Applied Mathematics, Vologda State University, IPI FRC CSC RAS, VolSC RAS, 160000 Vologda, Russia
2 Department of Mathematics, Vologda State University, 160000 Vologda, Russia; yacovi@mail.ru
3 Department of Applied Mathematics, Vologda State University, 160000 Vologda, Russia; ksushakiseleva@mail.ru
4 Faculty of Computational Mathematics and Cybernetics, Lomonosov Moscow State University, 119991 Moscow, Russia; vkorolev@cs.msu.ru
5 Department of Mathematics, School of Science, Hangzhou Dianzi University, Hangzhou 310018, China
* Correspondence: a_zeifman@mail.ru

Received: 15 April 2019; Accepted: 21 May 2019; Published: 26 May 2019

Abstract: We consider a multidimensional inhomogeneous birth-death process. In this paper, a general situation is studied in which the intensity of birth and death for each coordinate ("each type of particle") depends on the state vector of the whole process. A one-dimensional projection of this process on one of the coordinate axes is considered. In this case, a non-Markov process is obtained, in which the transitions to neighboring states are possible in small periods of time. For this one-dimensional process, by modifying the method previously developed by the authors of the note, estimates of the rate of convergence in weakly ergodic and null-ergodic cases are obtained. The simplest example of a two-dimensional process of this type is considered.

Keywords: multidimensional birth-death process; inhomogeneous continuous-time Markov chain; rate of convergence; one dimensional projection

1. Introduction and Preliminaries

Multidimensional birth-death processes (BDP) were objects of a number of studies in queueing theory and other applied fields. The authors of these papers studied different special classes of homogeneous multidimensional BDPs under some restrictions and considered fluid approximations [1], simulations [2–6], large deviations [7], stability [8,9], and other features. The problem of the product from solutions for such models was considered, for instance, in [10,11] (also, see the references therein). If the process is inhomogeneous and the transition intensities have a more general form, then the problem of computation of any probabilistic characteristics of the queueing model is much more difficult.

In the general case, it is impossible to obtain explicit solutions and their characteristics, as well as to construct any significant characteristics of the processes, as can be seen from the above list of works. This paper fills this gap and proposes a method of research and evaluation allowing one to estimate the rate of convergence for a one-dimensional projection of the multidimensional birth-death process. The approach also makes it possible to evaluate the main characteristics of the projection, as is demonstrated by the simplest example of an inhomogeneous two-dimensional process.

The background of our approach is the method of investigation of inhomogeneous BDP, see the detailed discussion and some preliminary results in [12–15]. Estimates for the state probabilities of one-dimensional projections of a multidimensional BDP were studied in [16,17]. However, within that methodology, it was impossible to obtain estimates of the rate of convergence, since the logarithmic norm of the operator cannot be applied to the corresponding nonlinear systems.

Here, we substantially modify that approach so that it can be used for estimation and construction of some explicit bounds on the rate of convergence for one-dimensional projection of a multidimensional BDP. Namely, in Section 2, we develop a simple but efficient method for bounding the rate of convergence for an arbitrary (which may be nonlinear, depending on the number of parameters and so on) differential equation in the space of sequences l_1, and in Section 3, we apply this method to bounding the rate of convergence for one-dimensional projections of BDP.

Let $\mathbf{X}(t) = (X_1(t), ..., X_d(t))$ be a d-dimensional BDP such that in the interval $(t, t+h)$, the following transitions are possible with order h: birth of a particle of type j, death of a particle of type j.

Let $\lambda_{j,\mathbf{m}}(t)$ be the corresponding birth rate (from the state $\mathbf{m} = (m_1, ..., m_d) = \sum_{i=1}^d m_i \mathbf{e}_i$ to the state $\mathbf{m} + \mathbf{e}_j$) and let $\mu_{j,\mathbf{m}}(t)$ be the corresponding death intensity (from the state $\mathbf{m} = (m_1, ..., m_d) = \sum_{i=1}^d m_i \mathbf{e}_i$ to the state $\mathbf{m} - \mathbf{e}_j$). Denote $p_{\mathbf{m}}(t) = \Pr(\mathbf{X}(t) = \mathbf{m})$.

To consider the existence and uniqueness, we renumber the states (only in this section), transforming the process into a one-dimensional one. Now, let the (finite or countable) state space of the vector process under consideration be arranged in a special order, say $0, 1, \ldots$. Denote by $p_i(t)$, the corresponding state probabilities, and by $\mathbf{p}(t)$, the corresponding column vector of state probabilities. Applying our standard approach (see details in [12,14,15]), we suppose in addition that all intensities are nonnegative functions locally integrable on $[0, \infty)$, and, moreover, in new enumeration,

$$\Pr\left(X(t+h) = j / X(t) = i\right) = \begin{cases} q_{ij}(t)h + \alpha_{ij}(t,h), & j \neq i, \\ 1 - \sum\limits_{k \neq i} q_{ik}(t)h + \alpha_i(t,h), & j = i, \end{cases} \tag{1}$$

where $q_{ij}(t)$ are the corresponding transition intensities and all $\alpha_i(t,h)$ are $o(h)$ uniformly in i, that is, $\lim_{h \to 0} \frac{1}{h} \sup_i |\alpha_i(t,h)| = 0$, for any $t \geq 0$.

We suppose that $\lambda_{j,\mathbf{m}}(t) \leq L < \infty$, $\mu_{j,\mathbf{m}}(t) \leq M < \infty$, for any j, $\mathbf{m}$ and almost all $t \geq 0$.

The probabilistic dynamics of the process is represented by the forward Kolmogorov system:

$$\frac{d\mathbf{p}}{dt} = A(t)\mathbf{p}(t), \tag{2}$$

where $A(t)$ is the corresponding infinitesimal (intensity) matrix.

Throughout the paper, we denote the l_1-norm by $\|\cdot\|$, i.e., $\|\mathbf{x}\| = \sum |x_i|$, and $\|B\| = \sup_j \sum_i |b_{ij}|$ for $B = (b_{ij})_{i,j=0}^{\infty}$.

Let Ω be the set all stochastic vectors, i.e., l_1-vectors with nonnegative coordinates and unit norm. We have the inequality $\|A(t)\| \leq 2d\,(L + M) < \infty$, for almost all $t \geq 0$. Hence, the operator function $A(t)$ from l_1 into itself is bounded for almost all $t \geq 0$ and is locally integrable on $[0; \infty)$. Therefore, we can consider (2) as a differential equation in the space l_1 with bounded operator.

It is well known, see [18], that the Cauchy problem for differential Equation (2) has unique solution for an arbitrary initial condition, and $\mathbf{p}(s) \in \Omega$ implies $\mathbf{p}(t) \in \Omega$ for $t \geq s \geq 0$.

We recall that a Markov chain $X(t)$ is called null-ergodic, if all $p_i(t) \to 0$ as $t \to \infty$ for any initial condition, and it is called weakly ergodic, if $\|\mathbf{p}^*(t) - \mathbf{p}^{**}(t)\| \to 0$ as $t \to \infty$ for any initial condition $\mathbf{p}^*(0), \mathbf{p}^{**}(0)$, see for instance [12,14].

2. Bounds on the Rate of Convergence for a Differential Equation

Consider a general (linear or nonlinear) differential equation

$$\frac{d\mathbf{y}}{dt} = H\mathbf{y}(t), \tag{3}$$

in the space of sequences l_1 under the assumption of existence and uniqueness of a solution for any initial condition $\mathbf{y}(0)$.

Let $H = (h_{ij})$, where all h_{ij} depend on some parameters (for instance, on $y, t, \ldots$).
We have

$$\frac{dy_i}{dt} = h_{ii}y_i + \sum_{j\neq i} h_{ij}y_j.$$

Now, if $y_i > 0$, then

$$\frac{d|y_i|}{dt} = \frac{dy_i}{dt} = h_{ii}|y_i| + \sum_{j\neq i} h_{ij}y_j \leq h_{ii}|y_i| + \sum_{j\neq i} |h_{ij}||y_j|,$$

and if $y_i < 0$, then we also have

$$\frac{d|y_i|}{dt} = -\frac{dy_i}{dt} = -h_{ii}y_i - \sum_{j\neq i} h_{ij}y_j \leq h_{ii}|y_i| + \sum_{j\neq i} |h_{ij}||y_j|.$$

Finally, using the continuity of all coordinates of the solution and the absolute convergence of all series, we obtain the estimate

$$\frac{d\|y\|}{dt} = \sum_i \frac{d|y_i|}{dt} \leq \sum_i \left(h_{ii}|y_i| + \sum_{j\neq i} |h_{ij}||y_j| \right) \leq \beta^* \|y\|, \tag{4}$$

where

$$\beta^* = \sup_i \left(h_{ii} + \sum_{j\neq i} |h_{ji}| \right). \tag{5}$$

Remark 1. *One can see that inequality (4) implies the bound*

$$\|y(t)\| \leq e^{\int_0^t \beta^*\, du} \|y(0)\|. \tag{6}$$

Moreover, if H is bounded for any t linear operator function from l_1 to itself, then $\beta^*(t) = \gamma(H(t))$ is the corresponding logarithmic norm of $H(t)$, see [12–15].

On the other hand, in a nonlinear situation, $\beta^*(t)$ yields a generalization of this notion.

3. Bounds on the Rate of Convergence for a Projection of Multidimensional BDP

Again, consider the forward Kolmogorov system (2) in the original vector form. Then, we have

$$\frac{dp_{\mathbf{m}}}{dt} = \sum_l \lambda_{l,\mathbf{m}-\mathbf{e}_l}(t) p_{\mathbf{m}-\mathbf{e}_l} + \tag{7}$$
$$\sum_l \mu_{l,\mathbf{m}+\mathbf{e}_l}(t) p_{\mathbf{m}+\mathbf{e}_l} - \sum_l \left(\lambda_{l,\mathbf{m}} + \mu_{l,\mathbf{m}}\right)(t) p_{\mathbf{m}},$$

for any $\mathbf{m}$.

In this section, we consider the one-dimensional process $X_j(t)$ for a fixed j. Denote $x_k(t) = \Pr\left(X_j(t) = k\right)$. Then, $x_k(t) = \sum_{\mathbf{m},m_j=k} p_{\mathbf{m}}(t)$. The process $X_j(t)$ has nonzero jump rates only for unit jumps (± 1), namely, if $X_j(t) = k$, then for small positive h only the jumps $X_j(t+h) = k \pm 1$ are possible with positive intensities, say $\tilde{\lambda}_k$ and $\tilde{\mu}_k$, respectively. Moreover, (7) implies the equalities

$$\tilde{\lambda}_k x_k(t) = \sum_{\mathbf{m},m_j=k} \lambda_{j,\mathbf{m}}(t) p_{\mathbf{m}}(t), \tag{8}$$

$$\tilde{\mu}_k x_k(t) = \sum_{\mathbf{m}, m_j = k} \mu_{j,\mathbf{m}}(t) p_{\mathbf{m}}(t), \tag{9}$$

and hence

$$\tilde{\lambda}_k = \frac{\sum_{\mathbf{m}, m_j = k} \lambda_{j,\mathbf{m}}(t) p_{\mathbf{m}}(t)}{\sum_{\mathbf{m}, m_j = k} p_{\mathbf{m}}(t)}, \tag{10}$$

and

$$\tilde{\mu}_k = \frac{\sum_{\mathbf{m}, m_j = k} \mu_{j,\mathbf{m}}(t) p_{\mathbf{m}}(t)}{\sum_{\mathbf{m}, m_j = k} p_{\mathbf{m}(t)}}. \tag{11}$$

Then, $X_j(t)$ is a (in general, non-Markovian) birth and death process with birth and death intensities $\tilde{\lambda}_k$ and $\tilde{\mu}_k$, respectively, (that is, it is a process with possible infinitesimal jumps ± 1, the intensities of which depend on t and on the initial condition for the original multidimensional process $X(t)$.)

For any fixed initial distribution $\mathbf{p}(0)$ and any $t > 0$, the probability distribution $\mathbf{p}(t)$ is unique. Hence, $\tilde{\lambda}_k = \lambda_k(\mathbf{p}(0), t)$ and $\tilde{\mu}_k = \mu_k(\mathbf{p}(0), t)$ uniquely define the system

$$\frac{d\mathbf{x}}{dt} = \tilde{A}\mathbf{x}(t), \tag{12}$$

for the vector $\mathbf{x}(t)$ of state probabilities of the projection $X_j(t)$ under the given initial condition. Obviously, different initial conditions specify different systems.

Here, $\tilde{A}$ is the corresponding three-diagonal "birth-death" transposed intensity matrix such that all off-diagonal elements are nonnegative and all column-wise sums are equal to zero.

Let for all $\mathbf{m}$ and any $t \geq 0$

$$l_j \leq \lambda_{j,\mathbf{m}}(t) \leq L_j, \quad m_j \leq \mu_{j,\mathbf{m}}(t) \leq M_j. \tag{13}$$

Then, from (10) and (11), we obtain the two-sided bounds

$$l_j \leq \tilde{\lambda}_k \leq L_j, \quad m_j \leq \tilde{\mu}_k \leq M_j, \tag{14}$$

for any k, any t, and any initial conditions.

1. Let the state space of $X_j(t)$ be countable and

$$M_j < l_j. \tag{15}$$

Put $\sigma = \sqrt{M_j/l_j} < 1$, $\delta_n = \sigma^n$, $n \geq 0$, $\tilde{x}_n = \delta_n x_n$, and $\tilde{\mathbf{x}} = (\tilde{x}_0, \tilde{x}_1, \dots)$. Let Λ be a diagonal matrix, $\Lambda = diag(\delta_0, \delta_1, \dots)$.

Note that in this situation, $\|\tilde{\mathbf{x}}(t)\| = \sum_{i=0}^{\infty} \delta_k x_k(t)$, and $\|\tilde{\mathbf{x}}(t)\| \to 0$ as $t \to \infty$ implying null ergodicity of $X_j(t)$, that is $p_k(t) = \Pr(X_j(t) = k) \to 0$ as $t \to \infty$ for any k.

Then,

$$\frac{d\tilde{\mathbf{x}}}{dt} = \Lambda \tilde{A} \Lambda^{-1} \tilde{\mathbf{x}}(t). \tag{16}$$

Then, we have

$$\tilde{\lambda}_k + \tilde{\mu}_k - \frac{\delta_{k+1}}{\delta_k}\tilde{\lambda}_k - \frac{\delta_{k-1}}{\delta_k}\tilde{\mu}_k \geq \tilde{\lambda}_k(1-\sigma) - \tilde{\mu}_k(1/\sigma - 1) \geq \tag{17}$$
$$l_j(1-\sigma) - M_j(1/\sigma - 1) = \left(\sqrt{l_j} - \sqrt{M_j}\right)^2 = \alpha^*,$$

implying the estimate

$$\frac{d\|\tilde{\mathbf{x}}\|}{dt} \leq \sup_k \left(\frac{\delta_{k+1}}{\delta_k}\tilde{\lambda}_k + \frac{\delta_{k-1}}{\delta_k}\tilde{\mu}_k - \tilde{\lambda}_k - \tilde{\mu}_k \right) \|\tilde{\mathbf{x}}\| =$$

$$- \inf_k \left(\tilde{\lambda}_k + \tilde{\mu}_k - \frac{\delta_{k+1}}{\delta_k}\tilde{\lambda}_k - \frac{\delta_{k-1}}{\delta_k}\tilde{\mu}_k \right) \|\tilde{\mathbf{x}}\| \leq -\alpha^* \|\tilde{\mathbf{x}}\|, \quad (18)$$

and the following statement.

Theorem 1. *Let (15) hold for some j. Then, $X_j(t)$ is null-ergodic and the following bounds hold:*

$$\|\tilde{\mathbf{x}}(t)\| \leq e^{-\alpha^* t} \|\tilde{\mathbf{x}}(0)\|, \quad (19)$$

and

$$\Pr\left(X_j(t) \leq n / X_j(0) = k\right) \leq \sigma^{k-n} \cdot e^{-\alpha^* t}. \quad (20)$$

Hence,

$$\Pr\left(X_j(t) > n / X_j(0) = k\right) > 1 - \sigma^{k-n} \cdot e^{-\alpha^* t}, \quad (21)$$

and $\Pr\left(X_j(t) > n / X_j(0) = k\right) \to 1$ *as* $t \to \infty$, *for any* $n,\ k$.

Remark 2. *It should be noted that the above requirements are imposed only on this one coordinate.*

2. Let

$$L_j < m_j, \quad \alpha_* = l_j + m_j - 2\sqrt{L_j M_j} > 0. \quad (22)$$

We have $\mathbf{x}(t) \in \Omega$ for any $t \geq 0$. Set $x_0(t) = 1 - \sum_{i \geq 1} x_i(t)$. Then, from (12), we obtain the system

$$\frac{d\mathbf{z}}{dt} = \tilde{B}\mathbf{z} + \tilde{\mathbf{f}}, \quad (23)$$

where $\mathbf{z} = (x_1, x_2, \dots)^\top$, $\tilde{\mathbf{f}} = \left(\tilde{\lambda}_0, 0, 0, \dots\right)^\top$, and the corresponding matrix $\tilde{B} = \left(\tilde{b}_{ij}\right)_{i,j=1}^{\infty}$, where $\tilde{b}_{ij} = \tilde{a}_{ij} - \tilde{a}_{i0}$ for the corresponding elements of the matrix $\tilde{A}$.

For the solutions of system (23), the rate of convergence is determined by the system

$$\frac{d\mathbf{w}}{dt} = \tilde{B}\mathbf{w}, \quad (24)$$

where all elements of $\tilde{B}$ depend on t and the initial condition of the original process.

Now, let $\beta = \sqrt{\frac{M_j}{L_j}} > 1$ in accordance with (22). Let $d_{k+1} = \beta^k$, $k \geq 0$. Denote by D, the upper triangular matrix

$$D = \begin{pmatrix} d_1 & d_1 & d_1 & \cdots \\ 0 & d_2 & d_2 & \cdots \\ 0 & 0 & d_3 & \cdots \\ & \ddots & \ddots & \ddots \end{pmatrix}. \quad (25)$$

Let $\tilde{\mathbf{w}} = D\mathbf{w}$. Then, the following bound holds:

$$\frac{d\|\tilde{\mathbf{w}}\|}{dt} \leq \sup_{i \geq 0} \left(\frac{d_{i+1}}{d_i}\tilde{\lambda}_{i+1} + \frac{d_{i-1}}{d_i}\tilde{\mu}_i - \left(\tilde{\lambda}_i + \tilde{\mu}_{i+1}\right) \right) \|\tilde{\mathbf{w}}\| =$$

$$- \inf_{i \geq 0} \left(\left(\tilde{\lambda}_i + \tilde{\mu}_{i+1} - \beta\tilde{\lambda}_{i+1} - \tilde{\mu}_i / \beta\right) \right) \|\tilde{\mathbf{w}}\| \leq -\alpha_* \|\tilde{\mathbf{w}}\|. \quad (26)$$

Note that $\|\tilde{\mathbf{w}}\| = \|D\mathbf{w}\| \geq \frac{1}{2}\|\mathbf{w}\|$, see detailed discussion in [15], therefore, if $\|\tilde{\mathbf{w}}(t)\| \to 0$ as $t \to \infty$, then $X_j(t)$ is weakly ergodic.

Thus, we obtain the following statement.

Theorem 2. *Let (22) hold for some j. Then, $X_j(t)$ is weakly ergodic and the following bound holds:*

$$\|D\mathbf{w}(t)\| \leq e^{-\alpha_* t}\|D\mathbf{w}(0)\|, \tag{27}$$

for any $t \geq 0$ and any corresponding initial conditions.

4. Example

Consider a simple two-dimensional BDP with finite state space $\{i,j\}$, $0 \leq i \leq 10$, $0 \leq j \leq 10$ and the following transition intensities:

(i) $\lambda_{1,i,0}(t) = \frac{i+1}{11}\lambda_1(t)$ from $(i,0)$ to $(i+1,0)$;
(ii) $\lambda_{1,i,j}(t) = \lambda_1(t)$ from (i,j) to $(i+1,j)$ if $j \neq 0$;
(iii) $\lambda_{2,i,j}(t) = \lambda_2(t)$ from (i,j) to $(i+1,j)$;
(iv) $\mu_{1,i,j}(t) = \mu_1(t)$ from (i,j) to $(i-1,j)$;
(v) $\mu_{2,i,j}(t) = \mu_2(t)$ from (i,j) to $(i,j-1)$;

where $\lambda_1(t) = 1 + \cos(2\pi t)$, $\lambda_2(t) = 5 + \sin(2\pi t)$, $\mu_1(t) = 11 + \sin(2\pi t)$, $\mu_2 = 3$.

Then, $\beta = \sqrt{\frac{M_1}{L_1}} = \sqrt{6}$, and Theorem 2 gives bound (27) with $\alpha_* = 10 - 4\sqrt{6}$.

We computed some important characteristics for the original process and its projection $X_1(t)$, namely:

Figures 1–3 show the behaviour of the state probabilities for $X_1(t)$, namely $\Pr(X_1(t) = 0)$, $\Pr(X_1(t) = 1)$, and $\Pr(X_1(t) = 2)$ under two initial conditions for the original BDP:

(i) $p_{i,j}(0) = \frac{1}{121}$, for any i,j (blue); and
(i) $p_{0,0}(0) = \frac{109}{121}$, $p_{i,j}(0) = \frac{1}{1210}$, for any i,j such that $i+j > 0$ (green).

Note that the corresponding initial conditions for the projection are $\mathbf{x}(0) = \left(\frac{1}{11}, \ldots, \frac{1}{11}\right)^T$, and $\mathbf{x}(0) = (10/11, 1/110, \ldots, 1/110)^T$.

These Figures illustrate the rate of convergence in a weak ergodic situation.

Figures 4 and 5 show the 'birth intensities' $\tilde{\lambda}_0$ and $\tilde{\lambda}_1$ for $X_1(t)$ under the same initial conditions.

Note that all the quantities are found by numerically solving the Cauchy problem for the forward Kolmogorov system (2) and the corresponding system (12) for its projection on the corresponding interval.

As can be seen from Theorem 2 and the figures below, to construct all the characteristics of interest with good accuracy, it suffices to carry out the numerical solution on the interval $[0,5]$.

As was already noted, the projection of the original process is not a Markov process, and all probabilistic characteristics depend on the initial conditions of the original process.

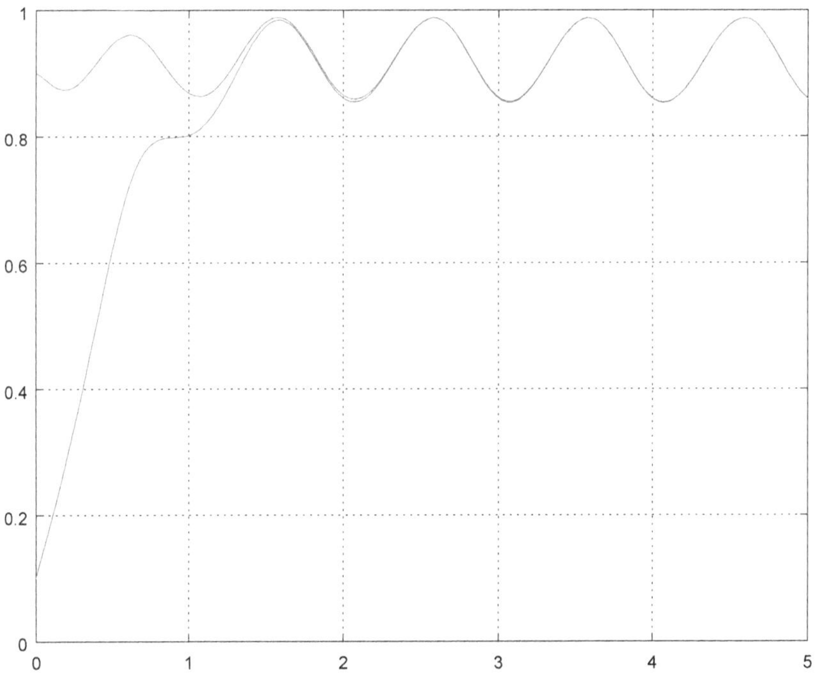

Figure 1. $\Pr(X_1(t) = 0)$ under initial conditions (i) and (ii).

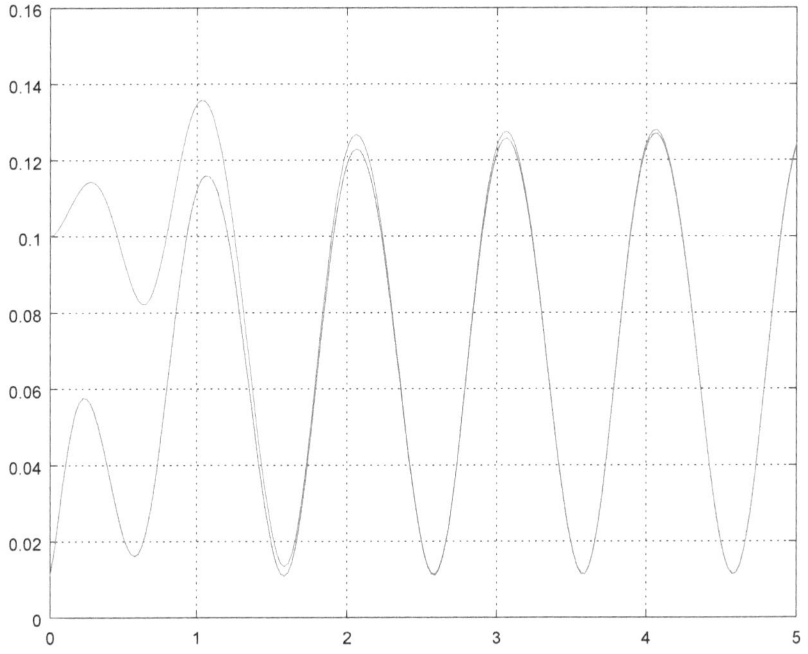

Figure 2. $\Pr(X_1(t) = 1)$ under initial conditions (i) and (ii).

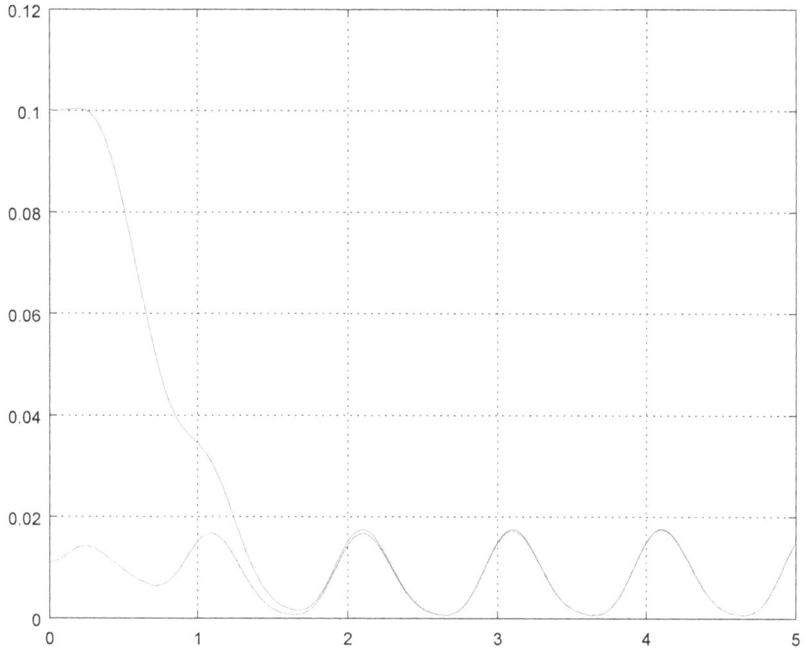

Figure 3. $\Pr(X_1(t) = 2)$ under initial conditions (i) and (ii).

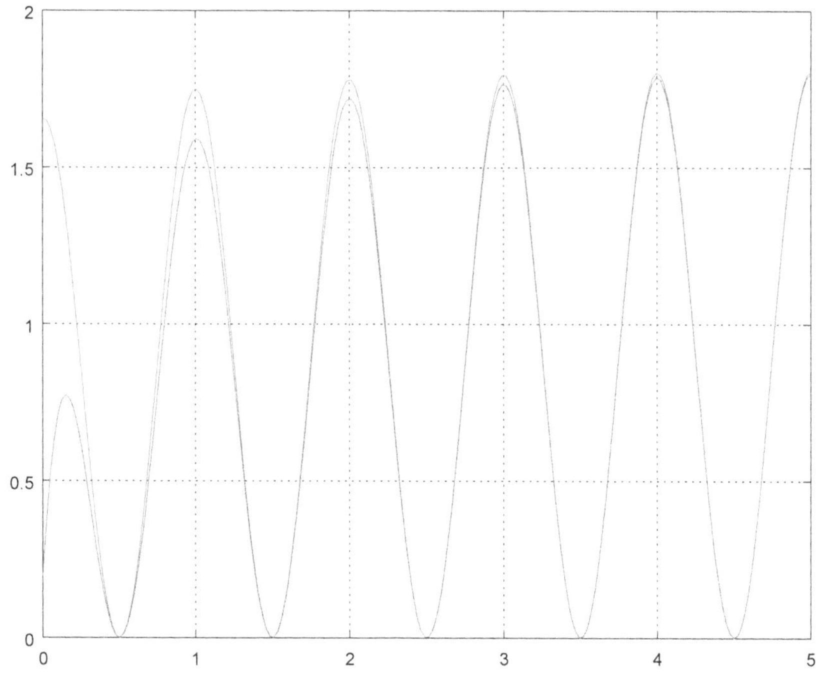

Figure 4. $\tilde{\lambda}_0$ under initial conditions (i) and (ii).

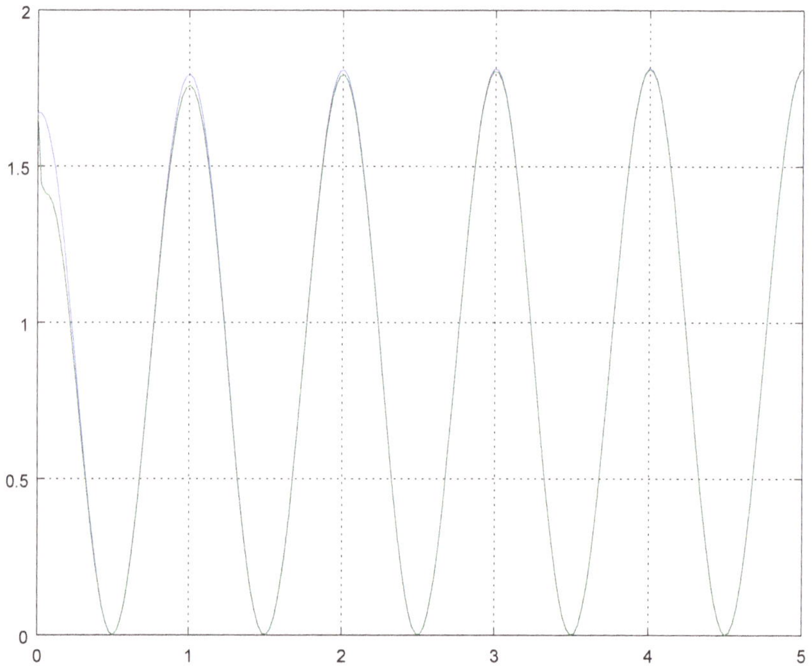

Figure 5. $\tilde{\lambda}_1$ under initial conditions (i) and (ii).

It can be seen that these characteristics present comprehensive information concerning the behavior of the projection of the original process.

5. Conclusions

In the paper, some estimates of the rate of convergence were discussed for one-dimensional projections of multidimensional inhomogeneous birth and death processes. Some specific queueing models were considered. The applied approach allows one to use an analogue of the logarithmic norm of an operator function for a nonlinear system of differential equations, as was shown in Section 2. In addition, similar estimates can be obtained for other one-dimensional processes related to the original one. For example, the total number of "particles" of all types can be studied. Moreover, it is possible to study multidimensional processes with possible transformations of particles from one type to another. Such processes play a very important role in stochastic models of epidemics, see, for example, References [19,20] and the references therein.

Author Contributions: Conceptualization, A.Z., Y.S.; Methodology, A.Z., Y.S.; Software, Y.S.; Validation, A.Z., Y.S., K.K., V.K.; Investigation, A.Z., V.K.; Writing—Original Draft Preparation, A.Z., K.K., Y.S., V.K.; Writing—Review and Editing, A.Z., K.K., V.K.; Supervision, A.Z.; Project Administration, K.K.

Funding: The work of A.Z. and V.K. was supported by the Russian Science Foundation under grant 18-11-00155.

Acknowledgments: The authors thank the referees for useful comments that improved the paper.

Conflicts of Interest: The authors declare no conflict of interest.

References

1. Michaelides, M.; Hillston, J.; Sanguinetti, G. Geometric fluid approximation for general continuous-time Markov chains. *arXiv* **2019**, arXiv:1901.11417.
2. Goswami, B.B.; Khouider, B.; Phani, R.; Mukhopadhyay, P.; Majda, A. Improving synoptic and intraseasonal variability in CFSv2 via stochastic representation of organized convection. *Geophys. Res. Lett.* **2017**, *44*, 1104–1113. [CrossRef]
3. Goswami, B.B.; Khouider, B.; Phani, R.; Mukhopadhyay, P.; Majda, A.J. The Stochastic Multi-cloud Model (SMCM) Convective Parameterization in the CFSv2: Scopes and Opportunities. In *Current Trends in the Representation of Physical Processes in Weather and Climate Models*; Springer: Singapore, 2019; pp. 157–181.
4. Deng, Q.; Khouider, B.; Majda, A.J. The MJO in a coarse-resolution GCM with a stochastic multicloud parameterization. *J. Atmos. Sci.* **2015**, *72*, 55–74. [CrossRef]
5. Marsan, M.A.; Marano, S.; Mastroianni, C.; Meo, M. Performance analysis of cellular mobile communication networks supporting multimedia services. *Mob. Netw. Appl.* **2000**, *5*, 167–177. [CrossRef]
6. Stevens-Navarro, E.; Mohsenian-Rad, A.H.; Wong, V.W. Connection admission control for multiservice integrated cellular/WLAN system. *IEEE Trans. Veh. Technol.* **2008**, *57*, 3789–3800. [CrossRef]
7. Jonckheere, M.; Lypez, S. Large deviations for the stationary measure of networks under proportional fair allocations. *Math. Oper. Res.* **2013**, *39*, 418–431. [CrossRef]
8. Jonckheere, M.; Shneer, S. Stability of multi-dimensional birth-and-death processes with state-dependent 0-homogeneous jumps. *Adv. Appl. Probab.* **2014**, *46*, 59–75. [CrossRef]
9. Lee, C. On moment stability properties for a class of state-dependent stochastic networks. *J. Korean Stat. Soc.* **2011**, *40*, 325–336. [CrossRef]
10. Mather, W.H.; Hasty, J.; Tsimring, L.S.; Williams, R.J. Factorized time-dependent distributions for certain multiclass queueing networks and an application to enzymatic processing networks. *Queueing Syst.* **2011**, *69*, 313–328. [CrossRef] [PubMed]
11. Tsitsiashvili, G.S.; Osipova, M.A.; Koliev, N.V.; Baum, D. A product theorem for Markov chains with application to PF-queueing networks. *Ann. Oper. Res.* **2002**, *113*, 141–154. [CrossRef]
12. Granovsky, B.; Zeifman, A. Nonstationary queues: Estimation of the rate of convergence. *Queueing Syst.* **2004**, *46*, 363–388. [CrossRef]
13. Zeifman, A.I. On the estimation of probabilities for birth and death processes. *J. Appl. Probab.* **1995**, *32*, 623–634. [CrossRef] [PubMed]
14. Zeifman, A.I. Upper and lower bounds on the rate of convergence for nonhomogeneous birth and death processes. *Stoch. Proc. Appl.* **1995**, *59*, 157–173. [CrossRef]
15. Zeifman, A.; Leorato, S.; Orsingher, E.; Satin, Y.; Shilova, G. Some universal limits for nonhomogeneous birth and death processes. *Queueing Syst.* **2006**, *52*, 139–151. [CrossRef]
16. Zeifman, A.; Sipin, A.; Korolev, V.; Bening, V. Estimates of some characteristics of multidimensional birth-and-death processes. *Dokl. Math.* **2015**, *92*, 695–697. [CrossRef]
17. Zeifman, A.I.; Sipin, A.S.; Korotysheva, A.V.; Panfilova, T.L.; Satin, Y.A.; Shilova, G.N.; Korolev, V.Y. Estimation of Probabilities for Multidimensional Birth-Death Processes. *J. Math. Sci.* **2016**, *218*, 238–244. [CrossRef]
18. Daleckij, J.L.; Krein, M.G. Stability of solutions of differential equations in Banach space. *Am. Math. Soc. Transl.* **2002**, *43*, 1–386.
19. Britton, T. Stochastic epidemic models: A survey. *Math. Biosci.* **2010**, *225*, 24–35. [CrossRef] [PubMed]
20. Zeifman, A.I. Processes of birth and death and simple stochastic epidemic models. *Autom. Remote Control* **1985**, *6*, 128–135.

mathematics

MDPI

Article

Estimating the Expected Discounted Penalty Function in a Compound Poisson Insurance Risk Model with Mixed Premium Income

Yunyun Wang [1], Wenguang Yu [2,*], Yujuan Huang [3], Xinliang Yu [2] and Hongli Fan [2]

1 College of Mathematics and Statistics, Chongqing University, Chongqing 401331, China; yyw@cqu.edu.cn
2 School of Insurance, Shandong University of Finance and Economics, Jinan 250014, China; yuxinliang19870906@126.com (X.Y.); 20177686@sdufe.edu.cn (H.F.)
3 School of Science, Shandong Jiaotong University, Jinan 250357, China; yujuanh518@163.com
* Correspondence: yuwg@sdufe.edu.cn; Tel.: +86-152-5416-5648

Received: 24 February 2019; Accepted: 18 March 2019; Published: 26 March 2019

Abstract: In this paper, we consider an insurance risk model with mixed premium income, in which both constant premium income and stochastic premium income are considered. We assume that the stochastic premium income process follows a compound Poisson process and the premium sizes are exponentially distributed. A new method for estimating the expected discounted penalty function by Fourier-cosine series expansion is proposed. We show that the estimation is easily computed, and it has a fast convergence rate. Some numerical examples are also provided to show the good properties of the estimation when the sample size is finite.

Keywords: compound poisson insurance risk model; expected discounted penalty function; estimation; Fourier transform; Fourier-cosine series

1. Introduction

In this paper, we consider an insurance risk model with mixed premium income defined by

$$U(t) = u + ct + \sum_{i=1}^{M(t)} Y_i - \sum_{j=1}^{N(t)} X_j, \qquad t \geq 0, \tag{1}$$

where $u \geq 0$ is the initial surplus, $c \geq 0$ is the constant premium rate, and $U(t)$ denotes the surplus level of an insurance company at time t. The premium number process $M(t)$ and the claim number process $N(t)$ are homogenous Poisson processes with intensity $\mu > 0$ and $\lambda > 0$, respectively. The individual claim sizes, $X_1, X_2, \ldots$, are positive independent and identically distributed (i.i.d) continuous random variables with density function f. The premium sizes, $Y_1, Y_2, \ldots$ are positive i.i.d. continuous random variables with exponential distribution function $g(y) = \beta e^{-\beta y}, y, \beta > 0$, where β is unknown. Throughout this paper, we assume that $\{M(t)\}_{t\geq 0}$, $\{N(t)\}_{t\geq 0}$, $\{X_i\}_{i\geq 1}$ and $\{Y_j\}_{j\geq 1}$ are mutually independent.

Whenever the surplus process becomes negative, we say that ruin occurs. The ruin time is defined by

$$\tau = \inf\{t \geq 0 : U(t) < 0\},$$

if for all $t \geq 0$, $U(t) > 0$, we denote $\tau = \infty$. To avoid that ruin is a certain event, suppose that the following condition holds throughout this paper.

Net profit condition

$$c + \frac{\mu}{\beta} - \lambda \mathbb{E}[X] > 0.$$

The above condition guarantees that the expectation of the surplus process will always be positive at any time $t > 0$.

Let $\delta \geq 0$ be the interest force, defining the expected discounted penalty function by

$$\Phi(u) = \mathbb{E}[e^{-\delta\tau} \mathrm{w}(U(\tau-), |U(\tau)|)\mathbf{1}_{(\tau<\infty)}|U(0) = u], \qquad u \geq 0, \tag{2}$$

where, $\mathrm{w} : [0, \infty) \times [0, \infty) \rightarrow [0, \infty)$ is a measurable penalty function of the surplus before ruin and the deficit at ruin, and $\mathbf{1}_A$ is an indicator function of the event A. This function was first introduced by Gerber and Shiu [1], and is called Gerber-Shiu function in the literature. It has become an important and standard risk measure in ruin theory since various quantities of interests in ruin theory can be obtained for different values of the discount factor δ and different penalty functions w. For recent research progress on the Gerber-Shiu function, we can refer to work by Lin et al. [2], Yuen et al. [3,4], Zhao and Yin [5], Chi [6], Yin and Wang [7], Chi and Lin [8], Shen et al. [9], Yin and Yuen [10], Zhao and Yao [11], Li et al. [12,13], Dong et al. [14], and Wang and Zhang [15], among others.

For the classical insurance risk model, the premium rate is usually set to a constant. Extensive research has been done on the ruin measures of the model and its extension under the constant premium rate, such as ruin probability, Gerber-Shiu function (see, e.g., Gerber and Shiu [1], Willmot and Dickson [16], Wang et al. [17], Yin and Yuen [18], Dong and Yin [19], Yu [20], Yin et al. [21], Zeng et al. [22], Zhao et al. [23] and Yu et al. [24]). However, in the actual insurance business, the premium income of an insurance company, especially a small one, is sometimes volatile and random. In view of this situation, Boucherie et al. [25] first extended the classical risk model to the risk model with stochastic premium income by replacing the constant premium income with a compound Poisson process. Since then, the risk model with stochastic premium income has been widely studied by many scholars. The non-ruin probability and ruin probability were, respectively, studied by Boikov [26] and Temnov [27]. Bao [28], Bao and Ye [29] and Yang and Zhang [30] studied the Gerber-Shiu function in the classical risk model, delayed renewal risk model and Sparre Andersen risk model by assuming the premium process are Poisson process, respectively. Supposing premium and claim process follow compound Poisson processes, a defective renewal equation satisfied by the Gerber-Shiu function was established by Labbé and Sendova [31]. Yang and Zhang [32] further studied the above model by assuming that there exists a specific dependence structure among the claim sizes, inter-claim times and premium sizes and the individual premium sizes are exponentially distributed. Zhao and Yin [33] considered a renewal risk model where the premiums follow a compound Poisson risk process and the claims follow a generalized Erlang(n) process. The Laplace transform and a defective renewal equation for the Gerber-Shiu function are derived when the premium sizes are exponentially distributed. For more study on the risk model with stochastic premium income, the interested readers are referred to the work by Xu et al. [34], Yu [35,36], Zhou et al. [37], Gao and Wu [38], Zhou et al. [39], Deng et al. [40] and Zeng et al. [41] and the references therein.

The above-mentioned papers assume that some probability characteristics of the surplus process are known, however, which are usually unknown for an insurance company. Actually, some data information on the surplus process, such as surplus levels, claim and premium numbers, and claim and premium sizes, can be obtained. Thus, some semi-parametric and parametric estimations of the ruin probability and Gerber-Shiu function for different risk models are presented in recent literature (e.g., Politis [42], Wang and Yin [43], Yuen and Yin [44], Zhang et al. [45], Wang et al. [46], Zhang [47], Zhang and Yang [48,49] and Shimizu and Zhang [50], Peng and Wang [51,52], Yang et al. [53,54]). Besides, some effective methods, such as Laplace transform, Fourier-Sinc series expansion and Laguerre series expansion, have been applied in estimating the ruin probability and Gerber-Shiu function based on the observed data information. We refer the interested readers to Shimizu [55], Zhang [56], Zhang and Su [57,58] and Su et al. [59], among others.

In recent years, Fang and Oosterlee [60] proposed a novel method for pricing European options by Fourier-cosine series expansion, which is also called the COS method in the literature. It can be easily applied to approximate an integrable function as long as the corresponding Fourier transform has closed-form expression. Now, the COS method has been widely used for pricing options and other financial derivatives (see, e.g., Fang and Oosterlee [61] and Zhang and Oosterlee [62]). Recently, the COS method has also been used in risk theory to compute and estimate some risk measures by some actuarial researchers. For example, Chau et al. [63,64] used the COS method to compute the ruin probability and Gerber-Shiu function in the Lévy risk models. Zhang [65] applied the COS method to compute the density of the time to ruin in the classical risk model. Yang et al. [66] used a two-dimensional COS method to estimate the discounted density function of the deficit at ruin in the classical risk model with stochastic income where the premiums are only described by a compound Poisson process and the premium sizes are exponential distributed. Inspired by Yang et al. [66], in this paper, we extend the risk model of Yang et al. [66] by considering that there is also constant rate premium income process. Then, we use the COS method to estimate the expected discounted penalty function in this risk model with mixed premiums income process. The remainder of this paper is organized as follows. In Section 2, we first briefly introduce the Fourier-cosine series expansion method, and then derive the Fourier transform of the expected discounted penalty function. In Section 3, an estimator of the expected discounted penalty function is proposed by the observed sample of the surplus process. The consistent property is studied in Section 4 under large sample size setting. Finally, in Section 5, we present some simulation results to show that the estimator behaves well under finite sample size setting.

2. Preliminaries on Expected Discounted Penalty Function

2.1. Fourier-Cosine Series Expansion

In this subsection, we present some known results on the Fourier-cosine series expansion method. Let $L^1(\mathbb{R}_+)$ denote the class of integrable functions on the positive axis, and let $\mathcal{F}f$ and $\mathcal{L}f$ denote the Fourier transform and Laplace transform of a $f \in L^1(\mathbb{R}_+)$, respectively. For any complex number z, we denote its real part and imaginary part by $Re(z)$ and $Im(z)$, respectively.

It is known that an integrable function f defined on $[a_1, a_2]$ has the following cosine series expansion,

$$f(x) = \sum_{k=0}^{\infty}{}' \left\{ \frac{2}{a_2 - a_1} \int_{a_1}^{a_2} f(x) \cos\left(k\pi \frac{x - a_1}{a_2 - a_1}\right) dx \right\} \cos\left(k\pi \frac{x - a_1}{a_2 - a_1}\right), \tag{3}$$

where $\sum'$ means the first term of the summation has half weight. For a function $f \in L^1(\mathbb{R}_+)$, we introduce an auxiliary function

$$f_a(x) = f(x) \cdot \mathbf{1}_{(0 \le x \le a)}, \qquad a > 0,$$

then f_a has a finite domain $[0, a]$ and $f(x) = f_a(x)$ when $x \in [0, a]$. By Equation (3), we have

$$f(x) = f_a(x) = \sum_{k=0}^{\infty}{}' \left\{ \frac{2}{a} \int_0^a f(x) \cos\left(k\pi \frac{x}{a}\right) dx \right\} \cos\left(k\pi \frac{x}{a}\right), \qquad 0 \le x \le a. \tag{4}$$

For large a, we have

$$\frac{2}{a} \int_0^a f(x) \cos\left(k\pi \frac{x}{a}\right) dx = \frac{2}{a} Re\left\{ \int_0^a f(x) e^{i\frac{k\pi}{a}x} dx \right\} \approx \frac{2}{a} Re\left\{ \mathcal{F}f\left(\frac{k\pi}{a}\right) \right\},$$

thus

$$f(x) \approx \sum_{k=0}^{\infty}{}' \frac{2}{a} Re\left\{ \mathcal{F}f\left(\frac{k\pi}{a}\right) \right\} \cos\left(k\pi \frac{x}{a}\right), \qquad 0 \le x \le a.$$

Furthermore, for a large integer K, the above summation can be truncated as follows:

$$f(x) \approx \sum_{k=0}^{K-1}{}' \frac{2}{a} Re\left\{\mathcal{F}f\left(\frac{k\pi}{a}\right)\right\} \cos\left(k\pi\frac{x}{a}\right), \qquad 0 \leq x \leq a. \tag{5}$$

As for the expected discounted penalty function, it follows from Equation (5) that for $0 \leq u \leq a$,

$$\Phi(u) \approx \Phi_{K,a}(u) := \sum_{k=0}^{K-1}{}' \frac{2}{a} Re\left\{\mathcal{F}\Phi\left(\frac{k\pi}{a}\right)\right\} \cos\left(k\pi\frac{x}{a}\right). \tag{6}$$

It can be easily seen that the key of approximating the expected discounted penalty function by Fourier-cosine series expansion method is to calculate the Fourier transform $\mathcal{F}\Phi(s)$ for $s \in \{\frac{k\pi}{a} : k = 0, 1, \ldots, K-1\}$. Therefore, we derive a specific expression of Fourier transform $\mathcal{F}\Phi(\cdot)$ in the next section.

2.2. The Fourier Transform of Expected Discounted Penalty Function

In this subsection, we derive the Fourier transform of the expected discounted penalty function. For convenience, we introduce the Dickson-Hipp operator T_s (see, e.g., Dickson and Hipp [67] and Li and Garrido [68]), which for any integrable function f on $(0, \infty)$ and any complex number s with $Re(s) \geq 0$ is defined as

$$T_s f(y) = \int_y^\infty e^{-s(x-y)} f(x)dx = \int_0^\infty e^{-sx} f(x+y)dx, \qquad y \geq 0.$$

The Dickson-Hipp operator has been widely used in ruin theory to simplify the expression of ruin related functions. For properties on this operator, we refer the interested readers to Li and Garrido [68].

We call the following equation (with respect to s) the Lundberg's fundamental equation for the risk model defined in Equation (1),

$$(\beta - s)\left[s - \frac{\delta + \mu + \lambda}{c} + \frac{\mu}{c} \cdot \frac{\beta}{\beta - s} + \frac{\lambda}{c}\mathcal{L}f(s)\right] = 0. \tag{7}$$

By Lemma 2.1 in Zhao and Yin [33], we known that for $\delta \geq 0$ and $c > 0$, the above equation has exactly two nonnegative roots, denoted as ρ_1, ρ_2 in this literature, and one of the roots ρ_1 equals 0 when $\delta = 0$.

Remark 1. *Let*

$$\chi(s) = \delta + \mu + \lambda - cs - \lambda\mathcal{L}f(s) + \frac{\mu\beta}{s - \beta}, \qquad s \geq 0.$$

It is obvious that $\chi(s) = 0$ *also has exactly two nonnegative roots* ρ_1, ρ_2. *Under net profit condition, we have*

$$\begin{aligned} \chi'(s) &= -c + \lambda \int_0^\infty x e^{-sx} f(x)dx - \frac{\mu\beta}{(s-\beta)^2} \\ &\leq -c - \lambda\mathbb{E}[X] - \frac{\mu}{\beta} < 0, \qquad 0 \leq s < \beta, \end{aligned}$$

which implies that $\chi(s)$ *is decreasing on* $[0, \beta)$. *In addition,* $\chi(s)$ *is continuous on* $[0, \beta)$, $\chi(0) = \delta \geq 0$ *and* $\chi(\beta - 0) = -\infty$. *Therefore, we conclude that the root* ρ_1 *of equation* $\chi(s) = 0$ *is located in the interval* $[0, \beta)$, *and, in particular,* $\rho_1 = 0$ *when* $\delta = 0$; *then,* ρ_2 *is located in the interval* (β, ∞).

By Theorem 3.1 in Zhao and Yin [33], when stochastic premium income follows a compound Poisson process, we known that the expected discounted penalty function satisfies the following renewal equation

$$\Phi(u) = \int_0^u \Phi(u-x)H(x)dx + K(u), \tag{8}$$

where

$$\begin{aligned} H(x) &= \frac{\lambda}{c}\left[\frac{\beta-\rho_1}{\rho_1-\rho_2}T_{\rho_1}f(x) + \frac{\beta-\rho_2}{\rho_2-\rho_1}T_{\rho_2}f(x)\right], \\ K(u) &= \frac{\lambda}{c}\left[\frac{\beta-\rho_1}{\rho_1-\rho_2}T_{\rho_1}\omega(u) + \frac{\beta-\rho_2}{\rho_2-\rho_1}T_{\rho_2}\omega(u)\right], \\ \omega(u) &= \int_u^\infty \mathrm{w}(u, x-u)f(x)dx. \end{aligned}$$

To derive the Fourier transform of $\Phi(u)$, we assume the following condition holds true.

Condition 1. The penalty function w satisfies

$$\int_0^\infty \int_0^\infty (1+x)\mathrm{w}(x,y)f(x+y)dydx < \infty.$$

This condition guarantees $\Phi \in L^1(\mathbb{R}_+)$.

Now, we compute the Fourier transform of the expected discounted penalty function. Applying the Fourier transform on both sides of Equation (8) gives

$$\begin{aligned} \mathcal{F}\Phi(s) &= \int_0^\infty e^{isu}\Phi(u)du \\ &= \int_0^\infty e^{isu}\int_0^u \Phi(u-x)H(x)dxdu + \int_0^\infty e^{isu}K(u)du \\ &= \int_0^\infty \int_0^\infty e^{is(z+x)}\Phi(z)dzdx + \mathcal{F}K(s) \\ &= \mathcal{F}\Phi(s)\mathcal{F}H(s) + \mathcal{F}K(s), \end{aligned}$$

leading to

$$\mathcal{F}\Phi(s) = \frac{\mathcal{F}K(s)}{1-\mathcal{F}H(s)}. \tag{9}$$

For the Fourier transform $\mathcal{F}H(s)$, we have

$$\mathcal{F}H(s) = \int_0^\infty e^{isu}H(x)dx = \frac{\lambda}{c}\left[\frac{\beta-\rho_1}{\rho_1-\rho_2}\int_0^\infty e^{isx}T_{\rho_1}f(x)dx + \frac{\beta-\rho_2}{\rho_2-\rho_1}\int_0^\infty e^{isx}T_{\rho_2}f(x)dx\right],$$

where

$$\begin{aligned} \int_0^\infty e^{isu}T_{\rho_j}f(x)dx &= \int_0^\infty e^{isx}\int_x^\infty e^{-\rho_j(y-x)}f(y)dydx \\ &= \int_0^\infty \int_x^\infty e^{(is+\rho_j)x}\cdot e^{-\rho_j y}f(y)dydx \\ &= \frac{1}{\rho_j+is}\left[\int_0^\infty e^{isy}f(y)dy - \int_0^\infty e^{-\rho_i y}f(y)dy\right] \\ &= \frac{1}{\rho_j+is}[\mathcal{F}f(s) - \mathcal{L}f(\rho_j)], \qquad j=1,2. \end{aligned}$$

Then, we obtain

$$\mathcal{F}H(s) = \frac{\lambda}{c(\rho_1 - \rho_2)} \left[\frac{(\beta - \rho_2)[\mathcal{F}f(s) - \mathcal{L}f(\rho_2)]}{\rho_2 + is} - \frac{(\beta - \rho_1)[\mathcal{F}f(s) - \mathcal{L}f(\rho_1)]}{\rho_1 + is} \right]. \tag{10}$$

Similarly, for $\mathcal{F}K(s)$, we obtain

$$\mathcal{F}K(s) = \frac{\lambda}{c(\rho_1 - \rho_2)} \left[\frac{(\beta - \rho_2)[\mathcal{F}\omega(s) - \mathcal{L}\omega(\rho_2)]}{\rho_2 + is} - \frac{(\beta - \rho_1)[\mathcal{F}\omega(s) - \mathcal{L}\omega(\rho_1)]}{\rho_1 + is} \right]. \tag{11}$$

3. Estimation Procedure

In this section, we study how to estimate the expected discounted penalty function by Fourier-cosine series expansion based on the discretely observed information of the surplus process, the aggregate claims and premiums processes. According to Equation (6), we know that the key is to construct an estimation of $\mathcal{F}\Phi(s)$ based on these discrete information.

Assume that we can observe the surplus process over a long time interval $[0, T]$. Let $\Delta > 0$ be a fixed inter-observation interval. Furthermore, without loss of generality, we assume T/Δ is an integer denoted as n.

Suppose that the insurer can get the following datasets.

(1) Dataset of surplus levels:

$$\{U_{j\Delta} : j = 0, 1, 2, \ldots, n\},$$

where $U_{j\Delta}$ is the observed surplus level at time $t = j\Delta$.

(2) Dataset of claim numbers and claim sizes:

$$\{N_{j\Delta}, X_1, X_2, \ldots, X_{N_{j\Delta}}\}, \qquad j = 1, \ldots, n,$$

where $N_{j\Delta}$ is the total claim number up to time $t = j\Delta$.

(3) Dataset of premium numbers and claim sizes:

$$\{M_{j\Delta}, Y_1, Y_2, \ldots, Y_{N_{j\Delta}}\}, \qquad j = 1, \ldots, n,$$

where $M_{j\Delta}$ is the total premium number up to time $t = j\Delta$.

Next, we study how to estimate the Fourier transform $\mathcal{F}\Phi(s)$ based on the above datasets. To estimate $\mathcal{F}\Phi(s)$ by Equations (9)–(11), we should first estimate the following characteristics:

$$\lambda,\ \mu,\ \beta,\ \mathcal{F}f,\ \mathcal{F}\omega,\ \rho_j,\ \mathcal{L}f(\rho_j),\ \mathcal{L}\omega(\rho_j), \qquad j = 1, 2.$$

First, we can estimate $\mathcal{F}f(s)$ by the empirical characteristic function

$$\widehat{\mathcal{F}f}(s) = \frac{1}{N_T} \sum_{j=1}^{N_T} e^{isX_j}.$$

Similarly, $\mathcal{L}f(s)$ can be estimated by

$$\widehat{\mathcal{L}f}(s) = \frac{1}{N_T} \sum_{j=1}^{N_T} e^{-sX_j}.$$

Next, for the function $\omega(u)$, we have

$$\mathcal{F}\omega(s) = E\left(\int_0^X e^{isu} w(u, X-u)du\right),$$
$$\mathcal{L}\omega(s) = E\left(\int_0^X e^{-su} w(u, X-u)du\right).$$

Then, $\mathcal{F}\omega(s)$ and $\mathcal{L}\omega(s)$ can be respectively estimated by

$$\widehat{\mathcal{F}\omega}(s) = \frac{1}{N_T}\sum_{j=1}^{N_T}\int_0^{X_j} e^{isu} w(u, X_j-u)du,$$
$$\widehat{\mathcal{L}\omega}(s) = \frac{1}{N_T}\sum_{j=1}^{N_T}\int_0^{X_j} e^{-su} w(u, X_j-u)du.$$

According to the property of Poisson distribution, λ and μ can be estimated by

$$\widehat{\lambda} = \frac{1}{T}N_T, \qquad \widehat{\mu} = \frac{1}{T}M_T.$$

It is easily seen that

$$\widehat{\lambda} - \lambda = O_p(T^{-\frac{1}{2}}), \qquad \widehat{\mu} - \mu = O_p(T^{-\frac{1}{2}}).$$

Since the premium size Y follows exponential distribution with parameter β, we have $\mathbb{E}[Y] = \frac{1}{\beta}$; then, we can estimate β by

$$\widehat{\beta} = \frac{1}{\frac{1}{M_T}\sum_{i=1}^{M_T} Y_i}.$$

It is also easily seen that $\widehat{\beta} - \beta = O_p(T^{-\frac{1}{2}})$. The estimation of ρ_j, $j = 1, 2$, denoted as $\widehat{\rho_j}$, $j = 1, 2$, are defined to be the nonnegative roots of the following equation (in s),

$$\delta + \widehat{\mu} + \widehat{\lambda} - cs - \widehat{\lambda}\widehat{\mathcal{L}f}(s) + \frac{\widehat{\mu}\widehat{\beta}}{s - \widehat{\beta}} = 0.$$

Furthermore, we can, respectively, estimate $\mathcal{L}f(\rho_j)$, $\mathcal{L}\omega(\rho_j)$, $j = 1, 2$ by $\widehat{\mathcal{L}f}(\widehat{\rho_j})$, $\widehat{\mathcal{L}\omega}(\widehat{\rho_j})$, $j = 1, 2$.

Remark 2. *Let*

$$\widehat{\chi}(s) = \delta + \widehat{\mu} + \widehat{\lambda} - cs - \widehat{\lambda}\widehat{\mathcal{L}f}(s) + \frac{\widehat{\mu}\widehat{\beta}}{s - \widehat{\beta}}, \qquad s \geq 0.$$

By the similar arguments of the Lunderg's fundamental equation in Zhao and Yin [33], we can obtain that equation $\widehat{\chi}(s) = 0$ has exactly two nonnegative roots, denoted as $\widehat{\rho}_1$, $\widehat{\rho}_2$ in this literature. It is clear that $\widehat{\chi}(0) = \delta \geq 0$ and $\widehat{\chi}(\widehat{\beta} - 0) = -\infty$. Under net profit condition, we have

$$\begin{aligned}\widehat{\chi}'(s) &= -c + \widehat{\lambda}\frac{1}{N_T}\sum_{j=1}^{N_T} X_j e^{-sX_j} - \frac{\widehat{\mu}\widehat{\beta}}{(s-\widehat{\beta})^2} \\ &\leq -c + \widehat{\lambda}\frac{1}{N_T}\sum_{j=1}^{N_T} X_j - \frac{\widehat{\mu}}{\widehat{\beta}} \xrightarrow{a.s.} -c + \lambda\mathbb{E}[X] - \frac{\mu}{\beta} < 0, \qquad 0 \leq s < \widehat{\beta}.\end{aligned}$$

Therefore, we obtain that the probability that $\widehat{\chi}(s) = 0$ has unique root $\widehat{\rho_1}$ on $[0, \widehat{\beta})$ tends to one as $T \to \infty$. We set $\widehat{\rho_1} = 0$ when $\delta = 0$ since $\rho_1 = 0$ in this case. Then, the other root $\widehat{\rho_2}$ is located in the interval $(\widehat{\beta}, \infty)$ with probability tends to one as $T \to \infty$.

Proposition 1. *Suppose that net profit condition holds true, then we have $\widehat{\rho_1} \xrightarrow{p} \rho_1$ and $\widehat{\rho_2} \xrightarrow{p} \rho_2$.*

Proof of Proposition 1. By Remark 2, when $0 \le s < \widehat{\beta}$, $\widehat{\chi}(s)$ is continuous and the probability that $\widehat{\chi}(s) = 0$ has unique nonnegative root tends to one as $T \to \infty$; and when $s > \widehat{\beta}$, $\widehat{\chi}(s)$ is continuous and the probability that $\widehat{\chi}(s) = 0$ has unique positive root tends to one as $T \to \infty$. By Remark 1, it is easily seen that, for every $\varepsilon > 0$, $\chi(\rho_1 - \varepsilon) > 0 > \chi(\rho_1 + \varepsilon)$ and $\chi(\rho_2 - \varepsilon) > 0 > \chi(\rho_2 + \varepsilon)$. Besides, we find that for any $s > 0$, $\widehat{\chi}(s) \xrightarrow{p} \chi(s)$. Thus, it follows from Lemma 5.10 in Van Der Vaart [69] that $\widehat{\rho_1} \xrightarrow{p} \rho_1$ and $\widehat{\rho_2} \xrightarrow{p} \rho_2$. □

Once we have obtained the estimation of the above characteristics, by Equations (9)–(11), the estimation of Fourier transform $\mathcal{F}\Phi(s)$, denoted as $\widehat{\mathcal{F}\Phi}(s)$, can be defined as

$$\widehat{\mathcal{F}\Phi}(s) = \frac{\widehat{\mathcal{F}K}(s)}{1 - \widehat{\mathcal{F}H}(s)}, \tag{12}$$

where

$$\widehat{\mathcal{F}K}(s) = \frac{\widehat{\lambda}}{c(\widehat{\rho_1} - \widehat{\rho_2})} \left[\frac{(\widehat{\beta} - \widehat{\rho_2})[\widehat{\mathcal{F}\omega}(s) - \widehat{\mathcal{L}\omega}(\widehat{\rho_2})]}{\widehat{\rho_2} + is} - \frac{(\widehat{\beta} - \widehat{\rho_1})[\widehat{\mathcal{F}\omega}(s) - \widehat{\mathcal{L}\omega}(\widehat{\rho_1})]}{\widehat{\rho_1} + is} \right],$$

$$\widehat{\mathcal{F}H}(s) = \frac{\widehat{\lambda}}{c(\widehat{\rho_1} - \widehat{\rho_2})} \left[\frac{(\widehat{\beta} - \widehat{\rho_2})[\widehat{\mathcal{F}f}(s) - \widehat{\mathcal{L}f}(\widehat{\rho_2})]}{\widehat{\rho_2} + is} - \frac{(\widehat{\beta} - \widehat{\rho_1})[\widehat{\mathcal{F}f}(s) - \widehat{\mathcal{L}f}(\widehat{\rho_1})]}{\widehat{\rho_1} + is} \right].$$

Finally, replacing $\mathcal{F}\Phi(\cdot)$ in Equation (6) by the estimation $\widehat{\mathcal{F}\Phi}(\cdot)$, the expected discounted penalty function can be estimated by

$$\widehat{\Phi}_{K,a}(u) := \sum_{k=0}^{K-1}{}' \frac{2}{a} Re \left\{ \widehat{\mathcal{F}\Phi}\left(\frac{k\pi}{a}\right) \right\} \cos\left(k\pi \frac{x}{a}\right), \qquad 0 \le u \le a. \tag{13}$$

4. Consistency Properties

In this section, we study the asymptotic properties of the estimation $\widehat{\Phi}_{K,a}$. Let $L^2(\mathbb{R}_+)$ denote the class of square integrable functions on the positive axis. For any function $f \in L^2(\mathbb{R}_+)$, its L^2-norm is defined by $\|f\| = \left(\int_0^\infty f^2(x)dx \right)^{\frac{1}{2}}$. Throughout this section, C represents a positive generic constant that may take different values at different steps. In addition, we define

$$H_j(x) = \int_0^x u^j w(u, x-u)du, \qquad j = 0, 1, 2.$$

It is easy to see that

$$\int_0^\infty u^j \omega(u)du = E\left[H_j(X)\right].$$

For reader's convenience, we introduce some definitions in empirical process theory, which are used to study the asymptotic properties. For any measurable function f, its $L^r(P)$-norm is defined by $\|f\|_{P,r} = \left(\int |f(\omega)|^r dP(\omega) \right)^{\frac{1}{r}}$. Given two functions l and u, the bracket $[l, u]$ is the set of all functions f with $l \le f \le u$. An ε-bracket in $L^r(P)$ is a bracket $[l, u]$ with $\|u - l\|_{P,r} < \varepsilon$. For a class $\mathcal{G} \subset L^r(P)$, the bracketing number $N_\diamond(\varepsilon,\ \mathcal{G},\ L^r(P))$ is the minimum number of ε-brackets needed to cover $\mathcal{G}$. For $\beta > 0$, the bracketing integral is defined by $J_\diamond(\beta, \mathcal{G}, L^r(P)) = \int_0^\beta \sqrt{N_\diamond(\varepsilon,\ \mathcal{G},\ L^r(P))}d\varepsilon$.

We use the L^2-norm to study the asymptotic properties of the estimator. The following condition is useful in our discussion, which ensures $\Phi \in L^2(\mathbb{R}_+)$.

Condition 2. For the penalty function w, there exist some integers $\alpha_1,\ \alpha_2$ and constant C such that

$$w(x,y) \leq C(1+x)^{\alpha_1}(1+y)^{\alpha_2}.$$

Put $\Phi_{K,a} = \widehat{\Phi}_{K,a} = 0$ when $u > a$. By triangle inequality, we have

$$\|\Phi - \widehat{\Phi}_{K,a}\| \leq \|\Phi - \Phi_{K,a}\| + \|\Phi_{K,a} - \widehat{\Phi}_{K,a}\|, \tag{14}$$

where the first term $\|\Phi - \Phi_{K,a}\|$ is the bias caused by Fourier cosine series approximation, and the second term $\|\Phi_{K,a} - \widehat{\Phi}_{K,a}\|$ is the bias caused by statistical estimation.

For the bias $\|\Phi - \Phi_{K,a}\|$, by similar arguments to those of Zhang [65], we obtain the following result.

Theorem 1. *Suppose that* $\int_0^\infty |\Phi'(u)|du < \infty$ *and for some integer* m, $\Phi(u) \leq Cu^{-(m+1)}$*; then, under net profit condition, Conditions 1 and 2, we have*

$$\|\Phi - \Phi_{K,a}\| \leq C\left\{\frac{K+1}{a^{2m+1}} + \frac{a}{K-1}\right\}.$$

Next, we study the error $\|\Phi_{K,a} - \widehat{\Phi}_{K,a}\|$. For $\widehat{\rho_1}$ and $\widehat{\rho_2}$, we derive the following result.

Theorem 2. *Supposing that net profit condition holds, we have* $\widehat{\rho_1} - \rho_1 = O_p(T^{-\frac{1}{2}})$. *Supposing that* $c > \lambda\mathbb{E}[X]$, *we have* $\widehat{\rho_2} - \rho_2 = O_p(T^{-\frac{1}{2}})$.

Proof of Theorem 2. By the mean value theorem,

$$\widehat{\chi}(\rho_j) = \widehat{\chi}(\widehat{\rho_j}) + \widehat{\chi}'(\rho_j^*)(\rho_j - \widehat{\rho_j}) = \widehat{\chi}'(\rho_j^*)(\rho_j - \widehat{\rho_j}), \qquad j = 1, 2,$$

where ρ_j^* $(j = 1,2)$ is a random number between ρ_j $(j = 1,2)$ and $\widehat{\rho_j}$ $(j = 1,2)$. Since $\chi(\rho_j) = 0,\ j = 1,2$, we obtain

$$\widehat{\rho_j} - \rho_j = \frac{\chi(\rho_j) - \widehat{\chi}(\rho_j)}{\widehat{\chi}'(\rho_j^*)}, \qquad j = 1, 2.$$

It is easily seen that $\chi(\rho_j) - \widehat{\chi}(\rho_j) = O_p(T^{-\frac{1}{2}}),\ j = 1,2$ for $\widehat{\lambda} - \lambda = O_p(T^{-\frac{1}{2}}),\ \widehat{\mu} - \mu = O_p(T^{-\frac{1}{2}})$, and $\widehat{\beta} - \beta = O_p(T^{-\frac{1}{2}})$.

For ρ_1, under net profit condition, we introduce the following set:

$$A_{T,1} = \left\{|\widehat{\chi}'(\rho_1^*)| > \frac{1}{2}\left(c + \frac{\mu}{\beta} - \lambda\mathbb{E}[X]\right)\right\}.$$

Since $c+\frac{\widehat{\mu}}{\widehat{\beta}}-\widehat{\lambda}\frac{1}{N_T}\sum_{j=1}^{N_T}X_j \xrightarrow{p} c+\frac{\mu}{\beta}-\lambda\mathbb{E}[X]$ and $|\widehat{\chi}'(\rho_1^*)|\geq c+\frac{\widehat{\mu}}{\widehat{\beta}}-\widehat{\lambda}\frac{1}{N_T}\sum_{j=1}^{N_T}X_j$, we have

$$\begin{aligned}\mathbb{P}(A_{T,1})\geq&\,\mathbb{P}\left(c+\frac{\widehat{\mu}}{\widehat{\beta}}-\widehat{\lambda}\frac{1}{N_T}\sum_{j=1}^{N_T}X_j\geq\frac{1}{2}(c+\frac{\mu}{\beta}-\lambda\mathbb{E}[X])\right)\\ =&1-\mathbb{P}\left(c+\frac{\widehat{\mu}}{\widehat{\beta}}-\widehat{\lambda}\frac{1}{N_T}\sum_{j=1}^{N_T}X_j<\frac{1}{2}(c+\frac{\mu}{\beta}-\lambda\mathbb{E}[X])\right)\\ =&1-\mathbb{P}\left(c+\frac{\mu}{\beta}-\lambda\mathbb{E}[X]-c-\frac{\widehat{\mu}}{\widehat{\beta}}+\widehat{\lambda}\frac{1}{N_T}\sum_{j=1}^{N_T}X_j\geq\frac{1}{2}(c+\frac{\mu}{\beta}-\lambda\mathbb{E}[X])\right)\to 1,\ \ as\ T\to\infty.\end{aligned}$$

For ρ_2, we have

$$|\widehat{\chi}'(\rho_2^*)|\geq c+\frac{\widehat{\mu}\widehat{\beta}}{(s-\widehat{\beta})^2}-\widehat{\lambda}\frac{1}{N_T}\sum_{j=1}^{N_T}X_j\geq c-\widehat{\lambda}\frac{1}{N_T}\sum_{j=1}^{N_T}X_j.$$

Under condition $c>\lambda\mathbb{E}[X]$, we introduce the following set:

$$A_{T,2}=\left\{|\widehat{\chi}'(\rho_2^*)|>\frac{1}{2}\left(c-\lambda\mathbb{E}[X]\right)\right\}.$$

Similarly, we obtain that $\mathbb{P}(A_{T,2})\to 1\ as\ T\to\infty$ in that $c-\widehat{\lambda}\frac{1}{N_T}\sum_{j=1}^{N_T}X_j\xrightarrow{p}c-\lambda\mathbb{E}[X]$.

Furthermore,

$$\begin{aligned}\mathbb{P}\left(|\widehat{\rho}_j-\rho_j|>CT^{-\frac{1}{2}}\right)=&\,\mathbb{P}\left(\frac{|\chi(\rho_j)-\widehat{\chi}(\rho_j)|}{|\widehat{\chi}'(\rho_j^*)|}>CT^{-\frac{1}{2}}\right)\\ \leq&\,\mathbb{P}\left(\left\{\frac{|\chi(\rho_j)-\widehat{\chi}(\rho_j)|}{|\widehat{\chi}'(\rho_j^*)|}>CT^{-\frac{1}{2}}\right\}\bigcap A_{T,j}\right)+\mathbb{P}(A_{T,j}^c)\\ =&\,\mathbb{P}\left(|\chi(\rho_j)-\widehat{\chi}(\rho_j)|>CT^{-\frac{1}{2}}\right)+\mathbb{P}(A_{T,j}^c),\qquad j=1,\,2.\end{aligned}$$

As a result, since $\mathbb{P}(A_{T,1}^c)\to 0$ under net profit condition, $\mathbb{P}(A_{T,2}^c)\to 0$ under $c>\lambda\mathbb{E}[X]$, and $\chi(\rho_j)-\widehat{\chi}(\rho_j)=O_p(T^{-\frac{1}{2}})$, $j=1,2$, we derive desired results. □

The following two theorems give the uniform convergence rates of $\mathcal{F}H$ and $\mathcal{F}K$.

Theorem 3. *Suppose that $c>\lambda\mathbb{E}[X]$, $\|H_j(X)\|_{P,1}<\infty$, $j=0,1$, and $\|H_j(X)\|_{P,2}<\infty$, $j=1,2$. Then, for large a, K and T, we have*

$$\sup_{s\in[0,K\pi/a]}\left|\mathcal{F}K(s)-\widehat{\mathcal{F}K}(s)\right|=O_p\left(\sqrt{\log\left(\frac{K}{a}\right)/T}\right).$$

Proof of Theorem 3. By Equations (11) and (12),

$$\begin{aligned}\widehat{\mathcal{F}K}(s)-\mathcal{F}K(s)=&\frac{\widehat{\lambda}}{c(\widehat{\rho}_1-\widehat{\rho}_2)}\frac{\widehat{\beta}-\widehat{\rho}_2}{\widehat{\rho}_2+is}\left[\widehat{\mathcal{F}\omega}(s)-\widehat{\mathcal{L}\omega}(\widehat{\rho}_2)\right]-\frac{\lambda}{c(\rho_1-\rho_2)}\frac{\beta-\rho_2}{\rho_2+is}\left[\mathcal{F}\omega(s)-\mathcal{L}\omega(\rho_2)\right]\\ &+\frac{\widehat{\lambda}}{c(\widehat{\rho}_1-\widehat{\rho}_2)}\frac{\widehat{\beta}-\widehat{\rho}_1}{\widehat{\rho}_1+is}\left[\widehat{\mathcal{F}\omega}(s)-\widehat{\mathcal{L}\omega}(\widehat{\rho}_1)\right]-\frac{\lambda}{c(\rho_1-\rho_2)}\frac{\beta-\rho_1}{\rho_1+is}\left[\mathcal{F}\omega(s)-\mathcal{L}\omega(\rho_1)\right]\\ :=&I_1+I_2,\end{aligned}\tag{15}$$

Since I_1 and I_2 have similar formations, we only study I_1 in detail.

$$\begin{aligned}
I_1 =& \frac{\widehat{\lambda}}{c(\widehat{\rho_1}-\widehat{\rho_2})}\frac{\widehat{\beta}-\widehat{\rho_2}}{\widehat{\rho_2}+is}\left[\widehat{\mathcal{F}\omega}(s)-\widehat{\mathcal{L}\omega}(\widehat{\rho_2})\right]-\frac{\lambda}{c(\rho_1-\rho_2)}\frac{\beta-\rho_2}{\rho_2+is}\left[\mathcal{F}\omega(s)-\mathcal{L}\omega(\rho_2)\right]\\
=& \frac{\widehat{\lambda}}{c(\widehat{\rho_1}-\widehat{\rho_2})}\frac{\widehat{\beta}-\widehat{\rho_2}}{\widehat{\rho_2}+is}\frac{1}{N_T}\sum_{j=1}^{N_T}\int_0^{X_j}(e^{isu}-e^{-\widehat{\rho_2}u})\mathrm{w}(u,X_j-u)du\\
&-\frac{\lambda}{c(\rho_1-\rho_2)}\frac{\beta-\rho_2}{\rho_2+is}\mathbb{E}\left[\int_0^{X}(e^{isu}-e^{-\rho_2u})\mathrm{w}(u,X-u)du\right]\\
=& \frac{\widehat{\lambda}}{c(\widehat{\rho_1}-\widehat{\rho_2})}\frac{\widehat{\beta}-\widehat{\rho_2}}{\widehat{\rho_2}+is}\frac{1}{N_T}\sum_{j=1}^{N_T}\int_0^{X_j}(e^{-\rho_2u}-e^{-\widehat{\rho_2}u})\mathrm{w}(u,X_j-u)du\\
&+\frac{\widehat{\lambda}}{c(\widehat{\rho_1}-\widehat{\rho_2})}\frac{\widehat{\beta}-\widehat{\rho_2}}{\widehat{\rho_2}+is}\frac{1}{N_T}\sum_{j=1}^{N_T}\int_0^{X_j}(e^{isu}-e^{-\widehat{\rho_2}u})\mathrm{w}(u,X_j-u)du\\
&-\frac{\lambda}{c(\rho_1-\rho_2)}\frac{\beta-\rho_2}{\rho_2+is}\mathbb{E}\left[\int_0^{X}(e^{isu}-e^{-\rho_2u})\mathrm{w}(u,X-u)du\right]\\
=& II_1+II_2.
\end{aligned}$$

where

$$\begin{aligned}
II_1 =& \frac{\widehat{\lambda}}{c(\widehat{\rho_1}-\widehat{\rho_2})}\frac{\widehat{\beta}-\widehat{\rho_2}}{\widehat{\rho_2}+is}\frac{1}{N_T}\sum_{j=1}^{N_T}\int_0^{X_j}(e^{-\rho_2u}-e^{-\widehat{\rho_2}u})\mathrm{w}(u,X_j-u)du,\\
II_2 =& \frac{\widehat{\lambda}}{c(\widehat{\rho_1}-\widehat{\rho_2})}\frac{\widehat{\beta}-\widehat{\rho_2}}{\widehat{\rho_2}+is}\frac{1}{N_T}\sum_{j=1}^{N_T}\int_0^{X_j}(e^{isu}-e^{-\widehat{\rho_2}u})\mathrm{w}(u,X_j-u)du\\
&-\frac{\lambda}{c(\rho_1-\rho_2)}\frac{\beta-\rho_2}{\rho_2+is}\mathbb{E}\left[\int_0^{X}(e^{isu}-e^{-\rho_2u})\mathrm{w}(u,X-u)du\right].
\end{aligned}$$

For II_1, we have

$$\begin{aligned}
|II_1| \leq& \frac{\widehat{\lambda}}{c(\widehat{\rho_1}-\widehat{\rho_2})}\frac{\widehat{\beta}-\widehat{\rho_2}}{\widehat{\rho_2}+is}\frac{1}{N_T}\sum_{j=1}^{N_T}\int_0^{X_j}u|\rho_2-\widehat{\rho_2}|\mathrm{w}(u,X_j-u)du\\
=& \frac{\widehat{\lambda}}{c(\widehat{\rho_1}-\widehat{\rho_2})}\frac{\widehat{\beta}-\widehat{\rho_2}}{\widehat{\rho_2}+is}|\rho_2-\widehat{\rho_2}|\frac{1}{N_T}\sum_{j=1}^{N_T}H_1(X_j).
\end{aligned}$$

Since $\frac{1}{N_T}\sum_{j=1}^{N_T}H_1(X_j)\xrightarrow{p}\|H_1(X)\|_{P,1}<\infty$, $\rho_2-\widehat{\rho_2}=O_p(T^{-\frac{1}{2}})$, we have

$$|II_1| = O_p(T^{-\frac{1}{2}}). \tag{16}$$

For II_2 we derive

$$\begin{aligned} II_2 =& \frac{\widehat{\lambda}(\widehat{\beta}-\widehat{\rho_2})}{c(\widehat{\rho_1}-\widehat{\rho_2})}\frac{\rho_2+is}{\widehat{\rho_2}+is}\frac{1}{N_T}\sum_{j=1}^{N_T}\int_0^{X_j}\frac{e^{isu}-e^{-\widehat{\rho_2}u}}{\rho_2+is}\mathrm{w}(u,X_j-u)du \\ &-\frac{\lambda(\beta-\rho_2)}{c(\rho_1-\rho_2)}\mathbb{E}\left[\int_0^X\frac{e^{isu}-e^{-\rho_2 u}}{\rho_2+is}\mathrm{w}(u,X-u)du\right] \\ =& \frac{\widehat{\lambda}(\widehat{\beta}-\widehat{\rho_2})}{c(\widehat{\rho_1}-\widehat{\rho_2})}\frac{\rho_2+is}{\widehat{\rho_2}+is}\frac{1}{N_T}\sum_{j=1}^{N_T}\left[g_{1,s}(X_j)-\mathbb{E}[g_{1,s}(X)]\right] \\ &+(\rho_2+is)\left(\frac{\widehat{\lambda}}{c(\widehat{\rho_1}-\widehat{\rho_2})}\frac{\widehat{\beta}-\widehat{\rho_2}}{\widehat{\rho_2}+is}-\frac{\lambda}{c(\rho_1-\rho_2)}\frac{(\beta-\rho_2)}{\rho_2+is}\right)\mathbb{E}[g_{1,s}(X)], \end{aligned}$$

where

$$\begin{aligned} g_{1,s}(x) &= \int_0^x\frac{e^{isu}-e^{-\rho_2 u}}{\rho_2+is}\mathrm{w}(u,x-u)du \\ &= \int_0^x\int_0^u e^{is(u-y)-\rho_2 y}dy\mathrm{w}(u,x-u)du, \qquad x\geq 0. \end{aligned}$$

For $g_{1,s}(x)$, we have

$$\begin{aligned} \sup_{s\in[0,K\pi/a]}|\mathbb{E}\left[g_{1,s}(X)\right]| &\leq \int_0^\infty\int_0^x\int_0^u e^{-\rho_2 y}dy\mathrm{w}(u,x-u)duf(x)dx \\ &\leq \int_0^\infty\int_0^x u\mathrm{w}(u,x-u)duf(x)dx = \|H_1(X)\|_{P,1}<\infty, \end{aligned}$$

then

$$\begin{aligned} &\sup_{s\in[0,K\pi/a]}\left|(\rho_2+is)\left(\frac{\widehat{\lambda}}{c(\widehat{\rho_1}-\widehat{\rho_2})}\frac{\widehat{\beta}-\widehat{\rho_2}}{\widehat{\rho_2}+is}-\frac{\lambda}{c(\rho_1-\rho_2)}\frac{(\beta-\rho_2)}{\rho_2+is}\right)\mathbb{E}[g_{1,s}(X)]\right| \\ \leq&\left|(\rho_2+is)\left(\frac{\widehat{\lambda}}{c(\widehat{\rho_1}-\widehat{\rho_2})}\frac{\widehat{\beta}-\widehat{\rho_2}}{\widehat{\rho_2}+is}-\frac{\lambda}{c(\rho_1-\rho_2)}\frac{(\beta-\rho_2)}{\rho_2+is}\right)\right|\cdot\|H_1(X)\|_{P,1}=O_p(T^{-\frac{1}{2}}) \end{aligned} \tag{17}$$

for $\widehat{\lambda}-\lambda=O_p(T^{-\frac{1}{2}}),\widehat{\mu}-\mu=O_p(T^{-\frac{1}{2}}),\widehat{\rho_1}-\rho_1=O_p(T^{-\frac{1}{2}})$ and $\widehat{\rho_2}-\rho_2=O_p(T^{-\frac{1}{2}})$.

Let us introduce the following two types of real-valued functions,

$$\begin{aligned} \mathcal{G}_{K,R} &= \{g:\ g=Re(g_{1,s}), s\in[0,K\pi/a]\}, \\ \mathcal{G}_{K,I} &= \{g:\ g=Im(g_{1,s}), s\in[0,K\pi/a]\}. \end{aligned}$$

Then, we have

$$\begin{aligned} &\sup_{s\in[0,K\pi/a]}\left|\frac{1}{N_T}\sum_{j=1}^{N_T}\left[g_{1,s}(X_j)-\mathbb{E}\left(g_{1,s}(X)\right)\right]\right| \\ \leq&\sup_{g\in\mathcal{G}_{K,R}}\left|\frac{1}{N_T}\sum_{j=1}^{N_T}\left[g_{1,s}(X_j)-\mathbb{E}\left(g_{1,s}(X)\right)\right]\right|+\sup_{g\in\mathcal{G}_{K,I}}\left|\frac{1}{N_T}\sum_{j=1}^{N_T}\left[g_{1,s}(X_j)-\mathbb{E}\left(g_{1,s}(X)\right)\right]\right|. \end{aligned} \tag{18}$$

We only study the convergence rate of the first term $\sup_{g\in\mathcal{G}_{K,R}}\left|\frac{1}{N_T}\sum_{j=1}^{N_T}\left[g_{1,s}(X_j)-\mathbb{E}\left(g_{1,s}(X)\right)\right]\right|$, since the second term follows similarly.

For any real-valued function $g \in \mathcal{G}_{K,R}$, we have

$$
\begin{aligned}
|g(x)| \le & \sup_{s\in[0,K\pi/a]} \left| \int_0^x \int_0^u e^{is(u-y)-\rho_2 y} dy w(u,x-u)du \right| \\
\le & \int_0^x \int_0^u e^{-\rho_2 y} dy w(u,x-u)du \le \int_0^x u w(u,x-u)du = H_1(x),
\end{aligned}
$$

which implies that $\mathcal{G}_{K,R}$ is contained in the single bracket $[-H_1, H_1]$. For two functions g_{1,s_1}, g_{1,s_2}, where $s_j \in [0, K\pi/a], j = 1,2$, the mean value theorem gives

$$
\begin{aligned}
|Re(g_{1,s_1}) - Re(g_{1,s_2})| = & \left| \int_0^x \int_0^u (cos(s_1(u-y)) - cos(s_2(u-y)))e^{-\rho_2 y} dy w(u,x-u)du \right| \\
= & \left| - \int_0^x \int_0^u sin(s^*(u-y)) \cdot (u-y) \cdot (s_1 - s_2) e^{-\rho_2 y} dy w(u,x-u)du \right| \\
\le & \int_0^x \int_0^u (u-y)e^{-\rho_2 y} dy w(u,x-u)du \cdot |s_1 - s_2| \le H_2(x)|s_1 - s_2|,
\end{aligned}
$$

where s^* is a number between s_1 and s_2. Under the condition $\|H_2(X)\|_{P,2} < \infty$, it follows from Example 19.7 in Van Der Vaart [69] that, for any $0 < \varepsilon < K\pi/a$, there exists a constant C such that the bracket number for $\mathcal{G}_{K,R}$ satisfies

$$
N_\diamond(\varepsilon, \mathcal{G}_{K,R}, L^2(P)) \le C \frac{K\pi}{\varepsilon a} \|H_2(x)\|_{P,2}.
$$

As a result, for every $\delta > 0$, the bracketing integral

$$
J_\diamond(\delta, \mathcal{G}_{K,R}, L^2(P)) \le \int_0^\delta \sqrt{\log\left(C\frac{K\pi}{\varepsilon a}\|H_2(x)\|_{P,2}\right) d\varepsilon} \lesssim \sqrt{\log\left(\frac{K}{a}\right)}.
$$

Furthermore, by Corollary 19.35 in Van Der Vaart [69], we have

$$
\begin{aligned}
& \mathbb{E}\left(\frac{1}{\sqrt{N_T}} \sup_{g\in\mathcal{G}_{K,R}} \left| \sum_{j=1}^{N_T} [g(X_j) - \mathbb{E}(g(X))] \right| |N_T \right) \\
\le & J_\diamond(\delta, \mathcal{G}_{K,R}, L^2(P)) \le \int_0^\delta \sqrt{\log\left(C\frac{K\pi}{\varepsilon a}\|H_2(x)\|_{P,2}\right) d\varepsilon} \lesssim \sqrt{\log\left(\frac{K}{a}\right)},
\end{aligned}
$$

then

$$
\begin{aligned}
& \mathbb{E}\left(\frac{1}{N_T} \sup_{g\in\mathcal{G}_{K,R}} \left| \sum_{j=1}^{N_T} [g(X_j) - \mathbb{E}(g(X))] \right| \right) \\
= & \mathbb{E}\left(\frac{1}{\sqrt{N_T}} \mathbb{E}\left(\frac{1}{\sqrt{N_T}} \sup_{g\in\mathcal{G}_{K,R}} \left| \sum_{j=1}^{N_T} [g(X_j) - \mathbb{E}(g(X))] \right| |N_T \right)\right) \\
\lesssim & \sqrt{\log\left(\frac{K}{a}\right)} / \mathbb{E}[\sqrt{N_T}] \le \sqrt{\log\left(\frac{K}{a}\right)} / \sqrt{\mathbb{E}[N_T]} = \sqrt{\log\left(\frac{K}{a}\right) / \lambda T}.
\end{aligned}
$$

Therefore,

$$
\sup_{g\in\mathcal{G}_{K,R}} \left| \frac{1}{N_T} \sum_{j=1}^{N_T} [g(X_j) - \mathbb{E}(g(X))] \right| = O_p\left(\sqrt{\log\left(\frac{K}{a}\right) / T}\right). \tag{19}
$$

Similarly,

$$\sup_{g\in\mathcal{G}_{K,l}}\left|\frac{1}{N_T}\sum_{j=1}^{N_T}\left[g(X_j)-\mathbb{E}\left(g(X)\right)\right]\right| = O_p\left(\sqrt{\log\left(\frac{K}{a}\right)/T}\right). \tag{20}$$

Combining Equations (18)–(20), we obtain

$$\sup_{s\in[0,K\pi/a]}\left|\frac{1}{N_T}\sum_{j=1}^{N_T}\left[g_{1,s}(X_j)-\mathbb{E}\left(g_{1,s}(X)\right)\right]\right| = O_p\left(\sqrt{\log\left(\frac{K}{a}\right)/T}\right). \tag{21}$$

As a result, Equations (17) and (21) give

$$\sup_{s\in[0,K\pi/a]}|II_2| = O_p\left(\sqrt{\log\left(\frac{K}{a}\right)/T}\right). \tag{22}$$

By Equations (16) and (22), we have

$$\sup_{s\in[0,K\pi/a]}|I_1| = O_p\left(\sqrt{\log\left(\frac{K}{a}\right)/T}\right); \tag{23}$$

then by the similar arguments of I_1, we have

$$\sup_{s\in[0,K\pi/a]}|I_2| = O_p\left(\sqrt{\log\left(\frac{K}{a}\right)/T}\right). \tag{24}$$

Finally, we can derive the desired result by Equations (23) and (24). □

Theorem 4. *Suppose that $c > \lambda\mathbb{E}[X]$, $\mathbb{E}[X^k] < \infty$, $k = 1, 2$. Then, for large a, K and T, we have*

$$\sup_{s\in[0,K\pi/a]}\left|\mathcal{F}H(s)-\widehat{\mathcal{F}H}(s)\right| = O_p\left(\sqrt{\log\left(\frac{K}{a}\right)/T}\right).$$

Proof of Theorem 4. By Equations (11) and (12),

$$\begin{aligned}\widehat{\mathcal{F}H}(s)-\mathcal{F}H(s) =&\frac{\widehat{\lambda}}{c(\widehat{\rho_1}-\widehat{\rho_2})}\frac{\widehat{\beta}-\widehat{\rho_2}}{\widehat{\rho_2}+is}\left[\widehat{\mathcal{F}f}(s)-\widehat{\mathcal{L}f}(\widehat{\rho_2})\right]-\frac{\lambda}{c(\rho_1-\rho_2)}\frac{\beta-\rho_2}{\rho_2+is}\left[\mathcal{F}f(s)-\mathcal{L}f(\rho_2)\right]\\ &+\frac{\widehat{\lambda}}{c(\widehat{\rho_1}-\widehat{\rho_2})}\frac{\widehat{\beta}-\widehat{\rho_1}}{\widehat{\rho_1}+is}\left[\widehat{\mathcal{F}f}(s)-\widehat{\mathcal{L}f}(\widehat{\rho_1})\right]-\frac{\lambda}{c(\rho_1-\rho_2)}\frac{\beta-\rho_1}{\rho_1+is}\left[\mathcal{F}f(s)-\mathcal{L}f(\rho_1)\right]\\ :=&l_1+l_2,\end{aligned} \tag{25}$$

Since l_1 and l_2 have similar formations, we only study l_1 in detail. For l_1, we have

$$\begin{aligned}l_1 =&\frac{\widehat{\lambda}}{c(\widehat{\rho_1}-\widehat{\rho_2})}\frac{\widehat{\beta}-\widehat{\rho_2}}{\widehat{\rho_2}+is}\frac{1}{N_T}\sum_{j=1}^{N_T}\left[e^{isX_j}-e^{-\widehat{\rho_2}X_j}\right]-\frac{\lambda}{c(\rho_1-\rho_2)}\frac{\beta-\rho_2}{\rho_2+is}\mathbb{E}\left[e^{isX}-e^{-\rho_2X}\right]\\ =&\frac{\widehat{\lambda}}{c(\widehat{\rho_1}-\widehat{\rho_2})}\frac{\widehat{\beta}-\widehat{\rho_2}}{\widehat{\rho_2}+is}\frac{1}{N_T}\sum_{j=1}^{N_T}\left[e^{-\rho_2X_j}-e^{-\widehat{\rho_2}X_j}\right]+\frac{\widehat{\lambda}}{c(\widehat{\rho_1}-\widehat{\rho_2})}\frac{\widehat{\beta}-\widehat{\rho_2}}{\widehat{\rho_2}+is}\frac{1}{N_T}\sum_{j=1}^{N_T}\left[e^{isX_j}-e^{-\rho_2X_j}\right]\\ &-\frac{\lambda}{c(\rho_1-\rho_2)}\frac{\beta-\rho_2}{\rho_2+is}\mathbb{E}\left[e^{isX}-e^{-\rho_2X}\right]\\ :=&ll_1+ll_2,\end{aligned}$$

where

$$ll_1 = \frac{\widehat{\lambda}}{c(\widehat{\rho_1}-\widehat{\rho_2})}\frac{\widehat{\beta}-\widehat{\rho_2}}{\widehat{\rho_2}+is}\frac{1}{N_T}\sum_{j=1}^{N_T}\left[e^{-\rho_2 X_j}-e^{-\widehat{\rho_2}X_j}\right],$$

$$ll_2 = \frac{\widehat{\lambda}}{c(\widehat{\rho_1}-\widehat{\rho_2})}\frac{\widehat{\beta}-\widehat{\rho_2}}{\widehat{\rho_2}+is}\frac{1}{N_T}\sum_{j=1}^{N_T}\left[e^{isX_j}-e^{-\rho_2 X_j}\right]-\frac{\lambda}{c(\rho_1-\rho_2)}\frac{\beta-\rho_2}{\rho_2+is}\mathbb{E}\left[e^{isX}-e^{-\rho_2 X}\right].$$

For ll_1, we have

$$|ll_1| \le \frac{\widehat{\lambda}}{c(\widehat{\rho_1}-\widehat{\rho_2})}\frac{\widehat{\beta}-\widehat{\rho_2}}{\widehat{\rho_2}+is}|\rho_2-\widehat{\rho_2}|\frac{1}{N_T}\sum_{j=1}^{N_T}X_j.$$

Since $\frac{1}{N_T}\sum_{j=1}^{N_T}X_j \xrightarrow{p} \mathbb{E}[X]<\infty$, $\rho_2-\widehat{\rho_2}=O_p(T^{-\frac{1}{2}})$, we have

$$|ll_1| = O_p(T^{-\frac{1}{2}}). \quad (26)$$

For ll_2, we obtain

$$\begin{aligned} ll_2 =& \frac{\widehat{\lambda}(\widehat{\beta}-\widehat{\rho_2})}{c(\widehat{\rho_1}-\widehat{\rho_2})}\frac{\rho_2+is}{\widehat{\rho_2}+is}\frac{1}{N_T}\sum_{j=1}^{N_T}\left[g_{2,s}(X_j)-\mathbb{E}[g_{2,s}(X)]\right] \\ &+(\rho_2+is)\left(\frac{\widehat{\lambda}}{c(\widehat{\rho_1}-\widehat{\rho_2})}\frac{\widehat{\beta}-\widehat{\rho_2}}{\widehat{\rho_2}+is}-\frac{\lambda}{c(\rho_1-\rho_2)}\frac{(\beta-\rho_2)}{\rho_2+is}\right)\mathbb{E}[g_{2,s}(X)], \end{aligned}$$

where

$$g_{2,s}(x) = e^{isx}, \qquad x \ge 0.$$

For $g_{2,s}(x)$, we have

$$\sup_{s\in[0,K\pi/a]}|\mathbb{E}[g_{2,s}(X)]| = \sup_{s\in[0,K\pi/a]}\left|\int_0^\infty e^{isx}f(x)dx\right| \le 1 < \infty,$$

Under condition $\mathbb{E}[X^k]<\infty$, $k=1,2$, by similar arguments of $g_{1,s}$, we conclude that

$$\sup_{s\in[0,K\pi/a]}\left|\frac{1}{N_T}\sum_{j=1}^{N_T}\left[g_{2,s}(X_j)-\mathbb{E}\left(g_{2,s}(X)\right)\right]\right| = O_p\left(\sqrt{\log\left(\frac{K}{a}\right)/T}\right). \quad (27)$$

In addition, it follows from a similar analysis of II_2 in the proof of Theorem 3 that

$$\sup_{s\in[0,K\pi/a]}|ll_2| = O_p\left(\sqrt{\log\left(\frac{K}{a}\right)/T}\right). \quad (28)$$

Then, Equations (26) and (28) give

$$\sup_{s\in[0,K\pi/a]}|l_1| \le \sup_{s\in[0,K\pi/a]}|ll_2| + \sup_{s\in[0,K\pi/a]}|ll_2| = O_p\left(\sqrt{\log\left(\frac{K}{a}\right)/T}\right). \quad (29)$$

Similarly, we derive

$$\sup_{s\in[0,K\pi/a]} |I_2| = O_p\left(\sqrt{\log\left(\frac{K}{a}\right)/T}\right). \tag{30}$$

Finally, we can derive the desired result by Equations (29) and (30). □

Based on the above conclusions, we have the following result.

Theorem 5. *Suppose that* $c > \lambda\mathbb{E}[X]$, $\mathbb{E}[X^k] < \infty$, $k = 1,2$, $\|H_j(X)\|_{P,1} < \infty$, $j = 0,1$, *and* $\|H_j(X)\|_{P,2} < \infty$, $j = 1,2$. *Then, for large* a, K *and* T, *we have*

$$\|\Phi_{K,a} - \widehat{\Phi}_{K,a}\|^2 = O_p\left(\frac{K}{a}\log\left(\frac{K}{a}\right)/T\right).$$

Proof of Theorem 5. First, we have

$$\begin{aligned}
\left\|\Phi_{K,a} - \widehat{\Phi}_{K,a}\right\|^2 &= \int_0^a \left|\Phi_{K,a}(u) - \widehat{\Phi}_{K,a}(u)\right|^2 du \\
&\le \sum_{k=0}^{K-1} \frac{4}{a^2}\left(\mathrm{Re}\left\{\mathcal{F}\Phi\left(\frac{k\pi}{a}\right) - \widehat{\mathcal{F}\Phi}\left(\frac{k\pi}{a}\right)\right\}\right)^2 \int_0^a \left(\cos\left(\frac{k\pi}{a}u\right)\right)^2 du \\
&= \frac{2}{a}\sum_{k=0}^{K-1}\left|\mathcal{F}\Phi\left(\frac{k\pi}{a}\right) - \widehat{\mathcal{F}\Phi}\left(\frac{k\pi}{a}\right)\right|^2 \\
&\le \frac{2K}{a}\sup_{s\in[0,K\pi/a]}\left|\mathcal{F}\Phi(s) - \widehat{\mathcal{F}\Phi}(s)\right|^2.
\end{aligned} \tag{31}$$

Then, by Equations (9) and (12), we obtain

$$\begin{aligned}
&\sup_{s\in[0,K\pi/a]}\left|\mathcal{F}\Phi(s) - \widehat{\mathcal{F}\Phi}(s)\right| \\
&= \sup_{s\in[0,K\pi/a]}\left|\frac{\mathcal{F}K(s)}{1-\mathcal{F}H(s)} - \frac{\widehat{\mathcal{F}K}(s)}{1-\widehat{\mathcal{F}H}(s)}\right| \\
&= \sup_{s\in[0,K\pi/a]}\left|\frac{(1-\mathcal{F}H(s))\left(\mathcal{F}K(s)-\widehat{\mathcal{F}K}(s)\right) + \mathcal{F}K(s)\left(\mathcal{F}H(s)-\widehat{\mathcal{F}H}(s)\right)}{(1-\mathcal{F}H(s))\left(1-\widehat{\mathcal{F}H}(s)\right)}\right|.
\end{aligned} \tag{32}$$

Combining Theorems 3 and 4 gives

$$\sup_{s\in[0,K\pi/a]}|\mathcal{F}\Phi(s) - \widehat{\mathcal{F}\Phi}(s)| = O_p\left(\sqrt{\log\left(\frac{K}{a}\right)/T}\right).$$

Finally, by Equation (31), we derive that

$$\|\Phi_{K,a} - \widehat{\Phi}_{K,a}\|^2 \le \frac{2K}{a}\sup_{s\in[0,K\pi/a]}|\mathcal{F}\Phi(s) - \widehat{\mathcal{F}\Phi}(s)|^2 = O_p\left(\frac{K}{a}\log\left(\frac{K}{a}\right)/T\right).$$

This completes the proof. □

Combing Theorems 1 and 5, we finally obtain the following convergence rate:

$$\|\Phi - \widehat{\Phi}_{K,a}\|^2 = O\left(\frac{K+1}{a^{2m+1}}\right) + O\left(\frac{a}{K-1}\right) + O_p\left(\frac{K}{a}\log\left(\frac{K}{a}\right)/T\right). \tag{33}$$

We first find the optimal truncation parameter $a^* = O\left(K^{-(n+1)}\right)$ to minimize the convergence rate $O\left(\frac{K+1}{a^{2m+1}}\right) + O\left(\frac{a}{K-1}\right)$. Replacing a by a^* in Equation (33) gives

$$\|\Phi - \widehat{\Phi}_{K,a}\|^2 = O\left(K^{-\frac{m}{m+1}}\right) + O_p\left(K^{\frac{m}{m+1}} \log(K)/T\right). \tag{34}$$

In addition, we find the optimal truncation $K^* = O\left(T^{\frac{m+1}{2m}}\right)$. Thus, we obtain the smallest convergence rate $\|\Phi - \widehat{\Phi}_{K,a}\|^2 = O_p\left(T^{-\frac{2m^2}{(m+1)^2}}\right)$.

5. Simulation Studies

In this section, we present some numerical examples to explain the excellent properties of our estimator when the observed sample is finite. We mainly studied the following three kinds of special expected discounted penalty function:

(1) Ruin probability (RP: $w \equiv 1,\ \delta = 0$);

(2) Laplace transform of ruin time (LT: $w \equiv 1,\ \delta = 0.1$);

(3) Expected discounted deficit at ruin (EDD) when ruin is due to a claim $(w(x,y) = y,\ \delta = 0.1)$.

Some simulation examples of the above functions are presented for different distributions of claim sizes:

(1) Exponential distribution: $f(x) = e^{-x},\ x > 0$;

(2) Erlang(2) distribution: $f(x) = 4xe^{-2x},\ x > 0$;

(3) Combined exponential distribution: $f(x) = 3e^{-1.5x} - 3e^{-3x},\ x > 0$;

(4) Mixed exponential distribution: $f(x) = \frac{2}{3}e^{-2x} + \frac{2}{3}e^{-x},\ x > 0$.

We set $\Delta = 1, \lambda = 2$, and $\mu = 5$, where $\Delta = 1$ can be explained as one week. That is to say, in a long time interval, the insurer will observe the data once a week, the expected claim number is 2 times per week, and the expected premium number is 5 times per week. Furthermore, since there are 52 business weeks every year, we assumed that $T = \frac{1}{4} \times 52 \times \Delta,\ T = \frac{1}{2} \times 52 \times \Delta,\ T = 1 \times 52 \times \Delta,\ T = 5 \times 52 \times \Delta$, which means that we observed the surplus process for one quarter, half a year, one year and five years. Then, we used Equation (13) to estimate the above Gerber-Shiu functions, and the corresponding true value obtained by Laplace inversion. In all simulations, we set $c = 5,\ a = 30,\ K = 2^{13}$, and we carried out the relevant analysis based on 300 simulation experiments. We first introduce several concepts used in this section. The mean value and the mean relative error, which are, respectively, defined by

$$\frac{1}{300}\sum_{j=1}^{300} \widehat{\Phi}_{K,a,j}(u), \qquad \frac{1}{300}\sum_{j=1}^{300} \left|\frac{\widehat{\Phi}_{K,a,j}(u)}{\Phi(u)} - 1\right|,$$

and the integrated mean square error (IMSE), which is defined by

$$\frac{1}{300}\sum_{j=1}^{300} \int_0^\infty (\widehat{\Phi}_{K,a,j}(u) - \Phi(u))^2 du \approx \frac{1}{300}\sum_{j=1}^{300} \int_0^{30} (\widehat{\Phi}_{K,a,j}(u) - \Phi(u))^2 du,$$

where $\widehat{\Phi}_{K,a,j}(u)$ is the estimate of expected discounted penalty function in the jth experiment. For IMSE, we computed the integral on the finite domain $[0, 30]$, as both the true value and the estimator will be very small when $u \geq 30$. In reality, both the true value and the estimated value are very close to zero when $u > 20$, thus we present all images for $u \in [0, 20]$ to illustrate the performance of our estimators better.

First, we plot the mean curves of the estimated expected discounted penalty functions and compare them with the corresponding true curves. For the above-mentioned four distributions of claim sizes, we show the mean curves and the true curves of RP, LT, EDD due to a claim in

Figures 1–3, respectively. It is easily observed that, even though we used the quarter book data, the performance of the estimation for each claim distribution was still good. We could hardly distinguish the true curves from the mean curves when we used the five-year book data. Since it is difficult to distinguish the mean curves for different T values when u becomes large, we further show the mean relative error curves of the estimated expected discounted penalty functions for different claim distributions in Figures 4–6, respectively. We observed that the mean relative errors became small as T increased, and they were very small when $T = 260$. Besides, we found that the mean relative errors were small when u was small, but became very large when u was large. This is because the true values of the expected discounted penalty functions were very small when u was large. All of the above results illustrate the performance of our estimations by images. Finally, we present some values of IMSE for the estimators of RP, LT, and EDD in Tables 1 and 2 to further show that our estimators perform well. Combining all of the simulation results, we conclude that our estimators could effectively approximate the true values, and they performed well for the large T.

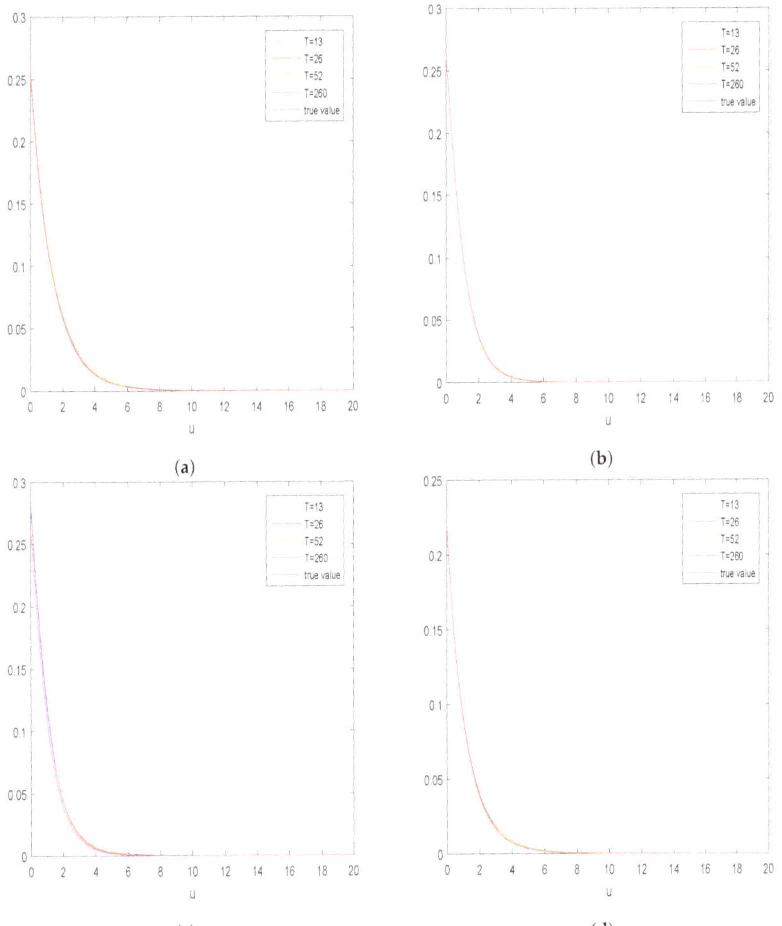

Figure 1. Estimation of the ruin probability for different claim sizes. Mean curves: (**a**) exponential claim sizes; (**b**) Erlang claim sizes; (**c**) combined-exponential claim sizes; and (**d**) mixed-exponential claim sizes.

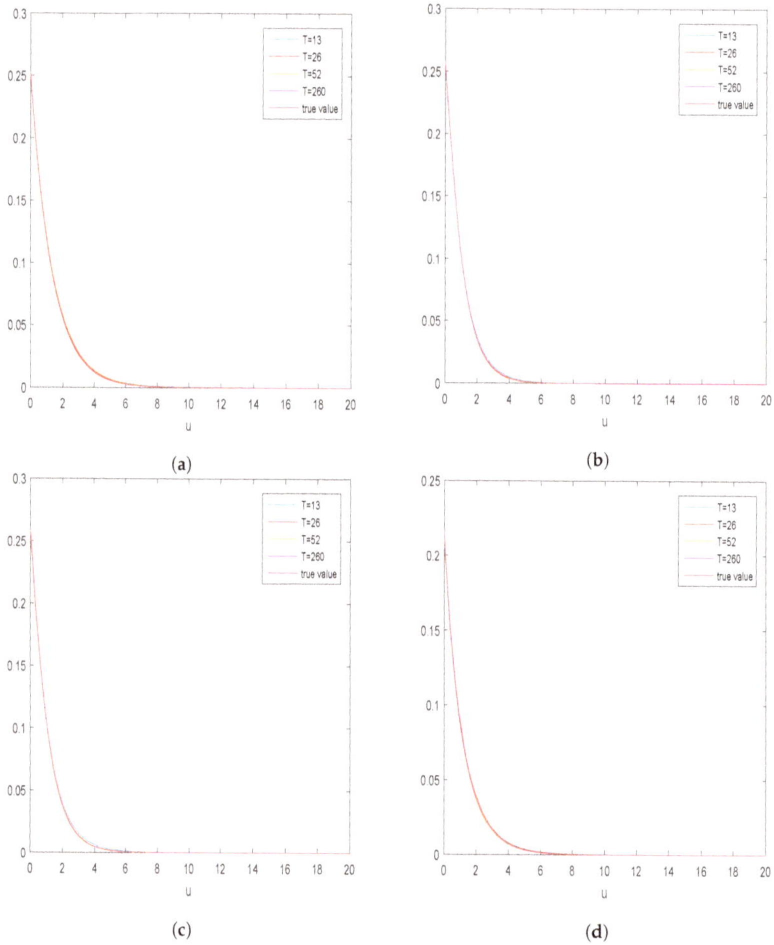

Figure 2. Estimation of the Laplace transform of ruin time for different claim sizes. Mean curves: (**a**) exponential claim sizes; (**b**) Erlang claim sizes; (**c**) combined-exponential claim sizes; and (**d**) mixed-exponential claim sizes.

Table 1. IMSE of $\widehat{\Phi}_{K,a}$.

T	Exp			Erlang(2)		
	RP	**LT**	**EDD**	**RP**	**LT**	**EDD**
13	0.00717	0.00644	0.01586	0.00351	0.00414	0.00412
26	0.00352	0.00316	0.00846	0.00186	0.00200	0.00214
52	0.00177	0.00199	0.00537	0.00108	0.00099	0.00095
260	0.00032	0.00034	0.00092	0.00018	0.00020	0.00019

Table 2. IMSE of $\widehat{\Phi}_{K,a}$.

T	Com-Exp			Mix-Exp		
	RP	LT	EDD	RP	LT	EDD
13	0.00719	0.00431	0.00527	0.00474	0.00511	0.01176
26	0.00271	0.00195	0.00277	0.00267	0.00233	0.00602
52	0.00160	0.00114	0.00124	0.00117	0.00107	0.00257
260	0.00046	0.00021	0.00024	0.00025	0.00018	0.00053

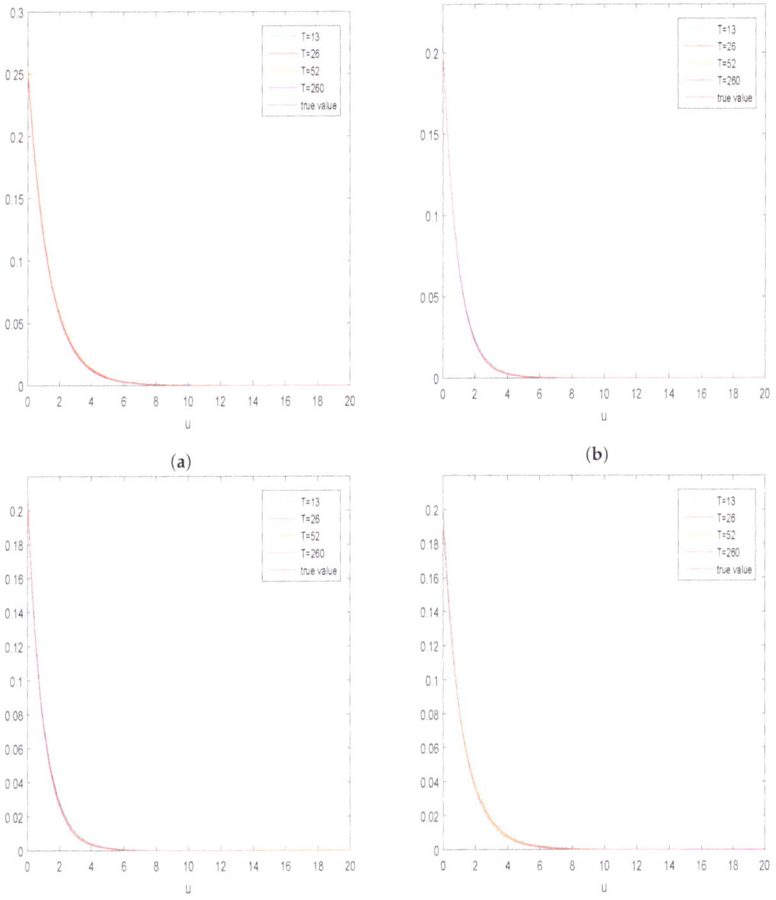

(a) (d)

Figure 3. Estimation of the expected discounted deficit at ruin due to a claim for different claim sizes. Mean curves: (**a**) exponential claim sizes; (**b**) Erlang claim sizes; (**c**) combined-exponential claim sizes; and (**d**) mixed-exponential claim sizes.

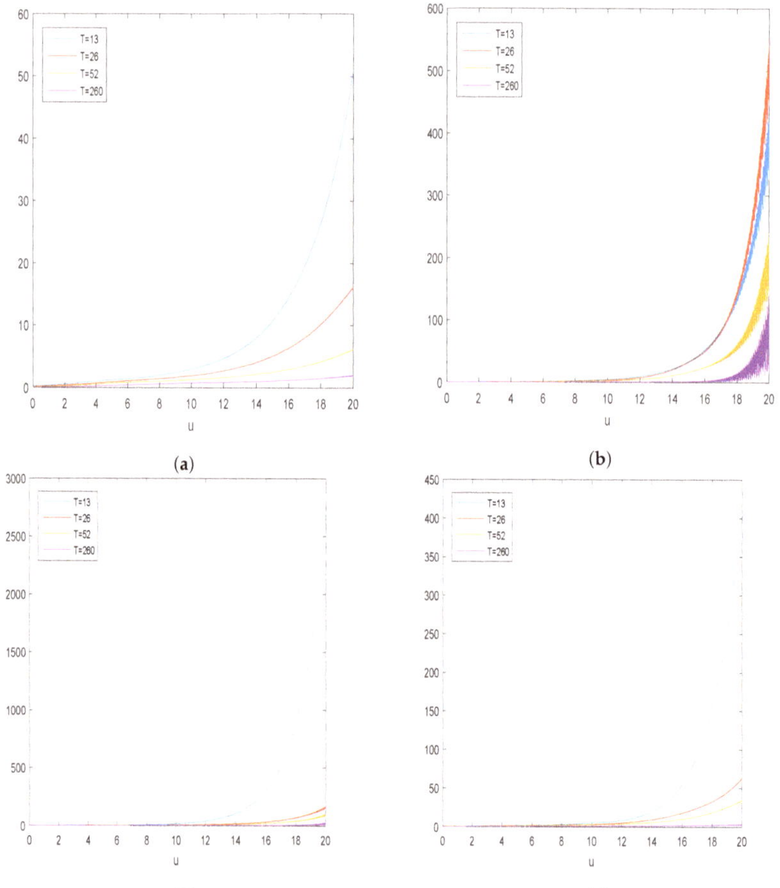

(a) (b) (c) (d)

Figure 4. Estimation of the ruin probability for different claim sizes. Mean relative error curves: (**a**) exponential claim sizes; (**b**) Erlang claim sizes; (**c**) combined-exponential claim sizes; and (**d**) mixed-exponential claim sizes.

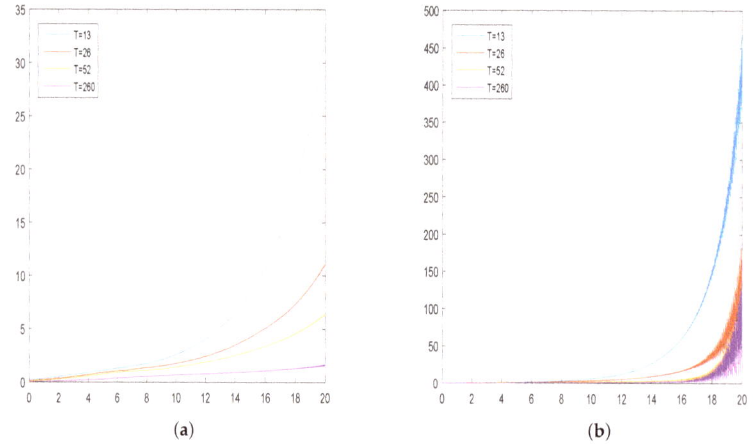

(a) (b)

Figure 5. *Cont.*

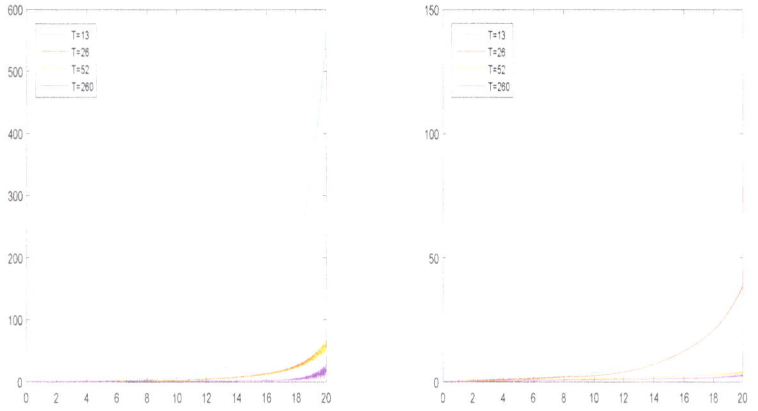

(c) (d)

Figure 5. Estimation of the Laplace transform of ruin time for different claim sizes. Mean relative error curves: (**a**) exponential claim sizes; (**b**) Erlang claim sizes; (**c**) combined-exponential claim sizes; and (**d**) mixed-exponential claim sizes.

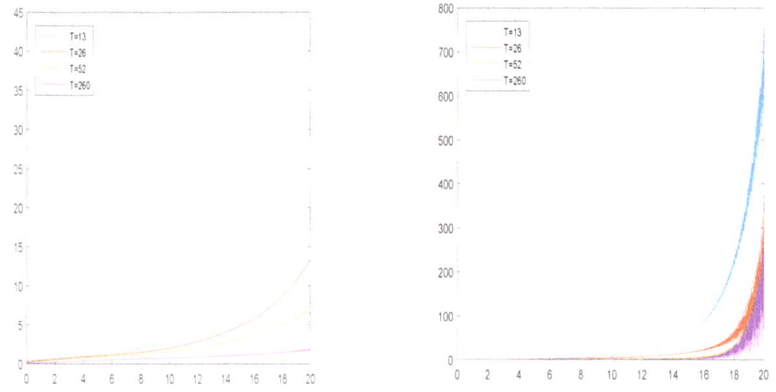

(a) (b)

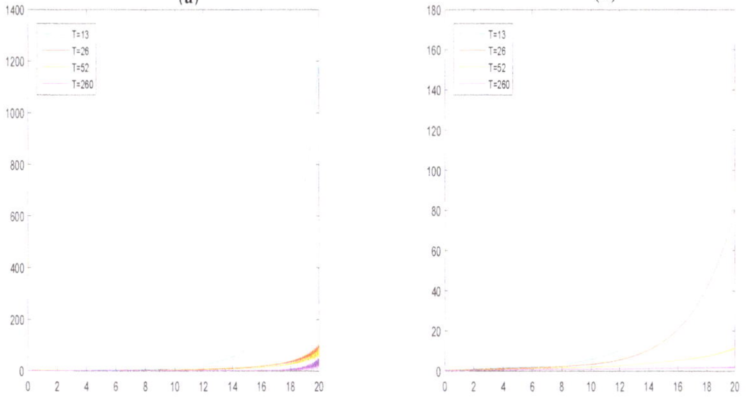

(c) (d)

Figure 6. Estimation of the expected discounted deficit at ruin due to a claim for different claim sizes. Mean relative error curves: (**a**) exponential claim sizes; (**b**) Erlang claim sizes; (**c**) combined-exponential claim sizes; and (**d**) mixed-exponential claim sizes.

Author Contributions: Data curation, Y.W.; Formal analysis, W.Y., Y.H. and X.Y.; Methodology, W.Y. and H.F.; and Writing—original draft, W.Y.

Funding: This research was partially supported by the National Natural Science Foundation of China (Grant Nos. 11301303 and 71804090), the National Social Science Foundation of China (Grant No. 15BJY007), the Taishan Scholars Program of Shandong Province (Grant No. tsqn20161041), the Humanities and Social Sciences Project of the Ministry Education of China (Grant Nos. 19YJA910002 and 16YJC630070), the Natural Science Foundation of Shandong Province (Grant No. ZR2018MG002), the Fostering Project of Dominant Discipline and Talent Team of Shandong Province Higher Education Institutions (Grant No. 1716009), the Social Science Foundation of Shandong Province (Grant Nos. 18DSHJ04 and 17CQXJ01), the Shandong Jiaotong University "Climbing" Research Innovation Team Program, the Risk Management and Insurance Research Team of Shandong University of Finance and Economics, the 1251 Talent Cultivation Project of Shandong Jiaotong University, and Collaborative Innovation Center Project of the Transformation of New and Old Kinetic Energy and Government Financial Allocation.

Acknowledgments: We thank two anonymous reviewers for their helpful comments on an earlier draft of this paper.

Conflicts of Interest: The authors declare no conflict of interest.

References

1. Gerber, H.U.; Shiu, E.S.W. On the time value of ruin. *N. Am. Actuar. J.* **1998**, *2*, 48–78. [CrossRef]
2. Lin, X.S.; Willmot, G.E.; Drekic, S. The classical risk model with a constant dividend barrier: Analysis of the Gerber-Shiu discounted penalty function. *Insur. Math. Econ.* **2003**, *33*, 391–408.
3. Yuen, K.C.; Wang, G.J.; Li, W.K. The Gerber-Shiu expected discounted penalty function for risk processes with interest and a constant dividend barrier. *Insur. Math. Econ.* **2007**, *40*, 104–112. [CrossRef]
4. Yuen, K.C.; Zhou, M.; Guo, J.Y. On a risk model with debit interest and dividend payments. *Stat. Probab. Lett.* **2008**, *78*, 2426–2432. [CrossRef]
5. Zhao, X.H.; Yin, C.C. The Gerber-Shiu expected discounted penalty function for Lévy insurance risk processes. *Acta Math. Sin.* **2010**, *26*, 575–586. [CrossRef]
6. Chi, Y.C. Analysis of expected discounted penalty function for a general jump diffusion risk model and applications in finance. *Insur. Math. Econ.* **2010**, *46*, 385–396. [CrossRef]
7. Yin, C.C.; Wang, C.W. The perturbed compound Poisson risk process with investment and debit Interest. *Methodol. Comput. Appl.* **2010**, *12*, 391–413. [CrossRef]
8. Chi, Y.C.; Lin, X.S. On the threshold dividend strategy for a generalized jump-diffusion risk model. *Insur. Math. Econ.* **2011**, 48, 326–337. [CrossRef]
9. Shen, Y.; Yin, C.C.; Yuen, K.C. Alternative approach to the optimality of the threshold strategy for spectrally negative Lévy processes. *Acta Math. Appl. Sin. Engl.* **2013**, *29*, 705–716. [CrossRef]
10. Yin, C.C.; Yuen, K.C. Exact joint laws associated with spectrally negative Lévy processes and applications to insurance risk theory. *Front. Math. China* **2014**, *9*, 1453–1471. [CrossRef]
11. Zhao, Y.X.; Yao, D.J. Optimal dividend and capital injection problem with a random time horizon and a ruin penalty in the dual model. *Appl. Math. Ser. B* **2015**, 30, 325–339. [CrossRef]
12. Li, S.M.; Lu, Y.; Sendova, K.P. The expected discounted penalty function: From infinite time to finite time. *Scand. Actuar. J.* **2019**. [CrossRef]
13. Li, Y.Q.; Yin, C.C.; Zhou, X.W. On the last exit times for spectrally negative Lévy processes. *J. Appl. Probab.* **2017**, *54*, 474–489. [CrossRef]
14. Dong, H.; Yin, C.C.; Dai, H.S. Spectrally negative Lévy risk model under Erlangized barrier strategy. *J. Comput. Appl. Math.* **2019**, *351*, 101–116. [CrossRef]
15. Wang, W.Y.; Zhang, Z.M. Computing the Gerber-Shiu function by frame duality projection. *Scand. Actuar. J.* **2019**. [CrossRef]
16. Willmot, G.E.; Dickson, D.C.M. The Gerber-Shiu discounted penalty function in the stationary renewal risk model. *Insur. Math. Econ.* **2003**, *32*, 403–411. [CrossRef]
17. Wang, C.W.; Yin, C.C.; Li, E.Q. On the classical risk model with credit and debit interests under absolute ruin. *Stat. Probab. Lett.* **2010**, *80*, 427–436. [CrossRef]
18. Yin, C.C.; Yuen, K.C. Optimality of the threshold dividend strategy for the compound Poisson model. *Stat. Probab. Lett.* **2011**, *81*, 1841–1846. [CrossRef]

19. Dong, H.; Yin, C.C. Complete monotonicity of the probability of ruin and DE Finetti's dividend problem. *J. Syst. Sci. Complex.* **2012**, *25*, 178–185. [CrossRef]
20. Yu, W.G. Some results on absolute ruin in the perturbed insurance risk model with investment and debit interests. *Econ. Model.* **2013**, *31*, 625–634. [CrossRef]
21. Yin, C.C.; Shen, Y.; Wen, Y.Z. Exit problems for jump processes with applications to dividend problems. *J. Comput. Appl. Math.* **2013**, *245*, 30–52. [CrossRef]
22. Zeng, Y.; Li, D.P.; Gu, A.L. Robust equilibrium reinsurance-investment strategy for a mean-variance insurer in a model with jumps. *Insur. Math. Econ.* **2016**, *66*, 138–152. [CrossRef]
23. Zhao, Y.X.; Wang, R.M.; Yao, D.J. Optimal dividend and equity issuance in the perturbed dual model under a penalty for ruin. *Commun. Stat. Theory Methods* **2016**, *45*, 365–384. [CrossRef]
24. Yu, W.G.; Huang, Y.J.; Cui, C.R. The absolute ruin insurance risk model with a threshold dividend strategy. *Symmetry* **2018**, *10*, 377. [CrossRef]
25. Boucherie, R.J.; Boxma, O.J.; Sigman, K. A note on negative customers, GI/G/1 workload, and risk process. *Probab. Eng. Inf. Sci.* **1997**, *11*, 305–311. [CrossRef]
26. Boikov, A.V. The Cramér-Lundberg model with stochastic premium process. *Theory Probab. Appl.* **2003**, *47*, 489–493. [CrossRef]
27. Temnov, G. Risk process with random income. *J. Math. Sci.* **2004**, *123*, 3780–3794. [CrossRef]
28. Bao, Z.H. The expected discounted penalty at ruin in the risk process with random income. *Appl. Math. Comput.* **2006**, *179*, 559–566. [CrossRef]
29. Bao, Z.H.; Ye, Z.X. The Gerber-Shiu discounted penalty function in the delayed renewal risk process with random income. *Appl. Math. Comput.* **2007**, *184*, 857–863. [CrossRef]
30. Yang, H.; Zhang, Z.M. On a class of renewal risk model with random income. *Appl. Stoch. Model. Bus.* **2009**, *25*, 678–695. [CrossRef]
31. Labbé, C.; Sendova, K.P. The expected discounted penalty function under a risk model with sochastic income. *Appl. Math. Comput.* **2009**, *215*, 1852–1867.
32. Zhang, Z.M.; Yang, H. On a risk model with stochastic premiums income and dependence between income and loss. *J. Comput. Appl. Math.* **2010**, *234*, 44–57. [CrossRef]
33. Zhao, Y.X.; Yin, C.C. The expected discounted penalty function under a renewal risk model with stochastic income. *Appl. Math. Comput.* **2012**, *218*, 6144–6154. [CrossRef]
34. Xu, L.; Wang, R.M.; Yao, D.J. On maximizing the expected terminal utility by investment and reinsurance. *J. Ind. Manag. Optim.* **2008**, *4*, 801–815.
35. Yu, W.G. On the expected discounted penalty function for a Markov regime switching risk model with stochastic premium income. *Discret. Dyn. Nat. Soc.* **2013**, *2013*, 1–9. [CrossRef]
36. Yu, W.G. Randomized dividends in a discrete insurance risk model with stochastic premium income. *Math. Probl. Eng.* **2013**, *2013*, 1–9. [CrossRef]
37. Zhou, J.M.; Mo, X.Y.; Ou, H.; Yang, X.Q. Expected present value of total dividends in the compound binomial model with delayed claims and random income. *Acta Math. Sci.* **2013**, *33*, 1639–1651. [CrossRef]
38. Gao, J.W.; Wu, L.Y. On the Gerber-Shiu discounted penalty function in a risk model with two types of delayed-claims and random income. *J. Comput. Appl. Math.* **2014**, *269*, 42–52. [CrossRef]
39. Zhou, M.; Yuen, K.C.; Yin, C.C. Optimal investment and premium control in a nonlinear diffusion model. *Acta Math. Appl. Sin. Engl.* **2017**, *33*, 945–958. [CrossRef]
40. Deng, Y.C.; Liu, J.; Huang, Y.; Li, M.; Zhou, J.M. On a discrete interaction risk model with delayed claims and stochastic incomes under random discount rates. *Commun. Stat. Theory Methods* **2018**, *47*, 5867–5883. [CrossRef]
41. Zeng, Y.; Li, D.P.; Chen, Z.; Yang, Z. Ambiguity aversion and optimal derivative-based pension investment with stochastic income and volatility. *J. Econ. Dyn. Control* **2018**, *88*, 70–103. [CrossRef]
42. Politis, K. Semiparametric estimation for non-ruin probabilities. *Scand. Actuar. J.* **2003**, *2003*, 75–96. [CrossRef]
43. Wang, Y.F.; Yin, C.C. Approximation for the ruin probabilities in a discrete time risk model with dependent risks. *Stat. Probab. Lett.* **2010**, *80*, 1335–1342. [CrossRef]
44. Yuen, K.C.; Yin, C.C. Asymptotic results for tail probabilities of sums of dependent and heavy-tailed random variables. *Chin. Ann. Math. B* **2012**, *33*, 557–568. [CrossRef]
45. Zhang, Z.M.; Yang, H.L.; Yang, H. On a nonparametric estimator for ruin probability in the classical risk model. *Scand. Actuar. J.* **2014**, *2014*, 309–338. [CrossRef]

46. Wang, Y.F.; Yin, C.C.; Zhang, X.S. Uniform estimate for the tail probabilities of randomly weighted sums. *Acta Math. Appl. Sin. Engl.* **2014**, *30*, 1063–1072. [CrossRef]
47. Zhang, Z.M. Nonparametric estimation of the finite time ruin probability in the classical risk model. *Scand. Actuar. J.* **2017**, *2017*, 452–469. [CrossRef]
48. Zhang, Z.M.; Yang, H.L. Nonparametric estimate of the ruin probability in a pure-jump Lévy risk model. *Insur. Math. Econ.* **2013**, *53*, 24–35. [CrossRef]
49. Zhang, Z.M.; Yang, H.L. Nonparametric estimation for the ruin probability in a Lévy risk model under low-frequency observation. *Insur. Math. Econ.* **2014**, *59*, 168–177. [CrossRef]
50. Shimizu, Y.; Zhang, Z.M. Estimating Gerber-Shiu functions from discretely observed Lévy driven surplus. *Insur. Math. Econ.* **2017**, *74*, 84–98. [CrossRef]
51. Peng, J.Y.; Wang, D.C. Asymptotics for ruin probabilities of a non-standard renewal risk model with dependence structures and exponential lévy process investment returns. *J. Ind. Manag. Optim.* **2017**, *13*, 155–185. [CrossRef]
52. Peng, J.Y.; Wang, D.C. Uniform asymptotics for ruin probabilities in a dependent renewal risk model with stochastic return on investments. *Stochastics* **2018**, *90*, 432–471. [CrossRef]
53. Yang, Y.; Wang, K.Y.; Liu, J.J.; Zhang, Z.M. Asymptotics for a bidimensional risk model with two geometric Lévy price processes, *J. Ind. Manag. Optim.* **2019**, *15*, 481–505 [CrossRef]
54. Yang, Y.; Yuen, K.C.; Liu, J.F. Asymptotics for ruin probabilities in Lévy-driven risk models with heavy-tailed claims. *J. Ind. Manag. Optim.* **2018**, *14*, 231–247. [CrossRef]
55. Shimizu, Y. Estimation of the expected discounted penalty function for Lévy insurance risks. *Math. Methods Stat.* **2011**, *20*, 125–149. [CrossRef]
56. Zhang, Z.M. Estimating the Gerber-Shiu function by Fourier-Sinc series expansion. *Scand. Actuar. J.* **2017**, *2017*, 898–919. [CrossRef]
57. Zhang, Z.M.; Su, W. A new efficient method for estimating the Gerber-Shiu function in the classical risk model. *Scand. Actuar. J.* **2018**, *2018*, 426–449. [CrossRef]
58. Zhang, Z.M.; Su, W. Estimating the Gerber-Shiu function in a Lévy risk model by Laguerre series expansion. *J. Comput. Appl. Math.* **2019**, *346*, 133–149. [CrossRef]
59. Su, W.; Yong, Y.D.; Zhang, Z.M. Estimating the Gerber-Shiu function in the perturbed compound Poisson model by Laguerre series expansion. *J. Math. Anal. Appl.* **2019**, *469*, 705–729. [CrossRef]
60. Fang, F.; Oosterlee, C.W. A novel option pricing method based on Fourier cosine series expansions. *SIAM J. Sci. Comput.* **2008**, *31*, 826–848. [CrossRef]
61. Fang, F.; Oosterlee, C.W. Pricing early-exercise and discrete barrier options by Fourier-cosine series expansions. *Numer. Math.* **2009**, *114*, 27–62. [CrossRef]
62. Zhang, B.; Oosterlee, C.W. Pricing of early-exercise Asian options under Lévy processes based on Fourier cosine expansions. *Appl. Numer. Math.* **2014**, *78*, 14–30. [CrossRef]
63. Chau, K.W.; Yam, S.C.P.; Yang, H.L. Fourier-cosine method for Gerber-Shiu functions. *Insur. Math. Econ.* **2015**, *61*, 170–180. [CrossRef]
64. Chau, K.W.; Yam, S.C.P.; Yang, H.L. Fourier-cosine method for ruin probabilities. *J. Comput. Appl. Math.* **2015**, *281*, 94–106. [CrossRef]
65. Zhang, Z.M. Approximating the density of the time to ruin via Fourier-cosine series expansion. *Astin Bull.* **2017**, *47*, 169–198. [CrossRef]
66. Yang, Y.; Su, W.; Zhang, Z.M. Estimating the discounted density of the deficit at ruin by Fourier cosine series expansion. *Stat. Probab. Lett.* **2019**, *146*, 147–155. [CrossRef]
67. Dickson, D.C.M.; Hipp, C. On the time to ruin for Erlang(2) risk processes. *Insur. Math. Econ.* **2001**, *29*, 333–344. [CrossRef]
68. Li, S.M.; Garrido, J. On ruin for the Erlang(n) risk process. *Insur. Math. Econ.* **2004**, *34*, 391–408. [CrossRef]
69. Van Der Vaart, A.W. *Asymptotic Statistics*; Cambridge University Press: Cambridge, UK, 1998.

mathematics

MDPI

Article

Monte Carlo Algorithms for the Parabolic Cauchy Problem

Alexander Sipin

Institute of Mathematics, Natural Science and Computer Science, Vologda State University, 160000 Vologda, Russia; cac1909@mail.ru

Received: 19 December 2018; Accepted: 12 February 2019; Published: 15 February 2019

Abstract: New Monte Carlo algorithms for solving the Cauchy problem for the second order parabolic equation with smooth coefficients are considered. Unbiased estimators for the solutions of this problem are constructed.

Keywords: parabolic equation; Cauchy problem; Monte Carlo method; unbiased estimator; von-Neumann–Ulam scheme

1. Introduction

Consider the *parabolic operator*

$$\begin{aligned} L &= L(x,t,\partial/\partial x,\partial/\partial t) \\ &= \frac{\partial}{\partial t} - \sum_{i,j=1}^{n} a_{ij}(x,t)\frac{\partial^2}{\partial x_i \partial x_j} + \sum_{i=1}^{n} a_i(x,t)\frac{\partial}{\partial x_i} + a_0(x,t) \end{aligned} \tag{1}$$

Let all coefficients of the operator L be defined in the domain $D_{n+1}^{(T)} = R^n \times (0,T)$. Denote by $A(x,t)$ the coefficient matrix of the highest derivatives of the operator L and suppose that $A(x,t)$ is symmetric matrix. Suppose that all eigenvalues of the matrix $A(x,t)$ belong to the fixed interval $[\nu,\mu]$, where $\nu > 0$.

Consider the *Cauchy problem* in the domain $D_{n+1}^{(T)}$

$$\begin{aligned} L(x,t,\partial/\partial x,\partial/\partial t)u(x,t) &= f(x,t), \\ u|_{t=0} &= \varphi(x). \end{aligned} \tag{2}$$

A Random variable $\xi(x,t)$ is called an *unbiased estimator* for a function $u(x,t)$ if mathematical expectation $E\xi(x,t)$ is equal to $u(x,t)$. Every unbiased estimator gives a stochastic numerical method for evaluation of the function $u(x,t)$. Now we briefly discuss some known stochastic methods for solving the Cauchy problem.

Let $0 < \alpha < 1$ and the coefficients of the parabolic operator are elements of the Hölder class $H^{\alpha,\frac{\alpha}{2}}(\overline{D_{n+1}^{(T)}})$, then Equation (2) has a *fundamental solution* $Z(x,y,t,\tau)$ [1]. Let the function $f(x,t)$ satisfy the Hölder condition with respect to all of its arguments, and let the function $\varphi(x)$ be continuous function. Let in addition $f(x,t)$ and $\varphi(x)$ grow no faster than $e^{a|x|^2}$, as $|x| \to \infty$. Then, the solution of the Cauchy problem can be written in the following form

$$u(x,t) = \int_0^t d\tau \int_{R^n} Z(x,y,t,\tau)f(y,\tau)dy + \int_{R^n} Z(x,y,t,0)\varphi(y)dy. \tag{3}$$

If $a_0(x,t) \equiv 0$ then the fundamental solution $Z(x,y,t,\tau)$ is a probability density (as a function of y). So, if the fundamental solution is known one can construct the corresponding unbiased estimator. Particularly, if the coefficients of the equation are constant, it is enought to generate a normally distributed random vector in R^n for the evaluation of $u(x,t)$. In the general case, $Z(x,y,t,\tau)$ is a transition density of a stochastic process X_t, which started from a point x at time $\tau = 0$. Hence,

$$u(x,t) = E\int_0^t f(X_s, t-s)ds + E\varphi(X_t), \tag{4}$$

and random variable $\eta = tf(X_{t\theta}, t(1-\theta)) + \varphi(X_t)$ is an unbiased estimator for $u(x,t)$, where the variable θ is uniformly distributed in $[0,1]$. Then we can use this estimator in the Monte Carlo procedure if we can generate the process X_t. The process X_t is a solution of the respective stochastic differential equation, and we can approximate it by another process Y_t, using, for example, the Euler scheme. Let $0 = t_0 < t_1 < \ldots < t_m = t$, and $Y_{t_0} = x, Y_{t_1}, \ldots, Y_{t_m}$ be the Euler approximation for the corresponding values of $X_s, s \in [0,t]$. After replacing X by Y, the estimator η became the biased one. Let $p_X(y_1,\ldots,y_m)$ and $p_Y(y_1,\ldots,y_m)$ be the densities of $m-$dimentional distributions for the X and Y processes, respectively. The estimator $p_X(Y_{t_1},\ldots,Y_{t_m})\varphi(Y_{t_m})/p_Y(Y_{t_1},\ldots,Y_{t_m})$ is an unbiased estimator for $E\varphi(X_t)$. Finally, if random variable ζ is an unbiased estimator for $p_X(Y_{t_1},\ldots,Y_{t_m})$, then

$$E\zeta\varphi(Y_{t_m})/p_Y(Y_{t_1},\ldots,Y_{t_m}) = E\varphi(X_t). \tag{5}$$

The first factor ζ in the formula (5) was constructed by W. Wagner in his papers [2]. It was shown that the fundamental solution is a functional of the solution of some integral Volterra equation. The von-Neumann–Ulam scheme [3] was applied for estimation of the fundamental solution. Monte Carlo algorithms for evaluation of some other functionals can be found in the works [4–6].

In paper [7], the von-Neumann–Ulam scheme was used for constructing another class of estimators for $u(x,t)$ without using a grid. A conjugate (dual) scheme of construction of unbiased estimators for functionals of the solutions of an integral equation, which is equivalent to the Cauchy problem, was considered in [8]. This scheme simplifies the modeling procedure, because boundaries of the spectrum for the matrix $A(x,t)$ are not required to be known.

Finally, if the operator L has differentiable coefficients, then we can obtain an integral equation for $u(x,t)$ by using the Green formula and solve this equation via the Monte Carlo method. Such algorithms were considered in [9,10] for equations whose principal part one is the Laplace operator. We obtain a Volterra equation for the Cauchy problem solution $u(x,t)$ in the general case. In this paper we investigate the von-Neumann–Ulam scheme for regular and conjugate cases.

It is necessary to note that the Multilevel Monte Carlo Method [11,12] is often used for evaluation of the functional $E\varphi(X_t)$, where process X_t is a solution of the respective stochastic differential equation. This approach is not covered in this paper.

This paper does not contain any results of numerical experiments. Numerical experiments and the efficiency of various stochastic algorithms for solving the Cauchy problem will be the subject of the separate paper.

2. Integral Representation

Let all coefficients of the operator L be elements of the Hölder class and let there exist continuous and bounded derivatives

$$\partial^2 a_{ij}(x,t)/\partial x_i \partial x_j, \quad \partial a_{ij}(x,t)/\partial x_j, \quad \partial a_i(x,t)/\partial x_i$$

for $i,j = 1,2,\ldots,n$.

We also suppose that the Cauchy problem solution is continuous and bounded. We define $\|u\|$ by equality

$$\|u\| = \sup_{(x,t)\in \overline{D_{n+1}^{(T)}}} |u(x,t)|.$$

Take a point (x,t). Let $A^{(i,j)}(x,t)$ be the elements of the inverse matrix $A^{-1}(x,t)$. Let us define a function $\sigma(y,x,t)$ by equality

$$\sigma(y,x,t)=\left(\sum_{i,j=1}^{n}A^{(i,j)}(x,t)(y_i-x_i)(y_j-x_j)\right)^{\frac{1}{2}}. \tag{6}$$

Define the function Z^0 for $t>\tau$ by equality

$$Z^0(x,y,t,\tau)=\frac{1}{[4\pi(t-\tau)]^{\frac{n}{2}}(\det A(x,t))^{\frac{1}{2}}}\cdot\exp\left(-\frac{\sigma^2(y,x,t)}{4(t-\tau)}\right). \tag{7}$$

For $t<\tau$ we set $Z^0(x,y,t,\tau)=0$. We denote $Z^0(x,y,t,\tau)$ by $v_0(y,\tau)$ if the point (x,t) is fixed.

For $\rho>0$, we define a function $v(y,\tau)$ by equality

$$v(y,\tau)=\frac{1}{[4\pi(t-\tau)]^{\frac{n}{2}}(\det A(x,t))^{\frac{1}{2}}}$$
$$\times\left(\exp\left(-\frac{\sigma^2(y,x,t)}{4(t-\tau)}\right)-\exp\left(-\frac{\rho^2}{4(t-\tau)}\right)\right).$$

Using a Green formula, it is easy to prove that

$$\begin{aligned}u(x,t)=&\int_0^t\int_{D_\rho}[v(y,\tau)Lu(y,\tau)-u(y,\tau)Mv(y,\tau)]\,dyd\tau\\&+\int_{D_\rho}u(y,0)v(y,0)dy\\&+\int_0^t\int_{\partial D_\rho}\frac{[(y-x)^TA^{-1}(x,t)A(y,\tau)A^{-1}(x,t)(y-x)]}{2(t-\tau)\|A^{-1}(x,t)(y-x)\|}\\&\times Z^0(x,y,t,\tau)u(y,\tau)d_ySd\tau,\end{aligned} \tag{8}$$

where the inner integral in the third term is a surface integral on the boundary of the domain $D_\rho\subset R^n$, which is defined by $D_\rho=\{y\in R^n|\sigma(y,x,t)<\rho\}$. Let M be a conjugate operator for $L=L(y,\tau,\partial/\partial y,\partial/\partial\tau)$:

$$\begin{aligned}Mv(y,\tau)=&-\frac{\partial v(y,\tau)}{\partial\tau}-\sum_{i,j=1}^{n}\frac{\partial^2}{\partial y_i\partial y_j}(a_{ij}(y,\tau)v(y,\tau))\\&-\sum_{i=1}^{n}\frac{\partial}{\partial y_i}(a_i(y,\tau)v(y,\tau))+a_0(y,\tau)v(y,\tau).\end{aligned} \tag{9}$$

Using the Cauchy inequality we have

$$[(y-x)^TA^{-1}(x,t)A(y,\tau)A^{-1}(x,t)(y-x)]\le\|A(y,\tau)\|\cdot\|A^{-1}(x,t)(y-x)\|^2.$$

Define a new scalar product $[v,w]$ by equality $[v,w]=v^TA^{-1}(x,t)w$. Then

$$[y-x,y-x]=\sigma^2(x,y,t),\quad [w,w]\le\|A^{-1}(x,t)\|\|w\|^2.$$

Using the Cauchy inequality we have

$$\|A^{-1}(x,t)(y-x)\|^4=[(y-x),A^{-1}(x,t)(y-x)]^2,$$

$$\|A^{-1}(x,t)(y-x)\|^4 \le [A^{-1}(x,t)(y-x), A^{-1}(x,t)(y-x)]\cdot \sigma^2(x,y,t).$$

Hence,

$$\|A^{-1}(x,t)(y-x)\|^2 \le \|A^{-1}(x,t)\|\cdot \sigma^2(x,y,t).$$

Now, we can evaluate the last integral in formula (8):

$$\int_0^t \int_{\partial D_\rho} \frac{[(y-x)^T A^{-1}(x,t)A(y,\tau)A^{-1}(x,t)(y-x)]}{2(t-\tau)\|A^{-1}(x,t)(y-x)\|}$$

$$\times Z^0(x,y,t,\tau)u(y,\tau)d_ySd\tau$$

$$\le \frac{\mu}{\nu}\int_0^t \int_{\partial D_\rho} \frac{\rho}{2(t-\tau)} \frac{1}{[4\pi(t-\tau)]^{\frac{n}{2}}(\det A(x,t))^{\frac{1}{2}}}$$

$$\times \exp\left(-\frac{\rho^2}{4(t-\tau)}\right) u(y,\tau)d_ySd\tau$$

$$\le \frac{\|u\|}{\Gamma(\frac{n}{2})}\frac{\mu}{\nu}\int_0^t \frac{\rho^n}{(t-\tau)}\frac{1}{[4(t-\tau)]^{\frac{n}{2}}}\cdot \exp\left(-\frac{\rho^2}{4(t-\tau)}\right)d\tau$$

$$\le \frac{\|u\|}{\Gamma(\frac{n}{2})}\frac{\mu}{\nu}\int_{\frac{\rho^2}{4t}}^{\infty} s^{\frac{n}{2}-1}\cdot \exp(-s)\,ds.$$

Hence, the last integral in formula (8) converges to zero as $\rho \to \infty$. For the function v we have inequality $v(y,\tau) \le v_0(y,\tau)$. Moreover, $v(y,\tau) \to v_0(y,\tau)$ as $\rho \to \infty$. So, using the equality

$$\int_{R^n} v_0(y,\tau)dy = 1,$$

we have

$$\int_{D_\rho} u(y,0)v(y,0)dy \to \int_{R^n} u(y,0)v_0(y,0)dy$$

and

$$\int_0^t \int_{D_\rho} v(y,\tau)Lu(y,\tau)dyd\tau \to \int_0^t \int_{R^n} v_0(y,\tau)Lu(y,\tau)dyd\tau.$$

It is easy to see that

$$Mv(y,\tau) - Mv_0(y,\tau) =$$

$$= (\frac{\partial}{\partial \tau} - d_0)\left(\frac{1}{[4\pi(t-\tau)]^{\frac{n}{2}}(\det A(x,t))^{\frac{1}{2}}}\exp\left(-\frac{\rho^2}{4(t-\tau)}\right)\right)$$

$$= \left(\frac{n}{2(t-\tau)} - \frac{\rho^2}{4(t-\tau)^2} - d_0\right)\frac{1}{[4\pi(t-\tau)]^{\frac{n}{2}}(\det A(x,t))^{\frac{1}{2}}}\exp\left(-\frac{\rho^2}{4(t-\tau)}\right),$$

where

$$d_0 = -\sum_{i,j=1}^n \frac{\partial^2 a_{ij}(y,\tau)}{\partial y_i \partial y_j} - \sum_{j=1}^n \frac{\partial a_j(y,\tau)}{\partial y_j} + a_0(y,\tau)$$

is a coefficient of the function u in the operator Mu. The inequalities

$$\int_0^t \int_{D_\rho} |u(y,\tau)Mv(y,\tau) - u(y,\tau)Mv_0(y,\tau)|\,dyd\tau$$

$$\le const\cdot \|u\|\int_0^t \left(\frac{n}{2(t-\tau)} + \frac{\rho^2}{4(t-\tau)^2} + \|d_0\|\right)$$

$$\times \frac{\rho^n}{\Gamma(\frac{n+1}{2})\,[4(t-\tau)]^{\frac{n}{2}}} \exp\left(-\frac{\rho^2}{4(t-\tau)}\right) d\tau$$

$$\le const \cdot \|u\| \left(1 + \frac{2t\|d_0\|}{n}\right) \frac{1}{\Gamma\left(\frac{n}{2}\right)} \int_{\frac{\rho^2}{4t}}^{\infty} s^{\frac{n}{2}-1} \cdot \exp(-s)\, ds$$

$$+const \cdot \|u\| \frac{1}{\Gamma\left(\frac{n+1}{2}\right)} \int_{\frac{\rho^2}{4t}}^{\infty} s^{\frac{n}{2}} \cdot \exp(-s)\, ds$$

show that

$$\int_0^t \int_{D_\rho} u(y,\tau) M v(y,\tau) dy d\tau \to \int_0^t \int_{R^n} u(y,\tau) M v_0(y,\tau) dy d\tau.$$

Putting $\rho \to \infty$ in the formula (8), we have the following integral representation of the Cauchy problem (2)

$$u(x,t) = \int_0^t \int_{R^n} [v_0(y,\tau) f(y,\tau) - u(y,\tau) M v_0(y,\tau)]\, dy d\tau + \int_{R^n} \varphi(y) v_0(y,0) dy. \tag{10}$$

3. Von-Neumann–Ulam Scheme

Now we investigate some properties of the integral operator

$$Ku(x,t) = -\int_0^t \int_{R^n} u(y,\tau) M v_0(y,\tau) dy d\tau \tag{11}$$

in Equation (10). The matrix $A(x,t)$ of the coefficients of higher derivatives is symmetric. So, from the equation

$$\sum_{i,j=1}^{n} a_{ij}(x,t) \frac{\partial^2}{\partial y_i \partial y_j} v_0(y,\tau) = -\frac{\partial v_0(y,\tau)}{\partial \tau},$$

we have

$$M v_0(y,\tau) = \sum_{i,j=1}^{n} [a_{ij}(x,t) - a_{ij}(y,\tau)] \frac{\partial^2}{\partial y_i \partial y_j} v_0(y,\tau) + \sum_{i=1}^{n} d_i(y,\tau) \frac{\partial}{\partial y_i} v_0(y,\tau) + d_0(y,\tau) v_0(y,\tau), \tag{12}$$

where $d_i(y,\tau) = -2\sum_{j=1}^{n} \partial a_{ij}(y,\tau)/\partial y_j - a_i(y,\tau)$ are bounded.

The expression (12) has the same structure and properties as the kernel $K(x,y,t,\lambda)$ in formula (11.12) in ([1], Sec. IV). It follows from inequalities (11.3) and (11.17) in ([1], Sec. IV) that there exist positive constants C and c, such that

$$|M v_0(y,\tau)| \le c(t-\tau)^{-\frac{n+2-\alpha}{2}} \exp\left(-C\frac{|y-x|^2}{t-\tau}\right), \tag{13}$$

for $0 \le \tau < t$.

Examples of constants c, C and further discussion can be found in [7]. In particular, it is shown in [7] that the inequality (13) implies uniform convergence of the von-Neumann series for Equation (10), if $f(x,t)$ and $\varphi(x)$ are bounded functions. We have

$$u(x,t) = \sum_{i=0}^{\infty} K^i F(x,t), \tag{14}$$

$$F(x,t) = F_1(x,t) + F_2(x,t)$$
$$= \int_0^t \int_{R^n} v_0(y,\tau) f(y,\tau) dy d\tau + \int_{R^n} \varphi(y) v_0(y,0) dy. \tag{15}$$

We can apply methods of [7] for constructing unbiased estimators for $u(x,t)$. To realize the von-Neumann–Ulam scheme, it is sufficient to choose a transition probability density for a Markov chain consistent with the kernel $K_1(x,y,t,\tau)$ of the operator K. For instance, we can take a density in the form

$$p((x,t) \to (y,\tau)) = \frac{\alpha(1-q)}{2t^{\frac{\alpha}{2}}}(t-\tau)^{\frac{\alpha}{2}-1} Z_1(x-y, t-\tau), \tag{16}$$

where $0 < q < 1$ is the probability of absorption at a current step and

$$Z_1(x-y, t-\tau) = \left(\frac{C}{\pi(t-\tau)}\right)^{\frac{n}{2}} \exp\left(-C\frac{|x-y|^2}{t-\tau}\right) \tag{17}$$

for $0 \le \tau < t$ and $Z_1(x-y, t-\tau) = 0$ for $\tau > t$.

The constant C in these formulas is the same as in inequality (13). We can take any constant such that $4\mu C < 1$. Hence, we have the compatibility of the density and the kernel of the integral equation. The probability of absorption at each step is a constant. Therefore, the time of the absorption (N) has a geometric probability distribution with a parameter q: $P(N = m) = q(1-q)^m$ for $m = 0, 1, 2, \ldots$. Random variable N and the trajectory are independent random elements and $EN = q^{-1}$. We can use procedure described in [7] for generating a Markov chain $\{(x_m, t_m)\}_{m=1}^{\infty}$ which starts at the point $(x_0, t_0) = (x, t)$.

For constructing unbiased estimators for the solution of Equation (10), we use the formulas

$$\eta(x,t) = \sum_{m=0}^{N} W^{(m)} F(x_m, t_m), \tag{18}$$

$$\zeta(x,t) = \frac{W^{(N)} F(x_N, t_N)}{q} \tag{19}$$

We define weight functions as $W^{(0)} = 1$,

$$W^{(m)} = W^{(m-1)} \frac{K_1(x_{m-1}, x_m, t_{m-1}, t_m)}{p((x_{m-1}, t_{m-1}) \to (x_m, t_m))}, \tag{20}$$

for $m = 1, 2, \ldots$. Final unbiased estimators for $u(x,t)$ are obtained after replacement of $F(x_m, t_m)$ by their unbiased estimators

$$\hat{F}_m = t_m f(x_m + \sqrt{2t_m(1-\theta)}Y, t_m\theta) + \varphi(x_m + \sqrt{2t_m}Y),$$

where the random variable θ is uniformly distributed on the interval $[0,1]$, and a random vector Y has a normal distribution with mean 0 and covariance matrix $A(x_m, t_m)$. They are independent.

It is proved in [7] that the estimators have finite variances.

Numerical Algorithm

The numerical algorithm is based on the Monte Carlo method for calculating the mathematical expectation of a random variable.

Consider as an example the following unbiased estimator $\hat{\zeta}(x,t) = W^{(N)} \hat{F}_N / q$ for $u(x,t)$.

Let $\hat{\zeta}_1, \hat{\zeta}_2, \ldots, \hat{\zeta}_k$, be independent realizations of the estimator $\hat{\zeta}(x,t)$. Then we can approximate $u(x,t)$ by the sample average $\bar{\zeta} = (\hat{\zeta}_1 + \hat{\zeta}_2 + \ldots + \hat{\zeta}_k)/k$. The approximation error is calculated as $3\sqrt{S^2/k}$, where $S^2 = (\hat{\zeta}_1^2 + \hat{\zeta}_2^2 + \ldots + \hat{\zeta}_k^2)/k - \bar{\zeta}^2$ is the sample variance.

For simulating a Markov chain $\{(x_m, t_m)\}_{m=0}^N$, we can use the formulas

$$x_0 = x, \quad t_0 = t, \quad x_{m+1} = x_m + \sqrt{t_m \vartheta_m/(2C)}Y_m, \quad t_{m+1} = t_m(1 - \vartheta_m) \tag{21}$$

where the random variables $\{\vartheta\}_{m=0}^\infty$ and the random vectors $\{Y_m\}_{m=0}^\infty$ are stochastically independent. The variables ϑ_m are distributed on the interval $(0,1)$ and have a distribution density $(\alpha/2)s^{\alpha/2-1}$. All the components of the vector Y_m are stochastically independent and have a standard normal distribution.

4. Conjugate Scheme

Now we apply the technique developed in [8] to Equation (10). Fix a number q $(0 < q < 1)$ and generate a random variable N having a geometric distribution $(P(N = m) = q(1-q)^m, m = 0, 1, \ldots)$. The random variables

$$\xi_1(x,t) = \frac{K^N F(x,t)}{q(1-q)^N}, \qquad \xi_2(x,t) = \sum_{m=0}^{N} \frac{K^m F(x,t)}{(1-q)^m}$$

are unbiased estimators for $u(x,t)$. We execute m times the procedure of evaluation of the integral in (11) to determine the unbiased estimator for $K^m F(x,t)$. This procedure is similar to the procedure of evaluation of the integral (3.8) in [8]. Namely, let $S_1^0(x,t) = \{\omega \in R^n | \omega' A^{-1}(x,t)\omega = 1\}$ be an ellipsoid centered at zero, and let $\sigma_n = 2\pi^{\frac{n}{2}}/\Gamma(\frac{n}{2})$ be an area of the sphere of radius 1 in R^n. The random vector Ω is distributed on $S_1^0(x,t)$ with density

$$p(x,t,\omega) = \frac{1}{\sigma_n \sqrt{\det(A(x,t))}|A^{-1}(x,t)\omega|}. \tag{22}$$

After the calculation of the kernel $K_1(x,y,t,\tau) = -Mv_0(y,\tau)$, we have

$$\begin{aligned}
Ku(x,t) = &-\int_0^t d\tau \int_0^\infty dr E \frac{n - \mathrm{Tr}\left(A(x+r\Omega,\tau)A^{-1}(x,t)\right)}{(t-\tau)\Gamma(\frac{n}{2})(4(t-\tau))^{\frac{n}{2}}} \\
&\times \exp\left(-\frac{r^2}{4(t-\tau)}\right) r^{n-1} u(x+r\Omega,\tau) \\
&+ \int_0^t d\tau \int_0^\infty dr E \frac{2r^2\left[\Omega' A^{-1}(x,t)A(x+r\Omega,\tau)A^{-1}(x,t)\Omega - 1\right]}{4(t-\tau)^2\Gamma(\frac{n}{2})(4(t-\tau))^{\frac{n}{2}}} \\
&\times \exp\left(-\frac{r^2}{4(t-\tau)}\right) r^{n-1} u(x+r\Omega,\tau) \\
&+ \int_0^t d\tau \int_0^\infty dr E\left(\frac{rd'(x+r\Omega,\tau)A^{-1}(x,t)\Omega}{(t-\tau)\Gamma(\frac{n}{2})(4(t-\tau))^{\frac{n}{2}}}\right. \\
\times \exp\left(-\frac{r^2}{4(t-\tau)}\right) &r^{n-1} u(x+r\Omega,\tau) - \int_0^t d\tau \int_0^\infty dr E d_0(x+r\Omega,\tau) \\
&\times u(x+r\Omega,\tau) \frac{2r^{n-1}}{\Gamma(\frac{n}{2})(4(t-\tau))^{\frac{n}{2}}} \exp\left(-\frac{r^2}{4(t-\tau)}\right),
\end{aligned} \tag{23}$$

where d' denotes the transposed vector $d^\top = (d_1, d_2, \ldots, d_n)$ and $\mathrm{Tr}(A)$ denotes the trace of the matrix A. E is the mathematical expectation of the function of random variable Ω.

All coefficients in the Equation (2) belong to the Hölder class. Hence, we can simplify the expressions in (23):

$$
\begin{aligned}
& n - \mathrm{Tr}\left(A(x+r\Omega,\tau)A^{-1}(x,t)\right) \\
&= \mathrm{Tr}\left(\left[A(x,t) - A(x+r\Omega,t)\right] A^{-1}(x,t)\right) \\
&+ \mathrm{Tr}\left(\left[A(x+r\Omega,t) - A(x+r\Omega,\tau)\right] A^{-1}(x,t)\right) \\
&= \tilde{g}_1(x+r\Omega,x,t)r^{\alpha} + \tilde{g}_2(x+r\Omega,x,\tau,t)(t-\tau)^{\frac{\alpha}{2}},
\end{aligned}
\tag{24}
$$

$$
\begin{aligned}
& \left[\Omega' A^{-1}(x,t)A(x+r\Omega,\tau)A^{-1}(x,t)\Omega - 1\right] \\
&= \Omega' A^{-1}(x,t)\left[A(x+r\Omega,t) - A(x,t)\right] A^{-1}(x,t)\Omega \\
&+ \Omega' A^{-1}(x,t)\left[A(x+r\Omega,\tau) - A(x+r\Omega,t)\right] A^{-1}(x,t)\Omega \\
&= \tilde{h}_1(x+r\Omega,x,t)r^{\alpha} + \tilde{h}_2(x+r\Omega,x,\tau,t)(t-\tau)^{\frac{\alpha}{2}},
\end{aligned}
\tag{25}
$$

where $\tilde{g}_1, \tilde{g}_2, \tilde{h}_1, \tilde{h}_2$ are bounded functions.

Substituting these expressions into (23) and putting $s = r^2/4(t-\tau)$, we obtain the following representation for $Ku(x,t)$:

$$
\begin{aligned}
Ku(x,t) &= \int_0^t d\tau (t-\tau)^{\frac{\alpha}{2}-1} \int_0^{\infty} ds \frac{2^{\alpha}}{2\Gamma(\frac{n}{2})} s^{\frac{n+\alpha}{2}-1} \exp(-s) \\
&\times E\left(\tilde{g}_1(x+2\sqrt{s(t-\tau)}\Omega,x,t)u(x+2\sqrt{s(t-\tau)}\Omega,\tau)\right) \\
&+ \int_0^t d\tau (t-\tau)^{\frac{\alpha}{2}-1} \int_0^{\infty} ds \frac{1}{2\Gamma(\frac{n}{2})} s^{\frac{n}{2}-1} \exp(-s) \\
&\times E\left(\tilde{g}_2(x+2\sqrt{s(t-\tau)}\Omega,x,\tau,t)u(x+2\sqrt{s(t-\tau)}\Omega,\tau)\right) \\
&+ \int_0^t d\tau (t-\tau)^{\frac{\alpha}{2}-1} \int_0^{\infty} ds \frac{2^{\alpha}}{\Gamma(\frac{n}{2})} s^{\frac{n+2+\alpha}{2}-1} \exp(-s) \\
&\times E\left(\tilde{h}_1(x+2\sqrt{s(t-\tau)}\Omega,x,t)u(x+2\sqrt{s(t-\tau)}\Omega,\tau)\right) \\
&+ \int_0^t d\tau (t-\tau)^{\frac{\alpha}{2}-1} \int_0^{\infty} ds \frac{1}{\Gamma(\frac{n}{2})} s^{\frac{n+2}{2}-1} \exp(-s) \\
&\times E\left(\tilde{h}_2(x+2\sqrt{s(t-\tau)}\Omega,x,\tau,t)u(x+2\sqrt{s(t-\tau)}\Omega,\tau)\right) \\
&+ \int_0^t d\tau (t-\tau)^{-\frac{1}{2}} \int_0^{\infty} ds \frac{1}{\Gamma(\frac{n}{2})} s^{\frac{n+1}{2}-1} \exp(-s) \\
&\times E\left(d'(x+2\sqrt{s(t-\tau)}\Omega,\tau)A^{-1}(x,t)\Omega u(x+2\sqrt{s(t-\tau)}\Omega,\tau)\right) \\
&- \int_0^t d\tau \int_0^{\infty} ds \frac{1}{\Gamma(\frac{n}{2})} s^{\frac{n}{2}-1} \exp(-s) \\
&\times E\left(d_0(x+2\sqrt{s(t-\tau)}\Omega,\tau)u(x+2\sqrt{s(t-\tau)}\Omega,\tau)\right).
\end{aligned}
\tag{26}
$$

The unbiased estimator $\tilde{\eta}(x,t)$ for $Ku(x,t)$ has the form:

$$
\begin{aligned}
\tilde{\eta}(x,t) = t^{\frac{\alpha}{2}} \frac{2^{\alpha}\Gamma(\frac{n+\alpha}{2})}{\alpha\Gamma(\frac{n}{2})} \tilde{g}_1 &\left(x + 2\sqrt{\gamma\left(\frac{n+\alpha}{2}\right)} t\vartheta\Omega, x, t \right) \\
\times u &\left(x + 2\sqrt{\gamma\left(\frac{n+\alpha}{2}\right)} t\vartheta\Omega, t - t\vartheta \right) \\
+ t^{\frac{\alpha}{2}} \frac{1}{\alpha} \tilde{g}_2 &\left(x + 2\sqrt{\gamma\left(\frac{n}{2}\right)} t\vartheta\Omega, x, t - t\vartheta, t \right) \\
\times u &\left(x + 2\sqrt{\gamma\left(\frac{n}{2}\right)} t\vartheta\Omega, t - t\vartheta \right) \\
+ t^{\frac{\alpha}{2}} \frac{2^{\alpha+1}\Gamma(\frac{n+2+\alpha}{2})}{\alpha\Gamma(\frac{n}{2})} \tilde{h}_1 &\left(x + 2\sqrt{\gamma\left(\frac{n+2+\alpha}{2}\right)} t\vartheta\Omega, x, t \right) \\
\times u &\left(x + 2\sqrt{\gamma\left(\frac{n+2+\alpha}{2}\right)} t\vartheta\Omega, t - t\vartheta \right) \\
+ t^{\frac{\alpha}{2}} \frac{n}{\alpha} \times \tilde{h}_2 &\left(x + 2\sqrt{\gamma\left(\frac{n+2}{2}\right)} t\vartheta\Omega, x, t - t\vartheta, t \right) \\
\times u &\left(x + 2\sqrt{\gamma\left(\frac{n+2}{2}\right)} t\vartheta\Omega, t - t\vartheta \right) \\
+ t^{\frac{1}{2}} \frac{2\Gamma(\frac{n+1}{2})}{\Gamma(\frac{n}{2})} d' &\left(x + 2\sqrt{\gamma\left(\frac{n+1}{2}\right)} t\delta\Omega, t - t\delta \right) A^{-1}(x,t)\Omega \\
\times u &\left(x + 2\sqrt{\gamma\left(\frac{n+1}{2}\right)} t\delta\Omega, t - t\delta \right) \\
- t d_0 \left(x + 2\sqrt{\gamma\left(\frac{n}{2}\right)} t\theta\Omega, t - t\theta \right) u &\left(x + 2\sqrt{\gamma\left(\frac{n}{2}\right)} t\theta\Omega, t - t\theta \right),
\end{aligned} \tag{27}
$$

where the random variables $\vartheta, \delta, \theta$ are distributed on the interval $[0,1]$. The variables ϑ and δ have densities $(\alpha/2)s^{\alpha/2-1}$ and $1/(2\sqrt{s})$, respectively, and θ is distributed uniformly. The variable $\gamma(m)$ has a gamma distribution with a density $s^{m-1}e^{-s}/\Gamma(m)$.

Choosing one of the summands in (27) with probability $\frac{1}{6}$ and multiplying it by 6, we obtain the final unbiased estimator $\tilde{\zeta}(x,t)$ for $Ku(x,t)$.

The unbiased estimators ψ_m for $K^m F(x,t)$ can be constructed on trajectories of the inhomogeneous Markov chain $\{(x_k, t_k)\}_{k=0}^{\infty}$ with initial point (x,t). Consider stochastically independent random elements $\{\vartheta_k\}_{k=0}^{\infty}$, $\{\delta_k\}_{k=0}^{\infty}$, $\{\theta_k\}_{k=0}^{\infty}$, $\{\Omega_k\}_{k=0}^{\infty}$. The initial value of the variable ψ_m is 1. At step k we consider $\tilde{\zeta}(x_{k-1}, t_{k-1})$ and multiply the variable ψ_m by the corresponding weight factor. The arguments of the function u determine the next state of the Markov chain. For example, if the first summand of the estimator (27) was chosen at step k, then we multiply variable ψ_m by

$$
6t_{k-1}^{\frac{\alpha}{2}} \frac{2^{\alpha}\Gamma(\frac{n+\alpha}{2})}{\alpha\Gamma(\frac{n}{2})} \tilde{g}_1 \left(x_{k-1} + 2\sqrt{\gamma_{k-1}\left(\frac{n+\alpha}{2}\right)} t_{k-1}\vartheta_{k-1}\Omega_{k-1}, x_{k-1}, t_{k-1} \right)
$$

and define the next point (x_k, t_k) by formulas:

$$
x_k = x_{k-1} + 2\sqrt{\gamma_{k-1}\left(\frac{n+\alpha}{2}\right)} t_{k-1}\vartheta_{k-1}\Omega_{k-1},
$$

$$
t_k = t_{k-1} - t_{k-1}\vartheta_{k-1}.
$$

After m steps, we multiply the variable ψ_m by an estimator for $F(x_m, t_m)$ which is equal to

$$t_m f\left(x_m + 2\sqrt{\gamma_m\left(\frac{n}{2}\right) t_m \theta_m} \Omega_m, t_m - t_m\theta_m\right) + \varphi\left(x_m + 2\sqrt{\gamma_m\left(\frac{n}{2}\right) t_m} \Omega_m\right).$$

So, the random variables

$$\tilde{\xi}_1(x,t) = \frac{\psi_N}{q(1-q)^N}, \qquad \tilde{\xi}_2(x,t) = \sum_{m=0}^{N} \frac{\psi_m}{(1-q)^m}$$

are unbiased estimators for $u(x,t)$. Repeating the arguments of the proof of Theorem 1 in [8], it is easy to prove that constructed estimators have finite variances.

Remark 1. *The unbiased estimators constructed above and the algorithm for calculating them can be used in the Monte Carlo method to find $u(x,t)$. This computational algorithm is more complex than the algorithm in Section 3. On the other hand this algorithm does not require an estimate of spectrum of matrix $A(x,t)$.*

Funding: The research is supported by the Russian Foundation for Basic Research, project No. 17-01-00267.

Conflicts of Interest: The authors declare no conflict of interest.

References

1. Ladyzhenskaya, O.A.; Solonnikov, V.A.; Uraltseva, N.N. *Linear and Quasilinear Equations of Parabolic Type*; Nauka: Moscow, Russia, 1967. (In Russian)
2. Wagner, W. Unbiased Monte Carlo estimators for functionals of weak solutions of stochastic diffretial equations. *Stoch. Stoch. Rep.* **1989**, *28*, 1–20 [CrossRef]
3. Ermakov, S.M.; Mikhailov, G.A. *Statistical Modeling*; Nauka: Moscow, Russia, 1982. (In Russian)
4. Wagner, W. Unbiased Monte Carlo evaluation of certain functional integrals. *J. Comput. Phys.* **1987**, *71*, 21–33. [CrossRef]
5. Wagner, W. Unbiased Multi-step Estimators for the Monte Carlo Evaluation of Certain Functional Integrals. *J. Comput. Phys.* **1988**, *79*, 336–352. [CrossRef]
6. Wagner, W. Monte Carlo evaluation of functionals of solutions of stochastic differential equations. Variance reduction and numerical examples. *Stoch. Anal. Appl.* **1988**, *6*, 447–468. [CrossRef]
7. Sipin, A.S. Statistical Algorithms for Solving the Cauchy Problem for Second-Order Parabolic Equations. *Vestn. Peterburg Univ. Math.* **2011**, *45*, 65–74. [CrossRef]
8. Sipin, A.S. Statistical Algorithms for Solving the Cauchy Problem for Second-Order Parabolic Equations: The "Dual" Scheme. *Vestn. Peterburg Univ. Math.* **2012**, *45*, 57–67. [CrossRef]
9. Sabelfeld, K.K. *Monte Carlo Methods in Boundary Value Problems*; Nauka: Novosibirsk, Russia, 1989. (In Russian)
10. Simonov, N.A. Stochastic iterative methods for solving equations of parabolic type. *Sib. Mat. Zhurnal* **1997**, *38*, 1146–1162.
11. Heinrich, S. *Multilevel Monte Carlo Methods*; Volume 2179 of Lecture Notes in Computer Science; Springer: Berlin, Germany, 2001; pp. 58–67.
12. Giles, M.B. Multi-Level Monte Carlo Path Simulation. *Oper. Res.* **2008**, *56*, 607–617. [CrossRef]

mathematics

MDPI

Article

Cumulative Measure of Inaccuracy and Mutual Information in k-th Lower Record Values

Maryam Eskandarzadeh [1], **Antonio Di Crescenzo** [2,*] and **Saeid Tahmasebi** [1]

1 Department of Statistics, Persian Gulf University, Bushehr 7516913817, Iran; eskandarymaryam@gmail.com (M.E.); tahmasebi@pgu.ac.ir (S.T.)
2 Dipartimento di Matematica, Università degli Studi di Salerno, Via Giovanni Paolo II n. 132, 84084 Fisciano (SA), Italy
* Correspondence: adicrescenzo@unisa.it

Received: 17 December 2018; Accepted: 5 February 2019; Published: 14 February 2019

Abstract: In this paper, we discuss the cumulative measure of inaccuracy in k-lower record values and study characterization results of dynamic cumulative inaccuracy. We also present some properties of the proposed measures, and the empirical cumulative measure of inaccuracy in k-lower record values. We prove a central limit theorem for the empirical cumulative measure of inaccuracy under exponentially distributed populations. Finally, we analyze the mutual information for measuring the degree of dependency between lower record values, and we show that it is distribution-free.

Keywords: measure of information; cumulative inaccuracy; mutual information; lower record values

MSC: 62N05; 94A17

---

1. Introduction and Background

The Information Theory provides various concepts of broad use in Probability and Statistics which are finalized to measure the information content of stochastic models. Apart from the classical differential entropy, which constitutes a useful tool for the analysis of absolutely continuous random variables, some information measures based on cumulative notions have been attracting an increasing amount of attention in the recent literature. Among such measures, in this paper we focus on the cumulative (past) inaccuracy of bivariate random lifetimes, which is a suitable extension of the cumulative entropy. We also deal with the mutual information, which is strictly related to the Shannon entropy, and is one of the most commonly adopted notions for bivariate random variables. Indeed, the mutual information is a measure of the mutual dependence of two random variables, and can be evaluated by means of the joint (and marginal) distributions.

We recall some recent papers dealing with stochastic models and information measures of interest in the reliability theory. Navarro et al. [1] presented some stochastic ordering and properties of aging classes of dynamic cumulative residual entropy, where Psarrakos and Navarro [2] generalized the concept of cumulative residual entropy by relating this concept to the mean time between record values, and also considered the dynamic version of this new measure. Moreover, Tahmasebi and Eskandarzadeh [3] proposed a new extension of the cumulative entropy based on k-th lower record values. Sordo and Psarrakos [4] provided comparison results for the cumulative residual entropy of systems and their dynamic versions.

Motivated by some of the articles mentioned above, in this paper we aim to investigate some applications of the previously mentioned information measures to the k-lower record values. Record values are widely studied in the literature as a suitable tool to convey essential information in stochastic models. More recently, they have also been attracting attention in applied contexts related to high-dimensional data, where it is computationally more convenient to determine the rank rather than the specific values of the observations under investigation.

More specifically, within the scope of this paper, we propose to study the cumulative measure of inaccuracy in k-lower record values and the parent random variable of a random sample. The context of dynamic observations will be also considered by analyzing the dynamic cumulative inaccuracy and related characterization results. We present some properties of the proposed measures, as well as the empirical cumulative measure of inaccuracy in k-lower record values. Moreover, the investigation focuses also on the mutual information measure, finalized to measure the degree of dependency between lower record values.

In the remaining part of this section, we recall the relevant notions that will be used in the following, and provide the plan of the paper. Specifically, we discuss the basic definitions and properties of the information measures mentioned above, i.e., the cumulative inaccuracy and the mutual information. Then, we mention the essential results on the lower record values and recall certain useful stochastic orders. We shall look into nonnegative random variables, with the case of truncated support being treatable in a similar way.

Throughout the paper, "log" means natural logarithm, prime denotes derivative, and the terms "increasing" and "decreasing" are used in a non-strict sense. Finally, as is customary, we assume that $0 \log 0$ is vanishing.

1.1. Notions of Information Theory

Consider an absolutely continuous random vector (X, Y) having nonnegative components. We denote by $f(x, y)$ the joint probability density function (PDF) by $f(x)$ and $g(x)$, the marginal PDFs, and by $F(x)$ and $G(x)$, the cumulative distribution functions (CDFs) of X and Y, respectively.

Bearing in mind the applications in the reliability theory, we assume that X and Y denote random lifetimes of suitable systems having support $(0, \infty)$. Let

$$H(X) = -\mathbb{E}[\log(f(X))] = -\int_0^\infty f(x)\log(f(x))\,dx \tag{1}$$

be the (Shannon) differential entropy of X. $H(Y)$ is defined similarly, and the bivariate entropy of (X, Y) is given by $H(X, Y) = -\mathbb{E}[\log(f(X, Y))] = -\int_0^\infty dx \int_0^\infty f(x, y)\log(f(x, y))\,dy$. Moreover, the conditional entropy of Y given by X is expressed by:

$$H(Y|X) = -\int_0^\infty dx \int_0^\infty f(x, y)\log\left(\frac{f(x, y)}{f(x)}\right) dy, \tag{2}$$

whereas the mutual information of (X, Y) is defined as:

$$M_{X,Y} = \int_0^\infty dx \int_0^\infty f(x, y)\log\left(\frac{f(x, y)}{f(x)\,g(y)}\right) dy. \tag{3}$$

We recall that $M_{X,Y}$ is a measure of dependence between X and Y, with $M_{X,Y} = 0$ if, and only if X and Y are independent. Furthermore, $M_{X,Y}$ is largely adopted to assess the information content in a variety of applied fields, such as signal processing and pattern recognition. Due to (3), the mutual information can be expressed in terms of suitable entropies as follows (see, for example, Ebrahimi et al. [5]):

$$M_{X,Y} = H(X) + H(Y) - H(X, Y) = H(Y) - H(Y|X). \tag{4}$$

See also Ebrahimi et al. [6] and Ahmadi et al. [7] for various results of interest in the reliability theory involving dynamic measures for multivariate distributions based on the mutual information.

Among the notions involving cumulative versions of information measures, we now recall the cumulative (past) inaccuracy of (X, Y), given by:

$$I(X,Y) = -\int_0^\infty F(x)\log(G(x))\,dx. \quad (5)$$

As specified in Section 1 of Kundu et al. [8], the measure given in (5) can be viewed as the cumulative analogue of the Kerridge inaccuracy measure of X and Y, which is expressed as $-\int_0^\infty f(x)\log(g(x))\,dx$ (cf. Kerridge [9]). The relevant difference is that Equation (5) involves the CDFs instead of the PDFs. In many real situations, it is more convenient to deal with distribution functions which carry information about the fact that an event occurs prior or after the current time. Moreover, the measure given in (5) provides information content when using $G(x)$, the CDF asserted by the experimenter due to missing or incorrect information in experiments, instead of the true distribution $F(x)$. Clearly, if F and G are identical, then $I(X,Y)$ identifies with the cumulative entropy studied by Di Crescenzo and Longobardi [10] and by Navarro et al. [1]. See Section 5 of Kumar and Taneja [11] for various results involving (5) and the related dynamic version, i.e., the dynamic cumulative past inaccuracy measure. We finally recall that the cumulative inaccuracy (5) and the cumulative entropy are also involved in the definition of other information measures of interest (see, for instance, Park et al. [12], and Di Crescenzo and Longobardi [13] for the cumulative Kullback-Leibler information).

1.2. Lower Record Values

Record values are often studied in various fields due to their relevance in specific applications. If statistical observations are difficult to obtain, or when experimental observations are destroyed and access to them is not available, then the researchers are forced to make inference about the distribution of the observations of used record amounts. Suppose, for example, that it is required to estimate the water level of a river solely based on the available records of previous flooding. Similarly, consider variables such as record rainfall, record temperature, wind speed record, and other quantities of interest in meteorology. In such cases, the analysis of statistical observations, if performed by resorting to lower or upper record values.

Let us now recall some basic notions about lower record values that will be used in this paper. Let X be an absolutely continuous nonnegative random variable with CDF $F(x)$ and PDF $f(x)$, and let $\{X_n, n \geq 1\}$ be a sequence of independent random variables, distributed identically as X. An observation X_j, $j \geq 1$, will be called a lower record value if its value is less than the values of all previous observations. Thus, X_j is a lower record value if $X_j < X_i$ for every $i < j$. For a fixed positive integer k, similarly to Dziubdziela and Kopociński [14], we define the sequence $\{T_{n(k)}, n \geq 1\}$ of k-th lower record times for the sequence $\{X_n, n \geq 1\}$ as follows:

$$T_{1(k)} = 1, \qquad T_{n+1(k)} = \min\left\{j > T_{n(k)} : X_{k:T_{n(k)}+k-1} > X_{k:k+j-1}\right\}, \quad (6)$$

where $X_{j:m}$ denotes the j-th order statistic in a sample of size m (see also Malinowska and Szynal [15]). Then,

$$L_{n(k)} := X_{k:T_{n(k)}+k-1},$$

is called a sequence of k-th lower record values of $\{X_n, n \geq 1\}$. Since the ordinary record values are contained in the k-records, the results for usual records can be obtained as a special case by setting $k = 1$. The PDF of $L_{n(k)}$, for $n \geq 1$ and $k \geq 1$, is given by:

$$f_{n(k)}(x) = \frac{k^n}{(n-1)!}[F(x)]^{k-1}[\tilde{\Lambda}(x)]^{n-1}f(x), \quad (7)$$

and the joint PDF of $(L_{m(k)}, L_{n(k)})$, for $1 \le m < n, k \ge 1$, is:

$$\begin{aligned} f_{m(k),n(k)}(x,y) = {} & \frac{k^n}{(m-1)!(n-m-1)!}[\widetilde{\Lambda}(y) - \widetilde{\Lambda}(x)]^{n-m-1} \\ & \times [\widetilde{\Lambda}(x)]^{m-1} h(x) [F(y)]^{k-1} f(y), \qquad x > y, \end{aligned} \tag{8}$$

where

$$\widetilde{\Lambda}(x) = -\log F(x) \quad \text{and} \quad h(x) = -\widetilde{\Lambda}'(x) = \frac{f(x)}{F(x)}, \qquad x > 0, \tag{9}$$

are the cumulative reversed hazard rate and the reversed hazard rate of X, respectively. Hence, the conditional PDF of $L_{n(k)}$ given $L_{m(k)}$, $1 \le m < n$, is given by:

$$f_{n|m}(y|x) = \frac{k^n [\widetilde{\Lambda}(y) - \widetilde{\Lambda}(x)]^{n-m-1} f(y) [F(y)]^{k-1}}{k^m (n-m-1)! [F(x)]^k}, \qquad x > y. \tag{10}$$

By using the well-known relation

$$\int_z^\infty \frac{\lambda^n}{(n-1)!} x^{n-1} e^{-\lambda x} dx = \sum_{i=0}^{n-1} \frac{[\lambda z]^i}{i!} e^{-\lambda z},$$

we see that the CDF corresponding to Equation (7) can be obtained as:

$$\begin{aligned} F_{n(k)}(x) &= \int_0^x \frac{k^n}{(n-1)!} [F(y)]^{k-1} [\widetilde{\Lambda}(y)]^{n-1} f(y) dy \\ &= [F(x)]^k \sum_{i=0}^{n-1} \frac{[k\widetilde{\Lambda}(x)]^i}{i!}. \end{aligned} \tag{11}$$

We note that the sequence of k-th *upper* record times is denoted as $\{U_{n(k)}, n \ge 1\}$. It is defined similarly to $T_{n(k)}$ by reverting the last inequality in (6) (see, for instance, Dziubdziela and Kopociński [14] or Tahmasebi et al. [16]). Record values apply in problems such as industrial stress testing, meteorological analysis, hydrology, sport, and economics. In reliability theory, record values are used to study things such as technical systems which are subject to shocks, e.g., peaks of voltages. For more details about records and their applications, one may refer to Arnold et al. [17]. Several authors investigated measures of inaccuracy for ordered random variables. Thapliyal and Taneja [18] proposed the measure of inaccuracy between the i-th order statistic and the parent random variable. Moreover, Thapliyal and Taneja [19] developed measures of dynamic cumulative residual and past inaccuracy. They studied characterization results of these dynamic measures under a proportional hazard model and proportional reversed hazard model. The same authors introduced the measure of residual inaccuracy of order statistics and proved a related characterization result (cf. Thapliyal and Taneja [20]). Equality of Rényi entropies of upper and lower k-records is also known to provide a characteristic property of symmetric distributions (see Fashandi and Ahmadi [21]). Furthermore, it is worth mentioning that the analysis of lower record values is related to the generalized cumulative entropy. For instance, its role in the study of a new measure of association based on the log-odds rate has recently been pinpointed by Asadi [22]. Finally, recent contributions on a measure of past entropy for nth upper k-record values can be found in Goel et al. [23].

1.3. Stochastic Orders and Related Notions

Aiming to use stochastic orders to perform suitable comparisons, here we recall some relevant definitions. Let X and Y be random variables, where X is said to be smaller than Y, according to the

- *usual stochastic ordering* (denoted by $X \le_{st} Y$) if $\mathbb{P}(X \ge x) \le \mathbb{P}(Y \ge x)$ for all $x \in \mathbb{R}$; it is known that $X \le_{st} Y \Leftrightarrow \mathbb{E}(\phi(X)) \le \mathbb{E}(\phi(Y))$ for all increasing functions ϕ;

- *likelihood ratio ordering* (denoted by $X \leq_{lr} Y$) if $\frac{g(x)}{f(x)}$ is increasing in x;
- *decreasing convex order*, denoted by $X \leq_{dcx} Y$, if $\mathbb{E}(\phi(X)) \leq \mathbb{E}(\phi(Y))$ for all decreasing convex functions ϕ, such that the expectations exist.

Moreover, we say that X has a *decreasing reversed hazard rate* (DRHR) if $h(x) = \frac{f(x)}{F(x)}$ is decreasing in x. For specific details on these notions, see, for instance, Shaked and Shanthikumar [24], and for applications of the decreasing convex order, see Ma [25].

1.4. Plan of the Paper

In this investigation, we propose the cumulative measure of inaccuracy and study characterization results of a dynamic cumulative inaccuracy measure. Also, we study the degree of dependency among the sequence of k-th lower record values through the mutual information of record values.

The paper is organized as follows: In Section 2, we consider a measure of inaccuracy associated with $L_{n(k)}$ and X. We provide some results and properties of such a measure, including an application to the proportional reversed hazards model. In Section 3, we propose the dynamic version of inaccuracy associated with $L_{n(k)}$ and X, and provide a characterization result. In Section 4, we study the problem of estimating the cumulative measure of inaccuracy by means of the empirical cumulative inaccuracy in k-lower record values. The rest of the section is devoted to a simple application to real data, with the discussion of some special cases, and a central limit theorem for the empirical cumulative measure of inaccuracy in the case of exponentially distributed random samples. Finally, in Section 5 we investigate the mutual information between sequences of lower record values, aiming to measure their degree of dependency. Specifically, we show that this measure is distribution-free and can be computed by using the distribution of the k-th lower record values of the sequence from the uniform distribution.

2. Cumulative Measure of Inaccuracy

Let us now consider the cumulative measure of inaccuracy between $L_{n(k)}$ and the parent non-negative random variable, say X. Recalling (5) and (11), we have:

$$\begin{aligned} I(L_{n(k)}, X) &= -\int_0^\infty F_{n(k)}(x) \log(F(x))\, dx \\ &= \sum_{i=0}^{n-1} \frac{k^i}{i!} \int_0^\infty [F(x)]^k [\widetilde{\Lambda}(x)]^{i+1} dx, \end{aligned} \tag{12}$$

with $\widetilde{\Lambda}(x)$ given in (9). According to the comments given in Section 1.1, $I(L_{n(k)}, X)$ can be used to gain information concerning an experiment for which the distribution of the k-lower record values is compared with the parent distribution. Noting that, due to (7), $L_{i+2(k)}$ being a random variable with density function

$$f_{i+2(k)}(x) = \frac{k^{i+2}}{(i+1)!}[F(x)]^{k-1}[\widetilde{\Lambda}(x)]^{i+1} f(x),$$

and recalling that $h(x)$ is the reversed hazard rate of X (see (9)), from (12) we obtain:

$$\begin{aligned} I(L_{n(k)}, X) &= \sum_{i=0}^{n-1} \frac{i+1}{k^2} \int_0^\infty \frac{k^{i+2}}{(i+1)!}[F(x)]^k [\widetilde{\Lambda}(x)]^{i+1} dx \\ &= \sum_{i=0}^{n-1} \frac{i+1}{k^2} \mathbb{E}\left[\frac{1}{h(L_{i+2(k)})}\right]. \end{aligned} \tag{13}$$

Remark 1. *Making use of the generalized cumulative entropy introduced in Definition 1.1 of Tahmasebi and Eskandarzadeh [3], given by:*

$$\mathcal{CE}_{i+1,k}(X) = \int_0^\infty \frac{k^{i+2}}{(i+1)!}[F(x)]^k[\widetilde{\Lambda}(x)]^{i+1}dx \equiv \mathbb{E}\left[\frac{1}{h(L_{i+2(k)})}\right], \tag{14}$$

from (13) one has that the cumulative measure of inaccuracy between $L_{n(k)}$ and X can be expressed: as a size-biased combination of generalized cumulative entropies through the following weighted sum with linearly increasing weights:

$$I(L_{n(k)}, X) = \frac{1}{k^2}\sum_{i=1}^{n} i\,\mathcal{CE}_{i,k}(X). \tag{15}$$

Furthermore, from (11) and (12) we obtain an alternative expression, that is:

$$I(L_{n(k)}, X) = \sum_{i=0}^{n-1}\int_0^\infty \widetilde{\Lambda}(x)[F_{i+1(k)}(x) - F_{i(k)}(x)]dx.$$

In the following proposition we provide another form of $I(L_{n(k)}, X)$.

Proposition 1. *Let X be a nonnegative random variable with cdf F; for the cumulative measure of inaccuracy between $L_{n(k)}$ and X, we have:*

$$I(L_{n(k)}, X) = \sum_{i=0}^{n-1}\frac{k^i}{i!}\int_0^{+\infty} h(z)\left\{\int_0^z [F(x)]^k[\widetilde{\Lambda}(x)]^i dx\right\}dz. \tag{16}$$

Proof. By (12) and the relation $-\log F(x) = \int_x^\infty h(z)dz$, we have:

$$I(L_{n(k)}, X) = \sum_{i=0}^{n-1}\frac{k^i}{i!}\int_0^{+\infty}\int_x^\infty h(z)[F(x)]^k[\widetilde{\Lambda}(x)]^i dzdx.$$

By using Fubini's theorem, we get:

$$I(L_{n(k)}, X) = \sum_{i=0}^{n-1}\frac{k^i}{i!}\int_0^{+\infty}\int_0^z h(z)[F(x)]^k[\widetilde{\Lambda}(x)]^i dxdz,$$

and the result thus follows. □

Hereafter, we present some examples and properties of $I(L_{n(k)}, X)$.

Example 1.

(i) If X is uniformly distributed in $[0, \theta]$, then:

$$I(L_{n(k)}, X) = \frac{\theta}{k^2}\sum_{i=0}^{n-1}(i+1)\left(\frac{k}{k+1}\right)^{i+2}.$$

(ii) If X is exponentially distributed with mean $\frac{1}{\lambda}$, then:

$$I(L_{n(k)}, X) = \frac{1}{\lambda}\sum_{i=0}^{n-1}(i+1)k^i\sum_{j=0}^{\infty}\left(\frac{1}{j+k+1}\right)^{i+2}.$$

(iii) If X has an inverse Weibull distribution with cdf $F(x) = \exp\{-(\frac{\alpha}{x})^{\beta}\}$, $x > 0$, *with* $\alpha > 0$ *and* $\beta > 1$, *then:*

$$I(L_{n(k)}, X) = \frac{\alpha}{\beta} k^{\frac{1}{\beta}-1} \sum_{i=0}^{n-1} \frac{1}{i!} \Gamma\left(\frac{(i+1)\beta - 1}{\beta}\right).$$

For suitable choices of n and k, such inaccuracy measures are plotted in Figure 1, where the parameters are chosen so that the considered distributions have a unity mean, i.e., (i) $\theta = 2$, *(ii)* $\lambda = 1$, *and (iii)* $\alpha = 1/\Gamma(1-\beta^{-1})$, *recalling that the mean of an inverse Weibull distribution is* $\mathbb{E}(X) = \alpha\Gamma(1-\beta^{-1})$ *(see de Gusmão et al. [26]). In all cases,* $I(L_{n(k)}, X)$ *is decreasing in k and increasing in n.*

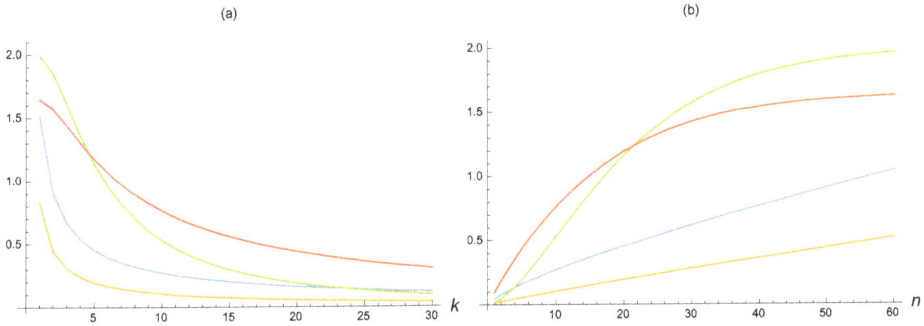

Figure 1. The values of $I(L_{n(k)}, X)$ related to Example 1; (**a**) for $1 \le k \le 30$ and $n = 10$, from top to bottom near the origin: cases (i), (ii), (iii) with $(\alpha, \beta) = (0.816049, 4)$, and (iii) with $(\alpha, \beta) = (0.935779, 10)$; (**b**) for $1 \le n \le 60$ and $k = 10$, from top to bottom for large n in the same cases as (**a**).

Let us now discuss the effect of identical linear transformations on $I(L_{n(k)}, X)$.

Proposition 2. *Let* $a > 0$ *and* $b \ge 0$; *for* $n \in \mathbb{N}$, *it holds that:*

$$I(aL_{n(k)} + b, aX + b) = aI(L_{n(k)}, X). \tag{17}$$

Proof. From (15), we have:

$$I(aL_{n(k)} + b, aX + b) = \frac{1}{k^2} \sum_{i=1}^{n} i\,\mathcal{CE}_{i,k}(aX + b).$$

Recalling (14), it is not hard to see that: $\mathcal{CE}_{i,k}(aX + b) = a\,\mathcal{CE}_{i,k}(X)$, so that the proof immediately follows. □

Remark 2. *Let X be a symmetric random variable with respect to the finite mean* $\mu = \mathbb{E}(X)$, *i.e.,* $F(x+\mu) = 1 - F(\mu - x)$ *for all* $x \in \mathbb{R}$. *Then the following relation holds:*

$$I(L_{n(k)}, X) = \overline{I}(R_{n(k)}, X) := -\int_0^{\infty} \overline{F}_{n(k)}(x) \log\left(\overline{F}(x)\right) dx,$$

where, similarly to (12), the latter term defines the cumulative residual measure of inaccuracy between the k-th upper record value and X. Here, as usual, $\overline{F}(x) = 1 - F(x)$ *denotes the survival function of X, and* $\overline{F}_{n(k)}(x)$ *denotes the survival function of the k-th upper record value* $R_{n(k)}$.

Hereafter, we obtain an upper bound for $I(L_{n(k)}, X)$.

Proposition 3. *Let X be an absolutely continuous non-negative random variable with a cumulative reversed hazard rate $\widetilde{\Lambda}(x)$, such that the following function is finite:*

$$h^{\star}_{i+1}(t) = \int_t^{\infty} [\widetilde{\Lambda}(x)]^{i+1} dx, \qquad t > 0. \tag{18}$$

Then, for $n \in \mathbb{N}$ we have:

$$I(L_{n(k)}, X) \leq \sum_{i=0}^{n-1} \frac{k^i}{i!} \mathbb{E}\left[h^{\star}_{i+1}(X)\right].$$

Proof. By (12), from $[F(x)]^k \leq F(x)]$ and Fubini's theorem, we obtain:

$$\begin{aligned} I(L_{n(k)}, X) &\leq \sum_{i=0}^{n-1} \frac{k^i}{i!} \int_0^{\infty} [\widetilde{\Lambda}(x)]^{i+1} \int_0^x f(t) dt dx \\ &= \sum_{i=0}^{n-1} \frac{k^i}{i!} \int_0^{\infty} f(t) \int_t^{\infty} [\widetilde{\Lambda}(x)]^{i+1} dx dt \\ &= \sum_{i=0}^{n-1} \frac{k^i}{i!} \mathbb{E}\left[h^{\star}_{i+1}(X)\right], \end{aligned}$$

thus completing the proof. □

Now we can prove a property of the considered inaccuracy measure by means of stochastic orderings. For that, we use the notions recalled in Section 1.3.

Theorem 1. *Suppose that the non-negative random variable X is DRHR. Then, for $n \in \mathbb{N}$, one has:*

$$I(L_{n+1(k)}, X) - I(L_{n(k)}, X) \leq \frac{1}{k^2} \sum_{m=1}^{n+1} \mathbb{E}\left[\frac{1}{h(L_{m(k)})}\right]. \tag{19}$$

Proof. Recalling that $f_{n(k)}(x)$ is the PDF of $L_{n(k)}$, provided in Equation (7), then the ratio $\frac{f_{n(k)}(x)}{f_{n+1(k)}(x)} = \frac{-n}{k \log F(x)}$ is increasing in x. Therefore, $L_{n+1(k)} \leq_{lr} L_{n(k)}$, and this implies that $L_{n+1(k)} \leq_{st} L_{n(k)}$, i.e., $\mathbb{E}[\phi(L_{n+1(k)})] \leq \mathbb{E}[\phi(L_{n(k)})]$ for all increasing functions ϕ such that these expectations exist. (For more details, see Shaked and Shanthikumar [24]). Thus, since X is DRHR and $h(x)$ is its reversed hazard rate, then $\frac{1}{h(x)}$ is increasing in x. As a consequence, from (13) we have:

$$\begin{aligned} I(L_{n+1(k)}, X) &= \sum_{i=0}^{n} \frac{i+1}{k^2} \mathbb{E}\left[\frac{1}{h(L_{i+2(k)})}\right] \\ &\leq \sum_{i=0}^{n} \frac{i+1}{k^2} \mathbb{E}\left[\frac{1}{h(L_{i+1(k)})}\right] \\ &= \sum_{j=-1}^{n-1} \frac{j+2}{k^2} \mathbb{E}\left[\frac{1}{h(L_{j+2(k)})}\right] \\ &= \sum_{j=0}^{n-1} \frac{j+2}{k^2} \mathbb{E}\left[\frac{1}{h(L_{j+2(k)})}\right] + \frac{1}{k^2} \mathbb{E}\left[\frac{1}{h(L_{1(k)})}\right] \\ &= I(L_{n(k)}, X) + \frac{1}{k^2} \sum_{i=1}^{n+1} \mathbb{E}\left[\frac{1}{h(L_{i(k)})}\right], \end{aligned}$$

where $I(L_{n(k)}, X)$ is expressed in (13). The proof is thus completed. □

Theorem 2. *Let X and Y be two non-negative random variables, such that $X \leq_{dcx} Y$; then we have:*

$$I(L_{n(k)}, X) \leq I(L^G_{n(k)}, Y) := -\int_0^\infty G_{n(k)}(y)\log(G(y))dy.$$

Here, similarly to (6), $L^G_{n(k)}$ denotes the k-th lower record times for the sequence $\{Y_n,\ n \geq 1\}$ with distribution function $G_{n(k)}(y)$, and $G(y)$ is the distribution function of Y.

Proof. Due to (18), $h^\star_{j+1}(x)$ is a decreasing convex function in x. The proof then immediately follows from Proposition 3. □

Let us now investigate the cumulative measure of inaccuracy within the proportional reversed hazards model (PRHM). We recall that two random variables X and X^*_θ satisfy the PRHM if their distribution functions are related by the following identity, for $\theta > 0$:

$$F^*_\theta(x) = [F(x)]^\theta, \qquad x \in \mathbb{R}. \tag{20}$$

For some properties of such a model associated with aging notions and the reversed relevation transform, see Gupta and Gupta [27] and Di Crescenzo and Toomaj [28], respectively.

In this case, we assume that X and X^*_θ are non-negative, absolutely continuous random variables. Due to Equation (20) and making use of (5) and (11), and noting that $\widetilde{\Lambda}^*_\theta(x) = \theta\widetilde{\Lambda}(x)$, we obtain the cumulative measure of inaccuracy between $L^*_{n(k)}$ and X^*_θ as follows, for $\theta > 0$:

$$\begin{aligned} I(L^*_{n(k)}, X^*_\theta) &= -\int_0^{+\infty} F^*_{n(k)}(x)\log\left(F^*_\theta(x)\right)dx \\ &= \sum_{i=0}^{n-1} k^i\theta^{i+1}\int_0^{+\infty}\frac{[\widetilde{\Lambda}(x)]^{i+1}}{i!}[F(x)]^{k\theta}dx, \end{aligned} \tag{21}$$

with $\widetilde{\Lambda}(x)$ expressed in (9). Moreover, if θ is a positive integer, then the last expression can be rewritten in terms of the generalized cumulative entropy (14) as follows:

$$I(L^*_{n(k)}, X^*_\theta) = \frac{1}{k^2\theta}\sum_{i=0}^{n-1}(i+1)\,\mathcal{CE}_{i+1,k\theta}(X).$$

We recall that in this case, i.e., when $\theta \in \mathbb{N}$, the PRHM expresses that X^*_θ is distributed as the first-order statistics of a random sample having size θ and taken from the distribution of X.

Let us now obtain suitable bounds under the PRHM.

Proposition 4. *Let X and X^*_θ be non-negative, absolutely continuous random variables satisfying the PRHM as specified in (20), with $\theta > 0$. If $\theta \geq (\leq)1$, then for any $n \in \mathbb{N}$ we have:*

$$I(L^*_{n(k)}, X^*_\theta) \leq (\geq)\frac{\theta^{i+1}}{k^2}\sum_{i=0}^{n-1}(i+1)\,\mathcal{CE}_{i+1,k}(X).$$

Proof. Clearly, for $\theta \geq (\leq)1$ it is $[F(x)]^{k\theta} \leq (\geq)[F(x)]^k$ for all $x \geq 0$, and then the thesis immediately follows from (14) and (21). □

We conclude this section with a remark on the cumulative measure of inaccuracy for bivariate first lower record values.

Remark 3. *Consider two identically distributed sequences of random variables* $\{X_n, n \geq 1\}$ *and* $\{Y_m, m \geq 1\}$, *and denote by* $L_{n,X} = L_{n(1),X}$ *and* $L_{m,Y} = L_{m(1),Y}$ *the corresponding first lower record times. Then, for* $k = 1$ *and making use of (11) it is not hard to see that the cumulative measure of inaccuracy between* $L_{n,X}$ *and* $L_{m,Y}$ *is given by:*

$$\begin{aligned} I(L_{n,X}, L_{m,Y}) &= -\int_0^{+\infty} F_n(x)\log(F_m(x)) \\ &= \sum_{i=0}^{n-1}(i+1)\mathbb{E}\left(\frac{1}{h(L_{i+1}(x))}\right) - \sum_{i=0}^{n-1}\int_0^{+\infty}\frac{F(x)[\widetilde{\Lambda}(x)]^i}{i!}\log\left(\sum_{i=0}^{m-1}\frac{[\widetilde{\Lambda}(x)]^i}{i!}\right), \end{aligned}$$

where $h(x)$ *is the reversed hazard rate of the underlying distribution.*

3. Dynamic Cumulative Measure of Inaccuracy

In this section, we study the dynamic version of the inaccuracy measure $I(L_{n(k)}, X)$. Let X be the random lifetime of a brand new system that begins to work at time 0 and is observed only at deterministic inspection times. Clearly, if the system is found failed at time t, then the conditional distribution function of $[X|X \leq t]$, known as the past lifetime, is given by $\dfrac{F(x)}{F(t)}$, $0 \leq x \leq t$. In a sequence of i.i.d. failure times having the same distribution as X, if the information about the k-th lower failure times is available, then $\dfrac{F_{n(k)}(x)}{F_{n(k)}(t)}$ is the conditional probability that the k-th lower failure time is smaller that x, given that it is smaller than t, for $0 \leq x \leq t$, where $F_{n(k)}(x)$ is the CDF given in (11). Hence, the dynamic cumulative measure of inaccuracy between $L_{n(k)}$ and X is expressed by the inaccuracy measure between the corresponding past lifetimes, i.e.,:

$$\begin{aligned} I(L_{n(k)}, X; t) &= -\int_0^t \frac{F_{n(k)}(x)}{F_{n(k)}(t)}\log\left(\frac{F(x)}{F(t)}\right)dx \\ &= \mu_{n(k)}(t)\log(F(t)) - \frac{1}{F_{n(k)}(t)}\int_0^t F_{n(k)}(x)\log(F(x))dx \\ &= \mu_{n(k)}(t)\log(F(t)) + \frac{1}{F_{n(k)}(t)}\sum_{i=0}^{n-1}\int_0^t \frac{k^i}{i!}[F(x)]^k[\widetilde{\Lambda}(x)]^{i+1}dx, \end{aligned} \tag{22}$$

where

$$\mu_{n(k)}(t) = \int_0^t \frac{F_{n(k)}(x)}{F_{n(k)}(t)}dx$$

is the mean inactivity time of the random variable $[t - L_{n(k)} \mid L_{n(k)} < t]$. Clearly, recalling (12) and assuming that X is a *bona fide* random variable, from (22) we have $\lim_{t\to\infty} I(L_{n(k)}, X; t) = I(L_{n(k)}, X)$. Moreover, if $F(t) > 0$ for all $t > 0$, since $\log F(t) \leq 0$, we immediately have:

$$I(L_{n(k)}, X; t) \leq \frac{I(L_{n(k)}, X)}{F_{n(k)}(t)}.$$

We can now obtain a characterization result for $I(L_{n(k)}, X; t)$.

Theorem 3. *Let* X *be a non-negative, absolutely continuous random variable with distribution function* $F(x)$. *If the dynamic cumulative inaccuracy (22) is finite for all* $t > 0$, *then* $I(L_{n(k)}, X; t)$ *characterizes the distribution function of* X.

Proof. Differentiating both sides of (22) with respect to t, we obtain:

$$\begin{aligned}\frac{d}{dt}I(L_{n(k)},X;t) &= h(t)\mu_{n(k)}(t) - h_{n(k)}(t)I(L_{n(k)},X;t)\\ &= h(t)\left[\mu_{n(k)}(t) - c(t)I(L_{n(k)},X;t)\right],\end{aligned}$$

where $h(t) = \dfrac{f(t)}{F(t)}$ and $h_{n(k)}(t) = \dfrac{f_{n(k)}(t)}{F_{n(k)}(t)}$ are the reversed hazard rates, and where we have set

$$c(t) = \frac{k^n[\widetilde{\Lambda}(t)]^{n-1}}{(n-1)!\sum_{i=0}^{n-1}\frac{k^i}{i!}[\widetilde{\Lambda}(t)]^i}.$$

Taking again the derivative with respect to t, we get:

$$h'(t) = \frac{(h(t))^2\left[c'(t)I(L_{n(k)},X;t) + c(t)\frac{d}{dt}I(L_{n(k)},X;t) - 1 + c(t)h(t)\mu_{n(k)}(t)\right]}{\frac{d}{dt}I(L_{n(k)},X;t)}. \tag{23}$$

Suppose that there are two CDFs, F and $\widehat{F}$, such that for all t,

$$I(L_{n(k)},X;t) = I(\widehat{L}_{n(k)},\widehat{X};t) = z(t),$$

and having reversed hazard rates $h(t)$ and $h_{\widehat{F}}(t)$, respectively. Then, from (23) we get, for all t:

$$h'(t) = \varphi(t,h(t)), \qquad h'_{\widehat{F}}(t) = \varphi(t,h_{\widehat{F}}(t)),$$

where

$$\varphi(t,y) := \frac{y^2\left[c'(t)z(t) + c(t)z'(t) - 1 + c(t)yw(t)\right]}{z'(t)},$$

for $w(t) := \mu_{n(k)}(t)$ and $y := h(t)$. By using Theorem 2.1 and Lemma 2.2 of Gupta and Kirmani [29], we obtain $h(t) = h_{\widehat{F}}(t)$, for all t. Since the reversed hazard rate function characterizes the distribution function uniquely, the proof is completed. □

4. Empirical Cumulative Measure of Inaccuracy

In this section, we address the problem of estimating the cumulative measure of inaccuracy by means of the empirical cumulative inaccuracy in lower record values. Let $X_1, X_2, \ldots, X_m$ be a random sample of size m from an absolutely continuous CDF $F(x)$. Then, according to (12), the *empirical cumulative measure of inaccuracy* is defined as:

$$\widehat{I}(L_{n(k)},X) = \sum_{i=0}^{n-1}\frac{k^i}{i!}\int_0^{\infty}[\widehat{F}_m(x)]^k\left[-\log\left(\widehat{F}_m(x)\right)\right]^{i+1}dx, \tag{24}$$

where

$$\widehat{F}_m(x) = \frac{1}{m}\sum_{i=1}^{m}\mathbf{1}_{\{X_i\le x\}}, \qquad x\in\mathbb{R}.$$

is the empirical distribution of the sample, $\mathbf{1}_{\{\cdot\}}$ being the indicator function. Let $X_{(1)} \le X_{(2)} \le \ldots \le X_{(m)}$ denote the order statistics of the sample. Then, (24) can be written as:

$$\widehat{I}(L_{n(k)},X) = \sum_{i=0}^{n-1}\frac{k^i}{i!}\sum_{j=1}^{m-1}\int_{X_{(j)}}^{X_{(j+1)}}[\widehat{F}_m(x)]^k\left[-\log\left(\widehat{F}_m(x)\right)\right]^{i+1}dx. \tag{25}$$

Finally, recalling that:

$$\widehat{F}_m(x) = \begin{cases} 0, & x < X_{(1)}, \\ \frac{j}{m}, & X_{(j)} \le x < X_{(j+1)}, \qquad j = 1, 2, \ldots, m-1 \\ 1, & x \ge X_{(m)}, \end{cases}$$

from Equation (25) we see that the empirical cumulative measure of inaccuracy can be expressed as:

$$\widehat{I}(L_{n(k)}, X) = \sum_{i=0}^{n-1} \frac{k^i}{i!} \sum_{j=1}^{m-1} U_{j+1} \left(\frac{j}{m}\right)^k \left[-\log\left(\frac{j}{m}\right)\right]^{i+1}, \tag{26}$$

where

$$U_{j+1} = X_{(j+1)} - X_{(j)}, \qquad j = 1, 2, \ldots, m-1 \tag{27}$$

are the sample spacings.

The following example provides an application of the empirical cumulative measure of inaccuracy to real data.

Example 2. *Consider the sample data by Abouammoh and Abdulghani [30] concerning the lifetimes (in days) of $m = 40$ patients suffering from blood cancer. The evaluation of the corresponding empirical cumulative measure of inaccuracy, obtained by means of Equation (26), shows that the values of $\widehat{I}(L_{n(k)}, X)$ are decreasing in k and increasing in n (see Figure 2).*

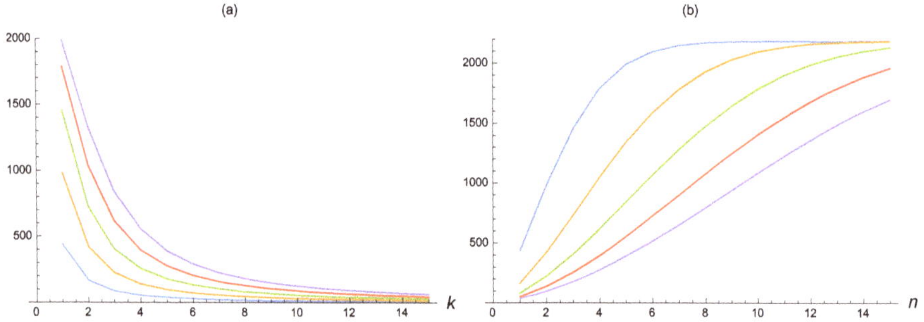

Figure 2. The values of $\widehat{I}(L_{n(k)}, X)$ concerning Example 2, (**a**) for $1 \le k \le 15$ and $n = 1, 2, 3, 4, 5$ (from bottom to top), and (**b**) for $1 \le n \le 15$ and $k = 1, 2, 3, 4, 5$ (from top to bottom).

Let us now discuss two special cases concerning populations from the uniform distribution and the exponential distribution.

Example 3. *Consider the random sample $X_1, X_2, \ldots, X_m$ from a population uniformly distributed in $[0, 1]$. In this case, the sample spacings (27) are independent and follow the beta distribution with parameters 1 and m (for more details, see Pyke [31]). Hence, making use of (26), the mean and the variance of the empirical cumulative measure of inaccuracy are, respectively:*

$$\mathbb{E}\left[\widehat{I}(L_{n(k)}, X)\right] = \sum_{i=0}^{n-1} \sum_{j=1}^{m-1} \frac{k^i}{i!(m+1)} \left(\frac{j}{m}\right)^k \left[-\log\left(\frac{j}{m}\right)\right]^{i+1}, \tag{28}$$

and

$$Var\left[\widehat{I}(L_{n(k)}, X)\right] = \sum_{i=0}^{n-1} \sum_{j=1}^{m-1} \frac{m}{m+2} \left[\frac{k^i}{i!(m+1)} \left(\frac{j}{m}\right)^k \left[-\log\left(\frac{j}{m}\right)\right]^{(i+1)}\right]^2. \tag{29}$$

Table 1 shows the values of the mean (28) and the variance (29) for $k = 2$, and for sample sizes $m = 10, 15, 20$, with $n = 2, 3, 4, 5$. We note that $\mathbb{E}[\widehat{I}(L_{n(2)}, X)]$ is increasing in m and n.

Table 1. Computed values of $\mathbb{E}[\widehat{I}(L_{n(2)}, X)]$ and $Var[\widehat{I}(L_{n(2)}, X)]$ for the uniform distribution.

	$\mathbb{E}[\widehat{I}(L_{n(2)}, X)]$				$Var[\widehat{I}(L_{n(2)}, X)]$			
m	$n=2$	$n=3$	$n=4$	$n=5$	$n=2$	$n=3$	$n=4$	$n=5$
10	0.23	0.37	0.48	0.57	0.003	0.006	0.008	0.010
15	0.24	0.38	0.50	0.59	0.002	0.004	0.006	0.008
20	0.25	0.39	0.51	0.61	0.002	0.003	0.005	0.006

Example 4. *Let $X_1, X_2, \ldots, X_m$ be a random sample drawn from the exponential distribution with parameter λ. Then, from (26) we see that the empirical cumulative measure of inaccuracy can be expressed as the following sum of independent and exponentially distributed random variables:*

$$\widehat{I}(L_{n(k)}, X) = \sum_{j=1}^{m-1} Y_j, \qquad \text{where} \quad Y_j := U_{j+1} \sum_{i=0}^{n-1} \frac{k^i}{i!} \left(\frac{j}{m}\right)^k \left[-\log\left(\frac{j}{m}\right)\right]^{i+1}. \tag{30}$$

Indeed, in this case, the sample spacings U_{j+1} defined in (27) are independent and exponentially distributed with mean $\dfrac{1}{\lambda(m-j)}$ (for more details, see Pyke [31]), so that the mean and the variance of $\widehat{I}(L_{n(k)}, X)$ are given by:

$$\mathbb{E}\left[\widehat{I}(L_{n(k)}, X)\right] = \sum_{j=1}^{m-1} \mu_j, \qquad Var\left[\widehat{I}(L_{n(k)}, X)\right] = \sum_{j=1}^{m-1} s_j^2, \tag{31}$$

where

$$\mu_j := \mathbb{E}[Y_j] = \frac{1}{\lambda} \sum_{i=0}^{n-1} \frac{k^i}{i!(m-j)} \left(\frac{j}{m}\right)^k \left[-\log\left(\frac{j}{m}\right)\right]^{i+1},$$

and

$$s_j^2 := Var\left[Y_j\right] = \frac{1}{\lambda^2} \sum_{i=0}^{n-1} \left\{\frac{k^i}{i!(m-j)} \left(\frac{j}{m}\right)^k \left[-\log\left(\frac{j}{m}\right)\right]^{i+1}\right\}^2.$$

In Table 2, for $k = 2$, we show the values of the mean and the variance (31) for sample sizes $m = 10, 15, 20$, with $\lambda = 0.5, 1, 2$ and $n = 2, 3, 4, 5$. One can easily see that $\mathbb{E}[\widehat{I}(L_{n(2)}, X)]$ is increasing in m, whereas $Var[\widehat{I}(L_{n(2)}, X)]$ is decreasing in m.

Table 2. Computed values of $\mathbb{E}[\widehat{I}(L_{n(2)}, X)]$ and $Var[\widehat{I}(L_{n(2)}, X)]$ for the exponential distribution.

	$\mathbb{E}[\widehat{I}(L_{n(2)}, X)]$											
λ	0.5	1	2	0.5	1	2	0.5	1	2	0.5	1	2
m		$n=2$			$n=3$			$n=4$			$n=5$	
10	1.30	0.65	0.33	1.77	0.89	0.44	2.12	1.063	0.53	2.37	1.19	0.59
15	1.33	0.67	0.33	1.81	0.91	0.45	2.17	1.086	0.54	2.43	1.22	0.61
20	1.35	0.68	0.34	1.83	0.92	0.46	2.19	1.096	0.55	2.46	1.23	0.62
	$Var[\widehat{I}(L_{n(2)}, X)]$											
λ	0.5	1	2	0.5	1	2	0.5	1	2	0.5	1	2
m		$n=2$			$n=3$			$n=4$			$n=5$	
10	0.13	0.032	0.008	0.16	0.04	0.009	0.18	0.046	0.011	0.20	0.050	0.012
15	0.09	0.022	0.005	0.11	0.028	0.007	0.12	0.031	0.008	0.14	0.035	0.008
20	0.07	0.017	0.004	0.08	0.021	0.005	0.09	0.024	0.006	0.11	0.026	0.007

Hereafter, we show a central limit theorem for the empirical cumulative measure of inaccuracy in the same case as Example 4.

Theorem 4. *If $X_1, X_2, \dots, X_m$ is a random sample drawn from the exponential distribution with parameter λ, then:*

$$\frac{\widehat{I}(L_{n(k)}, X) - \mathbb{E}[\widehat{I}(L_{n(k)}, X)]}{(Var[\widehat{I}(L_{n(k)}, X)])^{1/2}} \equiv \frac{\sum_{j=1}^{m-1}(Y_j - \mu_j)}{(\sum_{j=1}^{m-1} s_j^2)^{1/2}} \xrightarrow{d} \mathcal{N}(0,1).$$

Proof. With reference to the notation adopted in Example 4, by setting $\alpha_{j,r} = \mathbb{E}[|Y_j - \mathbb{E}(Y_j)|^r]$, $r = 2,3$, for large m, we have:

$$\begin{aligned}\sum_{j=1}^{m-1} \alpha_{j,2} = \sum_{j=1}^{m-1} s_j^2 &= \frac{1}{\lambda^2} \sum_{i=0}^{n-1} \left(\frac{k^i}{i!}\right)^2 \frac{1}{m^2} \sum_{j=1}^{m-1} \left\{ \frac{1}{(1-j/m)} \left(\frac{j}{m}\right)^k \left[-\log\left(\frac{j}{m}\right)\right]^{i+1} \right\}^2 \\ &\approx \frac{1}{\lambda^2 m} \sum_{i=0}^{n-1} \left(\frac{k^i}{i!}\right)^2 c_2^{(i,k)}\end{aligned}$$

and recalling that $\mathbb{E}[|Y_j - \mathbb{E}(Y_j)|^3] = 2(6-e)[\mathbb{E}(Y_j)]^3/e$ for the exponential distribution:

$$\begin{aligned}\sum_{j=1}^{m-1} \alpha_{j,3} = \frac{2(6-e)}{e} \sum_{j=1}^{m-1} \mu_j^3 &= \frac{2(6-e)}{e} \sum_{j=1}^{m-1} \left\{ \frac{1}{\lambda} \sum_{i=0}^{n-1} \frac{k^i}{i!(m-j)} \left(\frac{j}{m}\right)^k \left[-\log\left(\frac{j}{m}\right)\right]^{i+1} \right\}^3 \\ &\approx \frac{2(6-e)}{e\lambda^3 m^2} \left\{ \sum_{i=0}^{n-1} \frac{k^i}{i!} c_1^{(i,k)} \right\}^3,\end{aligned}$$

where

$$c_h^{(i,k)} := \int_0^1 \left[\frac{x^k}{1-x} (-\log x)^{i+1} \right]^h dx, \qquad h = 1,2.$$

Hence, for a suitable function $C(\lambda, i, k)$, for large m, it holds that:

$$\frac{\left(\sum_{j=1}^{m-1} \alpha_{j,3}\right)^{1/3}}{\left(\sum_{j=1}^{m-1} \alpha_{j,2}\right)^{1/2}} \approx C(\lambda, i, k)\, m^{-1/6} \to 0 \qquad \text{as } m \to \infty.$$

The Lyapunov's condition of the central limit theorem is thus fulfilled, this giving the proof. □

5. Mutual Information of Lower Record Values

In this section, we study the degree of dependency between the sequences of lower record values by means of the mutual information. Recall the basic relations given in Equations (3) and (4).

First, with reference to (1), in the following theorem we obtain the entropy of k-th lower record values.

Theorem 5. *Let $\{X_n, n \geq 1\}$ be a sequence of IID random variables having finite entropy. The entropy of $L_{n(k)}$ for all $n \geq 2$ is given by:*

$$H(L_{n(k)}) = -\log k - (n-1)\psi(n) + \log((n-1)!) + n\left(1 - \frac{1}{k}\right) - k^n \varphi_f(n-1), \tag{32}$$

where $\psi(n) = \Gamma'(n)/\Gamma(n)$ *is the digamma function, and:*

$$\varphi_f(n-1) := \int_0^{+\infty} \frac{z^{n-1}}{(n-1)!} e^{-zk} \log(f(F^{-1}(e^{-z})))dz. \tag{33}$$

Proof. As customary, we denote by $f(x)$ and $F(x)$ the PDF and CDF of X_1. By (7), we have:

$$\begin{aligned}
H(L_{n(k)}) &= -\int_{-\infty}^{+\infty} f_{n(k)}(x)\log(f_{n(k)}(x))dx \\
&= -\int_{-\infty}^{+\infty} \frac{k^n}{(n-1)!}[F(x)]^{k-1}[\widetilde{\Lambda}(x)]^{n-1} f(x) \log\left(\frac{k^n}{(n-1)!}[F(x)]^{k-1}[\widetilde{\Lambda}(x)]^{n-1} f(x)\right) dx \\
&= -\frac{k^n}{(n-1)!}\int_{-\infty}^{+\infty} \Big\{ n[\widetilde{\Lambda}(x)]^{n-1}[F(x)]^{k-1} f(x)\log k + (n-1)[\widetilde{\Lambda}(x)]^{n-1}[F(x)]^{k-1} f(x)\log(\widetilde{\Lambda}(x)) \\
&+ (k-1)[\widetilde{\Lambda}(x)]^{n-1}[F(x)]^{k-1} f(x)\log(F(x)) + [\widetilde{\Lambda}(x)]^{n-1}[F(x)]^{k-1} f(x)\log(f(x)) \\
&- [\widetilde{\Lambda}(x)]^{n-1}[F(x)]^{k-1} f(x)\log((n-1)!) \Big\} dx.
\end{aligned}$$

By taking $z = \widetilde{\Lambda}(x)$, we get:

$$\begin{aligned}
H(L_{n(k)}) = -k^n \int_0^{+\infty} \Big[& n\frac{z^{n-1}}{(n-1)!} e^{-zk}\log k + \frac{z^{n-1}}{(n-2)!} e^{-zk}\log z \\
& - (k-1)\frac{z^n}{(n-1)!} e^{-zk} + \frac{z^{n-1}}{(n-1)!} e^{-zk}\log(f(F^{-1}(e^{-z}))) - \frac{z^{n-1}}{(n-1)!} e^{-zk}\log((n-1)!) \Big] dz.
\end{aligned}$$

Hence, making use of Equation (A.8) of Zahedi and Shakil [32], after some calculations we finally get: Equation (32). □

It is worth noting the analogies between Equation (33) and the function $\phi_f(n)$ considered by Baratpour et al. [33] for the analysis of the upper record values.

In the following lemma we express the mutual information between $Z_{m(k)}$ and $Z_{n(k)}$ in terms of the digamma function.

Lemma 1. *Let* $Z_{n(k)}$ *be the k-th lower record values from the uniform distribution over interval* $(0,1)$. *Then, the mutual information between* $Z_{m(k)}$ *and* $Z_{n(k)}$, *for* $n > m$, *is given by:*

$$\begin{aligned}
M(Z_{m(k)}, Z_{n(k)}) &= m + \log((n-1)!) - \log((n-m-1)!) \\
&+ (n-m-1)\psi(n-m) - (n-1)\psi(n).
\end{aligned}$$

Proof. In this case, recalling (9), we have $\widetilde{\Lambda}(x) = -\log x$, $0 < x < 1$. Hence, making use of Equations (2), (4), (8) and (10), for $n > m$ we get:

$$\begin{aligned}
H(Z_{n(k)}|Z_{m(k)}) &= -\int_0^1 \int_0^x \frac{k^n[-\log(y)+\log(x)]^{n-m-1}[-\log(x)]^{m-1} y^{k-1}}{(m-1)!(n-m-1)!x} \\
&\times \log\left[\frac{k^n y^{k-1}}{k^m x^k}\frac{[-\log(y)+\log(x)]^{n-m-1}}{(n-m-1)!}\right] dydx \\
&= \sum_{r=1}^{5} I_r,
\end{aligned} \tag{34}$$

where

$$I_1 := -(n-m)\log(k)\int_0^1\int_0^x f_{m(k),n(k)}(x,y)dydx = (n-m)\log(k),$$
$$I_2 := \log[(n-m-1)!]\int_0^1\int_0^x f_{m(k),n(k)}(x,y)dydx = \log[(n-m-1)!],$$
$$I_3 := \int_0^1\int_0^x \frac{k^{n+1}[-\log(y)+\log(x)]^{n-m-1}[-\log(x)]^{m-1}y^{k-1}}{(m-1)!(n-m-1)!x}\log(x)dydx = -m,$$
$$I_4 := -\int_0^1\int_0^x \frac{k^{n}[-\log(y)+\log(x)]^{n-m-1}[-\log(x)]^{m-1}y^{k-1}}{(m-1)!(n-m-1)!x}\log[-\log(y)+\log(x)]^{n-m-1}dydx$$
$$= -(n-m-1)[\psi(n-m)-\log(k)],$$
$$I_5 := -\int_0^1\int_0^x \frac{(k-1)k^{n}[-\log(y)+\log(x)]^{n-m-1}[-\log(x)]^{m-1}y^{k-1}}{(m-1)!(n-m-1)!x}\log(y)dydx = n\left(1-\frac{1}{k}\right).$$

Hence, by straightforward calculations we obtain:

$$H(Z_{n(k)}|Z_{m(k)}) = \log((n-m-1)!) - (n-m)\log(k) - (n-m-1)[\psi(n-m)-\log k] + \frac{k-1}{k}(n-2m) - m, \quad (35)$$

where $\psi(n-m) = \int_0^{+\infty} t^{n-m}e^{-t}\log(t)\,dt$. Similarly, by (32) we have

$$H(Z_{n(k)}) = -\log k - (n-1)\psi(n) + \log((n-1)!) + n\left(1-\frac{1}{k}\right). \quad (36)$$

Recalling (4), the thesis thus follows from (35) and (36). □

Let us now come to the main result of this section. Recall that the mutual information of (X, Y) is defined in (3).

Theorem 6. *Under the assumptions of Theorem 5, the following result holds:*

(i) *The mutual information between the m-th and the n-th k-lower records is distribution-free, and is given by:*

$$\begin{aligned} M(L_{m(k)}, L_{n(k)}) &= m + \log((n-1)!) - \log((n-m-1)!) \\ &\quad + (n-m-1)\psi(n-m) - (n-1)\psi(n). \end{aligned} \quad (37)$$

(ii) *The mutual information $M(L_{m(k)}, L_{m+1(k)})$ is increasing in m.*

Proof.

(i) Let $Z_{m(k)} = F^{-1}(L_{m(k)})$ and $Z_{n(k)} = F^{-1}(L_{n(k)})$, where F^{-1} denotes the pseudo-inverse function of F, i.e., the quantile function of X. Then, $Z_{m(k)}$ and $Z_{n(k)}$ are the m-th and n-th k-lower records of the uniform distribution over the interval $(0, 1)$. By the invariance property of mutual information, we have: $M(L_{m(k)}, L_{n(k)}) = M(Z_{m(k)}, Z_{n(k)})$, thus the result follows from Lemma 1.

(ii) By taking $n = m+1$ in (37), we get:

$$M(L_{m(k)}, L_{m+1(k)}) = m + \log(m!) - m\psi(m+1),$$

so that

$$\begin{aligned} M(L_{m+1(k)}, L_{m+2(k)}) - M(L_{m(k)}, L_{m+1(k)}) &= \log((m+1)!) - \log(m!) \\ &\quad + m\psi(m+1) - (m+1)\psi(m+2) + 1. \end{aligned}$$

It is not hard to see that the right-hand-side is positive, and the proof thus follows. □

It is useful to assess the mutual information between the k-th lower record values, such as when such values correspond to successive failures of a repairable system. Hence, the information provided in Theorem 6 is useful for constructing suitable replacement criteria of components, in order to avoid failures and to improve system availability.

The mutual information between the m-th and the n-th k-lower records, determined in Equation (37), is shown in Figure 3 for different choices of m and n. For fixed values of m, such an information measure is decreasing in n, whereas for fixed n, it is increasing in m.

Finally, Figure 4 shows that $M(L_{m(k)}, L_{m+1(k)})$ is increasing in m, in agreement with the point *(ii)* of Theorem 6.

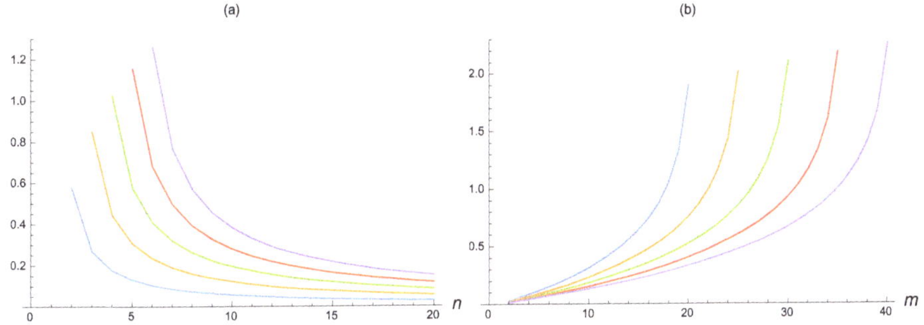

Figure 3. The values of $M(L_{m(k)}, L_{n(k)})$ are shown (**a**) for $m = 1, 2, 3, 4, 5$ (from bottom to top) and $m < n \leq 20$, and (**b**) for $n = 20, 25, 30, 35, 40$ (from top to bottom, near the origin) and $1 \leq m < n$.

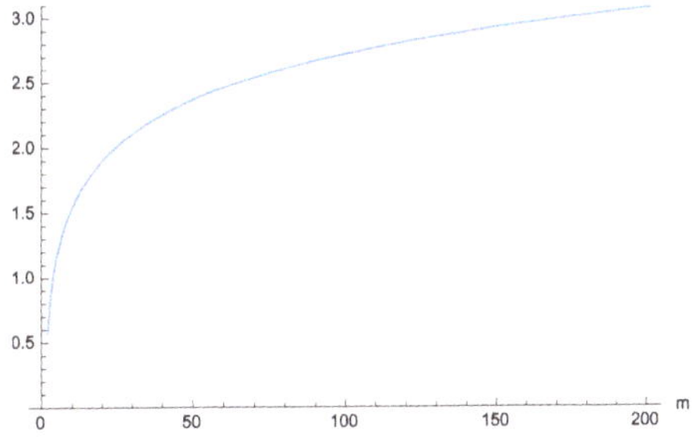

Figure 4. Values of $M(L_{m(k)}, L_{m+1(k)})$ for $1 \leq m \leq 200$.

6. Conclusions

In this paper, we discussed the concept of inaccuracy between $L_{n(k)}$ and X in a random sample generated by X. We proposed a dynamic version of cumulative inaccuracy and studied a related characterization result. We also proved that $I(L_{n(k)}, X; t)$ can uniquely determine the parent distribution F. Moreover, we constructed bounds for characterization results of $I(L_{n(k)}, X)$. Also, we estimated the cumulative measure of inaccuracy by means of the empirical cumulative inaccuracy in lower record values. These concepts can be applied in measuring the inaccuracy contained in the associated past lifetime. Finally, we studied the degree of dependency among the sequence of k-th lower record values in terms of mutual information. We showed that $M(L_{m(k)}, L_{n(k)})$ is distribution-free and can be computed by using the distribution of the k-th lower record values of the sequence from the uniform distribution.

Author Contributions: All the authors contributed equally to this work.

Funding: This research received no external funding.

Acknowledgments: The authors thank the referees for useful comments that improved the paper. Part of this research has been performed during a visit of Maryam Eskandarzadeh at Salerno University. A.D.C. is member of the group GNCS of INdAM.

Conflicts of Interest: The authors declare no conflict of interest.

References

1. Navarro, J.; del Aguila, Y.; Asadi, M. Some new results on the cumulative residual entropy. *J. Stat. Plan. Inference* **2010**, *140*, 310–322. [CrossRef]
2. Psarrakos, G.; Navarro, J. Generalized cumulative residual entropy and record values. *Metrika* **2013**, *76*, 623–640. [CrossRef]
3. Tahmasebi, S.; Eskandarzadeh, M. Generalized cumulative entropy based on *k*th lower record values. *Stat. Prob. Lett.* **2017**, *126*, 164–172. [CrossRef]
4. Sordo, M.A.; Psarrakos, G. Stochastic comparisons of interfailure times under a relevation replacement policy. *J. Appl. Prob.* **2017**, *54*, 134–145. [CrossRef]
5. Ebrahimi, N.; Soofi, E.S.; Soyer, R. Information measures in perspective. *Int. Stat. Rev.* **2010**, *78*, 383–412. [CrossRef]
6. Ebrahimi, N.; Kirmani, S.N.U.A.; Soofi, E.S. Multivariate dynamic information. *J. Multivar. Anal.* **2007**, *98*, 328–349. [CrossRef]
7. Ahmadi, J.; Di Crescenzo, A.; Longobardi, M. On dynamic mutual information for bivariate lifetimes. *Adv. Appl. Prob.* **2015**, *47*, 1157–1174. [CrossRef]
8. Kundu, C.; Di Crescenzo, A.; Longobardi, M. On cumulative residual (past) inaccuracy for truncated random variables. *Metrika* **2016**, *79*, 335–356. [CrossRef]
9. Kerridge, D.F. Inaccuracy and inference. *J. R. Stat. Soc. Ser. B Stat. Methodol.* **1961**, *23*, 184–194. [CrossRef]
10. Di Crescenzo, A.; Longobardi, M. On cumulative entropies. *J. Stat. Plan. Inference* **2009**, *139*, 4072–4087. [CrossRef]
11. Kumar, V.; Taneja, H.C. Dynamic cumulative residual and past inaccuracy measures. *J. Stat. Theory Appl.* **2015**, *14*, 399–412. [CrossRef]
12. Park, S.; Rao, M.; Shin, D.W. On cumulative residual Kullback-Leibler information. *Stat. Prob. Lett.* **2012**, *82*, 2025–2032. [CrossRef]
13. Di Crescenzo, A.; Longobardi, M. Some properties and applications of cumulative Kullback-Leibler information. *Appl. Stoch. Models Bus. Ind.* **2015**, *31*, 875–891. [CrossRef]
14. Dziubdziela, W.; Kopociński, B. Limiting properties of the *k*-th record values. *Appl. Math.* **1976**, *15*, 187–190. [CrossRef]
15. Malinowska, I.; Szynal, D. On characterization of certain distributions of *k*th lower (upper) record values. *Appl. Math. Comput.* **2008**, *202*, 338–347. [CrossRef]
16. Tahmasebi, S.; Eskandarzadeh, M.; Jafari, A.A. An extension of generalized cumulative residual entropy. *J. Stat. Theory Appl.* **2017**, *16*, 165–177. [CrossRef]
17. Arnold, B.C.; Balakrishnan, N.; Nagaraja, H.N. *A First Course in Order Statistics*; John Wiley and Sons: New York, NY, USA, 1992.
18. Thapliyal, R.; Taneja, H.C. A measure of inaccuracy in order statistics. *J. Stat. Theory Appl.* **2013**, *12*, 200–207. [CrossRef]
19. Thapliyal, R.; Taneja, H.C. On residual inaccuracy of order statistics. *Stat. Prob. Lett.* **2015**, *97*, 125–131. [CrossRef]
20. Thapliyal, R.; Taneja, H.C. On Rényi entropies of order statistics. *Int. J. Biomath.* **2015**, *8*, 1550080. [CrossRef]
21. Fashandi, M.; Ahmadi, J. Characterizations of symmetric distributions based on Rényi entropy. *Stat. Prob. Lett.* **2012**, *82*, 798–804. [CrossRef]
22. Asadi, M. A new measure of association between random variables. *Metrika* **2017**, *80*, 649–661. [CrossRef]
23. Goel, R.; Taneja, H.C.; Kumar, V. Measure of entropy for past lifetime and *k*-record statistics. *Physica A* **2018**, *503*, 623–631. [CrossRef]
24. Shaked, M.; Shanthikumar, J.G. *Stochastic Orders*; Springer: New York, NY, USA, 2007.

25. Ma, C. Convex orders for linear combinations of random variables. *J. Stat. Plan. Inference* **2000**, *84*, 11–25. [CrossRef]
26. De Gusmão, F.R.S.; Ortega, E.M.M.; Cordeiro, G.M. The generalized inverse Weibull distribution. *Stat. Pap.* **2011**, *52*, 591–619. [CrossRef]
27. Gupta, R.C.; Gupta, R.D. Proportional reversed hazard rate model and its applications. *J. Stat. Plan. Inference* **2007**, *137*, 3525–3536. [CrossRef]
28. Di Crescenzo, A.; Toomaj, A. Extension of the past lifetime and its connection to the cumulative entropy. *J. Appl. Prob.* **2015**, *52*, 1156–1174. [CrossRef]
29. Gupta, R.C.; Kirmani, S.N.U.A. Characterizations based on convex conditional mean function. *J. Stat. Plan. Inference* **2008**, *138*, 964–970. [CrossRef]
30. Abouammoh, A.M.; Abdulghani, S.A. On partial orderings and testing of new better than renewal used classes. *Reliab. Eng. Syst. Saf.* **1994**, *43*, 37–41. [CrossRef]
31. Pyke, R. Spacings. *J. R. Stat. Soc. Ser. B* **1965**, *27*, 395–449. [CrossRef]
32. Zahedi, H.; Shakil, M. Properties of entropies of record values in reliability and life testing context. *Commun. Stat. Theory Meth.* **2006**, *35*, 997–1010. [CrossRef]
33. Baratpour, S.; Ahmadi, J.; Arghami, N.R. Entropy properties of record statistics. *Stat. Pap* **2007**, *48*, 197–213. [CrossRef]

MDPI
St. Alban-Anlage 66
4052 Basel
Switzerland
Tel. +41 61 683 77 34
Fax +41 61 302 89 18
www.mdpi.com

Mathematics Editorial Office
E-mail: mathematics@mdpi.com
www.mdpi.com/journal/mathematics

www.ingramcontent.com/pod-product-compliance
Lightning Source LLC
Chambersburg PA
CBHW051845210326
41597CB00033B/5785